Radioactive Pollution and Biological Effects of Radioactivity

Radioactive Pollution and Biological Effects of Radioactivity

Editors

Fabrizio Ambrosino
Supitcha Chanyotha

MDPI • Basel • Beijing • Wuhan • Barcelona • Belgrade • Manchester • Tokyo • Cluj • Tianjin

Editors
Fabrizio Ambrosino
Department of Physics
University of Naples
Naples
Italy

Supitcha Chanyotha
Department of Nuclear
Engineering
Chulalongkorn University
Bangkok
Thailand

Editorial Office
MDPI
St. Alban-Anlage 66
4052 Basel, Switzerland

This is a reprint of articles from the Special Issue published online in the open access journal *Life* (ISSN 2075-1729) (available at: www.mdpi.com/journal/life/special_issues/Radioactive_Pollution).

For citation purposes, cite each article independently as indicated on the article page online and as indicated below:

LastName, A.A.; LastName, B.B.; LastName, C.C. Article Title. *Journal Name* **Year**, *Volume Number*, Page Range.

ISBN 978-3-0365-6767-9 (Hbk)
ISBN 978-3-0365-6766-2 (PDF)

Contents

About the Editors vii

Preface to "Radioactive Pollution and Biological Effects of Radioactivity" ix

Lucy A. Parker, Andrea Moreno-Garijo, Elisa Chilet-Rosell, Fermina Lorente and Blanca Lumbreras
Gender Differences in the Impact of Recommendations on Diagnostic Imaging Tests: A Retrospective Study 2007–2021
Reprinted from: *Life* **2023**, *13*, 289, doi:10.3390/life13020289 1

Narongchai Autsavapromporn, Chutima Krandrod, Pitchayaponne Klunklin, Rawiwan Kritsananuwat, Churdsak Jaikang, Kittikun Kittidachanan, et al.
Health Effects of Natural Environmental Radiation during Burning Season in Chiang Mai, Thailand
Reprinted from: *Life* **2022**, *12*, 853, doi:10.3390/life12060853 17

Akari Morita, Ko Sakauchi, Wataru Taira and Joji M. Otaki
Ingestional Toxicity of Radiation-Dependent Metabolites of the Host Plant for the Pale Grass Blue Butterfly: A Mechanism of Field Effects of Radioactive Pollution in Fukushima
Reprinted from: *Life* **2022**, *12*, 615, doi:10.3390/life12050615 31

Ko Sakauchi, Wataru Taira and Joji M. Otaki
Metabolomic Profiles of the Creeping Wood Sorrel *Oxalis corniculata* in Radioactively Contaminated Fields in Fukushima: Dose-Dependent Changes in Key Metabolites
Reprinted from: *Life* **2022**, *12*, 115, doi:10.3390/life12010115 51

Ondřej Harkut, Petr Alexa and Radim Uhlář
Radiocaesium Contamination of Mushrooms at High- and Low-Level Chernobyl Exposure Sites and Its Consequences for Public Health
Reprinted from: *Life* **2021**, *11*, 1370, doi:10.3390/life11121370 73

Narongchai Autsavapromporn, Pitchayaponne Klunklin, Imjai Chitapanarux, Churdsak Jaikang, Busyamas Chewaskulyong, Patumrat Sripan, et al.
A Potential Serum Biomarker for Screening Lung Cancer Risk in High Level Environmental Radon Areas: A Pilot Study
Reprinted from: *Life* **2021**, *11*, 1273, doi:10.3390/life11111273 83

Ko Sakauchi, Wataru Taira and Joji M. Otaki
Metabolomic Response of the Creeping Wood Sorrel *Oxalis corniculata* to Low-Dose Radiation Exposure from Fukushima's Contaminated Soil
Reprinted from: *Life* **2021**, *11*, 990, doi:10.3390/life11090990 93

Tanate Suksompong, Sirikanjana Thongmee and Wanwisa Sudprasert
Efficacy of a Graphene Oxide/Chitosan Sponge for Removal of Radioactive Iodine-131 from Aqueous Solutions
Reprinted from: *Life* **2021**, *11*, 721, doi:10.3390/life11070721 121

Ibrahim I. Suliman and Khalid Alsafi
Radiological Risk to Human and Non-Human Biota Due to Radioactivity in Coastal Sand and Marine Sediments, Gulf of Oman
Reprinted from: *Life* **2021**, *11*, 549, doi:10.3390/life11060549 137

Vittoria D'Avino, Mariagabriella Pugliese, Fabrizio Ambrosino, Mariateresa Bifulco, Marco La Commara, Vincenzo Roca, et al.
Radon Survey in Bank Buildings of Campania Region According to the Italian Transposition of Euratom 59/2013
Reprinted from: *Life* **2021**, *11*, 533, doi:10.3390/life11060533 . . . . **149**

Jayan D. M. Senevirathna and Shuichi Asakawa
Multi-Omics Approaches and Radiation on Lipid Metabolism in Toothed Whales
Reprinted from: *Life* **2021**, *11*, 364, doi:10.3390/life11040364 . . . . **161**

Nurul Absar, Jainal Abedin, M Mashiur Rahman, Md. Moazzem Hossain Miah, Naziba Siddique, Masud Kamal, et al.
Radionuclides Transfer from Soil to Tea Leaves and Estimation of Committed Effective Dose to the Bangladesh Populace
Reprinted from: *Life* **2021**, *11*, 282, doi:10.3390/life11040282 . . . . **183**

Filomena Loffredo, Federica Savino, Roberto Amato, Alfredo Irollo, Francesco Gargiulo, Giuseppe Sabatino, et al.
Indoor Radon Concentration and Risk Assessment in 27 Districts of a Public Healthcare Company in Naples,
South Italy
Reprinted from: *Life* **2021**, *11*, 178, doi:10.3390/life11030178 . . . . **199**

About the Editors

Fabrizio Ambrosino

He is currently working as scientific researcher at the Department of Physics of University of Naples in Italy. His scientific interests fall within the field of applied physics, mainly concerning the study, monitoring and analysis of radioactivity of environmental origin. His research interests are: radioactivity, environmental radiation, radon gas, ionizing radiation, radioactive pollution, nuclear power plant decommissioning, numerical analysis, time series analysis, neural networks, machine learning, Monte Carlo simulation, software development, data processing, geogenic phenomena, and Earth rotation (Length of the day).

Supitcha Chanyotha

She is currently working as associate professor at Nuclear Engineering Department of Chulalongkorn University in Bangkok (Thailand). Her research interests are: radioactive waste management, development of low level radioactive waste and mixed hazardous waste, inventory of low level radioactive waste in Thailand, naturally occurring radioactive materials (NORM), development of systematic measurements for characterization of NORM in Thailand industries, and nuclear power simulation development of the Thailand research reactor.

Preface to "Radioactive Pollution and Biological Effects of Radioactivity"

The Special Issue "Radioactive Pollution and Biological Effects of Radioactivity"of the Journal *Life* involved the participation of academic scientists, researchers, and scholars from all over the World that contributed with articles, reviews, and case reports, based on high-quality research works.

The common topic of the published works concerns the radioactive pollution, which occurs when radioactive elements enter the atmosphere and reach the Earth's surface, within solids, liquids, or gases, including the human body. This happens after natural and/or man-made activities, such as nuclear tests, industrial waste (e.g., radiodiagnostics), and excesses of naturally occurring radioactive sources. This kind of pollution entails risks of radiological contamination of the environment, with harmful effects on human health caused by the ionizing radiation. This theme has become even more current due to the increasing use of ionizing radiation for domestic, industrial, and medical purposes during the last century. Radiological monitoring is a primary objective of radiation protection in order to estimate and understand the impact of radionuclides on the environment and to assess the health risk for the population.

The aim of the Special Issue has been to provide an interdisciplinary platform for researchers to exchange and share their experiences and latest achievements on all aspects of radioactive pollution. Survey data analysis, original and unpublished results of conceptual, constructive, empirical, experimental, and theoretical work were welcome, as well as other concepts related to the radiation field.

Fabrizio Ambrosino and Supitcha Chanyotha
Editors

life

MDPI

Article

Gender Differences in the Impact of Recommendations on Diagnostic Imaging Tests: A Retrospective Study 2007–2021

Lucy A. Parker [1,2], **Andrea Moreno-Garijo** [3], **Elisa Chilet-Rosell** [1,2], **Fermina Lorente** [4] and **Blanca Lumbreras** [1,2,*]

1 Department of Public Health, University Miguel Hernández de Elche, 03550 Alicante, Spain
2 CIBER de Epidemiología y Salud Pública (CIBERESP), 28029 Madrid, Spain
3 Faculty of Pharmacy, University Miguel Hernández de Elche, 03550 Alicante, Spain
4 Radiology Department, University Hospital of San Juan de Alicante, Sant Joan d'Alacant, 03550 Alicante, Spain
* Correspondence: blumbreras@umh.es; Tel.: +34-965-919510

Abstract: (1) Background: The frequency of imaging tests grew exponentially in recent years. This increase may differ according to a patient's sex, age, or socioeconomic status. We aim to analyze the impact of the Council Directive 2013/59/Euratom to control exposure to radiation for men and women and explore the impact of patients' age and socioeconomic status; (2) Methods: The retrospective observational study that includes a catchment population of 234,424. We included data of CT, mammography, radiography (conventional radiography and fluoroscopy) and nuclear medicine between 2007–2021. We estimated the associated radiation effective dose per test according using previously published evidence. We calculated a deprivation index according to the postcode of their residence. We divided the study in 2007–2013, 2014–2019 and 2020–2021 (the pandemic period). (3) Results: There was an increase in the number of imaging tests received by men and women after 2013 ($p < 0.001$), and this increase was higher in women than in men. The frequency of imaging tests decreased during the pandemic period (2020–2021), but the frequency of CT and nuclear medicine tests increased even during these years ($p < 0.001$) and thus, the overall effective mean dose. Women and men living in the least deprived areas had a higher frequency of imaging test than those living in the most deprived areas. (4) Conclusions: The largest increase in the number of imaging tests is due to CTs, which account for the higher amount of effective dose. The difference in the increase of imaging tests carried out in men and women and according to the socioeconomic status could reflect different management strategies and barriers to access in clinical practice. Given the low impact of the available recommendations on the population exposure to radiation and the performance of high-dose procedures such as CT, deserve special attention when it comes to justification and optimization, especially in women.

Keywords: imaging tests; radiation exposure; recommendations; gender; socioeconomical status

Citation: Parker, L.A.; Moreno-Garijo, A.; Chilet-Rosell, E.; Lorente, F.; Lumbreras, B. Gender Differences in the Impact of Recommendations on Diagnostic Imaging Tests: A Retrospective Study 2007–2021. *Life* **2023**, *13*, 289. https://doi.org/10.3390/life13020289

Academic Editors: Fabrizio Ambrosino and Supitcha Chanyotha

Received: 20 December 2022
Revised: 13 January 2023
Accepted: 17 January 2023
Published: 20 January 2023

1. Introduction

In recent decades, diagnostic imaging tests using radiation, together with nuclear medicine, have been a major source of exposure to non-natural radiation in the general population in Western countries [1]. In most developed countries, it has been shown that the contribution of nuclear medicine diagnostic procedures is between 4% and 14% [2]. The widespread use of diagnostic techniques such as CT scans has also meant the number of doses of ionising radiation received has increased, at both individual and population level [3]. Over the last 22 years, radiation due to CT exposures were estimated to account for 0.7% of the cancer incidence and 1% of cancer mortality [4]. Moreover, the incidence of cancer in individuals who had been exposed to CT was found to be 24% higher compared to individuals who had not [4]. Other adverse effects of the increase in imaging tests are time and resource utilization and the presence of incidental findings that can lead to unnecessary clinical interventions [5].

The US Food and Drug Administration (FDA) and the European Union have developed strategies aimed at reducing unnecessary radiation exposure. In 2010, the FDA published an initiative to promote patient safety through the justification of each imaging test carried out in accordance with the patient's symptoms and medical history. In addition, they promoted the lowest possible dose based on each patient's anatomical and physiological factors [6]. The most relevant European initiative in this regard is the Directive 97/432/EURATOM [7] which has given rise to various projects such as the DOSE DATAMED. This project consisted of a survey to assess the radiation received by the European population. In accordance with this European directive other projects have been developed, which aim at unifying the training and knowledge of the professionals involved in radiation protection: MEDRAPET Project [8], European Medical ALARA Network [9] and Medical Physics Expert Project [10]. The Directive 2013/59/Euratom [11], which was an update of the previous directive, was published in 2013. It was supposed to be transposed in all member countries before the 6 February 2018. However, in Spain, this directive was not transposed until the 18 October 2019 (Royal Decree 601/2019 [12]) and has not yet been implemented.

Despite these recommendations, recent studies indicate that the frequency of imaging tests has grown exponentially in recent years. Even during the SARS-CoV-2 pandemic when access to healthcare was limited, the use of imaging tests remained widespread [13].

Moreover, this increase in the frequency of imaging tests may differ according to patient's characteristics such as sex, age, or socioeconomic status. It has been known for over 30 years that there are wide differences in the clinical management of men and women in many situations. Even though women use more health services than men [14], there is evidence that there is a diagnostic bias between men and women [15]. Research regarding appropriateness has shown that women are less likely to have an imaging test considered adequate than men [16]. Another study also found that more inappropriate and uncertain myocardial perfusion imaging was ordered for women compared to men [17]. Some studies estimate that radiation from imaging tests may be linked to 1% of cancers diagnosed in the United States, with adult women between the ages of 35 and 54 years being the largest population at risk [18,19] especially if they are under the age of 30 years [20,21]. Previous reports identified ionising radiation from CT as a contributing factor for both breast cancer [22] and a greater hazard of radiation-related solid cancer in women compared to men [23]. These data could have influenced the differential use of imaging diagnostic testing in women compared to men [24].

In addition, other aspects such as the socioeconomic context could also play a role in the different performance of imaging tests. Social determinants of health are known to play a large role in health outcomes. For example, they may impact an individual's ability to access nutritious food and healthcare resources, time, and space for physical activity [25,26]. A previous study found that the intersection of sex and social factors in influencing patient-relevant outcomes varies even among countries with similar healthcare and high gender equality [27]. High income countries have been associated with a higher frequency of CT examinations per inhabitant [28]. However, there is little data on the differences in the frequency of imaging tests according to the socioeconomic status of the population for men and women in the same country, including those with a public health system like Spain.

The purpose of this study was to analyze the impact of the establishment of new recommendations in 2013 to control exposure to radiation for men and women, and to explore the impact of patients' age and socioeconomic status on gender differences, in a single university hospital.

2. Materials and Methods

2.1. Study Design

We conducted a retrospective observational study to analyse the impact of the established recommendations in 2013 on tests performed in clinical practice according in men and women. We also calculated the per capita effective dose and the influence of patients' age and socioeconomic status.

2.2. Setting

The target population for the study were all residents in the catchment area of San Juan Hospital (Alicante), in the Valencian Community (Spain), a general centre, with an estimated catchment population of 234,424. This is a referral hospital for all individuals living in the catchment area who belong to the National Health Care System (NHS). Most of the Spanish population uses the NHS as the main medical service (the publicly funded insurance scheme covers 98.5% of the Spanish population).

2.3. Participants

We included utilization data of CT, radiography (including mammography, conventional radiography, and fluoroscopy) and nuclear medicine by the target population between 2007–2021 (in any care setting, inpatient, outpatient, or emergency department). We excluded imaging tests that did not involve radiation exposure (i.e., MRI and ultrasound) and patients who had an imaging test in this hospital but did not belong to its catchment area.

2.4. Imaging Test Frequency

For collecting data on imaging test frequency, we carried out procedures similar to those used in a previous study [29]. Briefly, we collected the following data from the Medical Image Bank of the Valencian Community from the Department of Universal Health and Public Health Service: sex and age at entry in the study, radiological examination, and date. Both the images and the patient data were anonymised and deidentified by the Health Informatics Department of the Hospital of San Juan using Research and Development (R&D) Cloud CEIB Architecture [30]. This digital register started in 2007 in our setting. Each imaging test received was classified as a single radiation exposure. However, abdomen and pelvis tests carried out in the same process were included as a single abdomen–pelvis test, while an abdomen or a pelvis test in a different process, even in the same patient, were included as two different tests. Thoracic and lumbar spine tests were included when they were performed alone but not when performed together with chest or abdominal tests.

2.5. Effective Dose Estimate

Given that it was impossible to get individual machine parameters for all imaging tests, we estimated the associated radiation effective dose per test according to its region of anatomical coverage by age and using previously published evidence [31]. This review provides values of the typical effective doses associated with the 20 most frequent imaging tests for adults and children and for the most widely used set of weights (ICRP60) as well as for the most recent (ICRP103). In addition, we estimated the effective dose of imaging tests different from the 20 most frequent imaging tests in Dose DataMed 2 project according to previous studies [32–34].

2.6. Socioeconomical Status

To represent the socioeconomic status of the individuals we calculated a deprivation index according to the postcode of their residence. The Spanish Society of Epidemiology (SEE) published a deprivation index for the entire country using the enumeration district [35] and the information from the 2011 census. We reconstructed the index within the catchment area of the hospital by assigning each enumeration district with the postcode and then estimated the six socioeconomic indicators used by the SEE at postcode level using the census data (percentage of manual working population, percentage of casual working population, percentage of unemployed population, percentage of population with insufficient education, percentage of young population with insufficient education, and percentage of main dwellings without internet access). We used principal components analysis in Stata SE to recalculate the deprivation index by postcode. As this is a standardized index (with mean 0 and standard deviation 1), values close to zero would indicate the average deprivation in the population area. We divided the population in tertials according to the deprivation index: least deprived between −2.579506 and −1.060423, medium deprived between −0.813536 and 0.276083, and most deprived between 0.399323 and 5.022990.

2.7. Calendar Time

We assessed the different values according to two periods of study: 2007–2013 and 2014–2019 according to the year of publication of the Council Directive 2013/59/Euratom. Although this recommendation was transposed in Spain on the 18 October 2019, we did not evaluate its impact because of the SARS-CoV-2 pandemic in 2020. However, we also assessed the pandemic period: 2020–2021.

2.8. Statistical Analysis

All analysis were stratified by sex. We evaluate the frequency of imaging tests performed by imaging modality, age (<18 years [children]; 18–64 years [adults], and >64 years [older adults]), deprivation index (grouped in terciles), and calendar year (2007–2013, 2014–2019, 2020–2021) using the Chi-Square test. The annual average frequency was assessed per 1000 persons per women and men (number of people who are administratively assigned to the university hospital in each year by sex and age group). We also estimated the effective radiation dose by imaging modality (median and interquartile range) age, deprivation, and calendar time using Mann–Whitney U test.

Statistical analyses of the data were performed with SPSS (V.25.0; SPSS). A p-value of 0.05 was considered significant.

3. Results

3.1. Population Included in the Study

In 2007, 232,446 people were administratively assigned to the selected hospital: 107,622 (46.3%) men and 124,824 (53.7%) women. The population in 2021 was 249,572 persons: 114,434 (45.9%) men and 135,138 (54.1%) women. There were not statistical differences according to the distribution of demographic variables (age and deprivation level) between men and women in the years 2007–2021: 16% were subjects < 18 years, 61% were subjects 18–64 years, and 23% >65 years. The deprivation level was divided in terciles, with the minimum value −2.84 (least deprived) and the maximum 5022 (mean −0.9514, sd 2.01).

3.2. Impact of Calendar Time in the Frequency of Imaging Tests According to Type of Imaging Test, Patients' Age and Deprivation Index for Men and Women (Table 1)

The frequency of imaging tests was higher in women than in men for the three periods of study. The increase in tests between the years 2007–2013 and 2014–2019 was higher in women than in men (from 575.3 tests per 1000 women to 634.3 tests per 1000 women, percentage of change of 10.3% vs. from 498.5 tests per 1000 men to 535.6 tests per 1000 men, percentage of change of 7.4%). The frequency of imaging tests decreased in 2020–2021 to 519.6 tests per 1000 women (percentage of change of −18.1%) ($p < 0.001$) and this decrease was lower in men during the years 2020–2021, 471 tests per 1000 men (percentage of change −12.1%) ($p < 0.001$). (Table 1).

Although the frequency of radiography in men was similar between 2007–2013 and 2014–2019, it decreased during the years 2020–2021. CT frequency in men, in contrast, increased throughout the period of study (67.1 tests per 1000 persons in 2007–2013, 88.8 tests per 1000 persons in 2014–2019, and 98.7 tests per 1000 persons in 2020–2021). Similarly, the frequency of nuclear medicine tests in men also increased during the period of study: 29.9 per 1000 men in 2007–2013, 38.4 per 1000 men in 2014–2019, and 41.2 per 1000 men in 2020–2021 ($p < 0.001$). In women, the frequency of mammography, radiography and nuclear medicine increased between 2007–2013 (26.3 per 1000 women, 460.4 per 1000 women, and 35.6 per 1000 women, respectively) and 2014–2019 (32.5 per 1000 women, 486.7 per 1000 women, and 43.4 per 1000 women, respectively) and these frequencies decreased in 2020–2021 (28.7 per 1000 women, 376.3 per 1000 women, and 37.7 per 1000 women, respectively). However, the frequency of CT also increased during the study ($p < 0.001$).

According to age, the frequency of tests decreased in 2014–2019 and 2020–2021 in comparison with 2007–2013 for men < 18 years and those aged 18–64 years; the frequency increased in men > 64 years between 2007–2013 and 2014–2019 (686.2 tests per 1000 men to 892.3 tests per 1000 men) and this frequency decreased in 2020–2021 (855.8 tests per 1000 men) ($p < 0.001$). Although the frequency of imaging tests in women < 18 years was lower than in men during the three periods of study, the frequency of tests in women > 64 years was higher than in men > 64 years during the period of study. As well as in men, the frequency of tests also decreased in women < 18 years and those 18–64 years in 2014–2019 and in 2020–2021 in comparison with 2007–2013; nevertheless, the frequency of imaging tests increased in women > 64 years in 2014–2019 in comparison with 2007–2013 (916.6 tests per 1000 women to 1180 tests per 1000 women) and this frequency decreased in 2020–2021 (1010.3 tests per 1000 women) in comparison with 2014–2019 ($p < 0.001$)

In Figure 1, we have shown that the amount of CT increased in 2014–2019 and in 2020–2021 in comparison with 2007–2013 in patients aged 18–64 years and in those older than 64 years, for men and women. The frequency of radiographies increased in 2014–2019 in comparison with 2007–2013 in patients > 64 years, although it decreased in patients < 18 years and in those aged 18–64 years in men and women. In addition, although the frequency of mammographies decreased during 2020–2021 in women aged 18–64 years, it increased in women older than 64 years. The frequency of nuclear medicine tests increased between 2007–2013 and 2014–2019 in men and women > 64 years and it decreased in 2020–2021.

Table 1. Description of the impact of calendar time in the frequency of imaging tests according to type of imaging test; patients' age and deprivation index for men and women.

Frequency per 1000 Persons	Men				Women			
	2007–2013	2014–2019	2020–2021	*p* Value	2007–2013	2014–2019	2020–2021	*p* Value
Imaging test				<0.001				<0.001
Mamography	0.9	0.9	0.8		26.3	32.5	28.7	
Radiography	400.5	407.6	330.4		460.4	486.7	376.3	
CT	67.1	88.8	98.7		53.0	71.7	76.9	
Nuclear medicine	29.9	38.4	41.2		35.6	43.4	37.7	
Age (years)				<0.001				<0.001
<18	409.4	327.3	223.4		310.5	255.6	168.1	
18–64	452.2	458.4	393.9		519.4	532.6	431	
>64	686.2	892.3	855.8		916.6	1180	1010.3	
Deprivation index				<0.001				<0.001
Least deprived	414.5	474.4	440.1		492.1	587.1	500.9	
Medium deprived	407.6	466.1	422.1		452.3	537.0	456.6	
Most deprived	323.1	346.1	310.4		400.1	448.5	376.2	
Total	498.5	535.6	471.0	<0.001	575.3	634.3	519.6	<0.001

The frequency of imaging tests was higher in those patients living in the least deprived areas than in those living in the most deprived areas for women and men (Figure 2). In men and women, the frequency of imaging tests increased between the years 2007–2013 and 2014–2019 for the three groups of patients according to their living area and decreased in the years 2020–2021 ($p < 0.001$). In Figure 2, the frequency of radiographies decreased for men and women in 2020–2021 in comparison with 2014–2019 and with 2007–2013 regardless the deprivation area; however, the amount of CT increased in the three periods of time for men and women in the three deprivation groups. In women, the frequency of mammographies decreased in 2020–2021 in comparison with 2014–2019, but it was higher than the frequency in 2007–2013. In men, the frequency of nuclear medicine tests increased in 2020–2021 in comparison with the previous years, mainly in men in the least deprived area; in women, the frequency of nuclear medicine tests decreased in 2020–2021 in comparison with the previous years for women regardless of the deprivation index.

The impact of calendar time in the mean dose (mSv), according to the type of imaging test, patients' age and deprivation index for men and women, is shown in Table 2.

Table 2. Description of the impact of calendar time in the mean dose (mSv) according to the type of imaging test; patients' age and deprivation index for men and women.

Mean Dose (mSv)	Men				Women			
	2007–2013	2014–2019	2020–2021	*p* Value	2007–2013	2014–2019	2020–2021	*p* Value
Imaging test				<0.001				<0.001
Mamography	0	0	0		0.01	0.01	0.01	
Radiography	0.14	0.20	0.17		0.17	0.24	0.20	
CT	0.57	0.71	0.80		0.44	0.56	0.63	
Nuclear medicine	0.32	0.36	0.30		0.37	0.43	0.32	
Age				<0.001				<0.001
<18	0.31	0.16	0.22		0.25	0.21	0.15	
18–64	0.80	1.00	1.05		0.83	1.04	1.05	
>64	2.20	3.18	5.59		1.98	2.80	2.88	
Deprivation index				<0.001				<0.001
Least deprived	0.77	1.02	1.23		0.70	0.97	1.02	
Medium deprived	0.71	0.96	1.07		0.63	0.85	0.91	
Most deprived	0.69	0.82	0.90		0.64	0.78	0.82	
Total	1.04	1.27	1.27	<0.001	0.99	1.24	1.16	<0.001

Men received higher effective doses of radiation than women in the three periods of time, and there was an increase in the effective mean dose received throughout the period of study for men and women.

In men, the mean dose associated with radiographs increased between 2007–2013 and 2014–2019 (0.14 mSv and 0.20 mSv, respectively), and decreased in 2020–2021 (0.17 mSv). This trend was the same for nuclear medicine tests: the frequency increased between 2007–2013 and 2014–2019 (0.32 mSv and 0.36, respectively), and decreased in 2020–2021 (0.30 mSv). In contrast, the mean dose associated with CT increased throughout the period of study ($p < 0.001$). The same trend was found in women throughout the period of study ($p < 0.001$).

Regarding age, the mean dose increased in men > 64 years throughout the study (2007–2013, 2.20 mSv; 2014–2019, 3.18 mSv, and 2020–2021, 5.59 mSv) (percentage of change between 2007–2013 and 2020–2021, 154%%) ($p < 0.001$). The mean dose received in women > 64 years also increased throughout the period of study and the percentage of change between 2007–2013 and 2020–2021 was lower than in men (45%) ($p < 0.001$).

Men living in the least deprived areas received the higher mean dose in comparison with men living in the medium and the most deprived areas. In addition, the mean dose received by men living in the least deprived areas increased 59.7% between 2007–2013 (0.77 mSv) and 2020–2021 (1.23 mSv). In men living in the most deprived areas, the mean dose received increased by 30.4% between 2007–2013 (0.60 mSv) and 2020–2021 (0.90mSv) ($p < 0.001$). Women living in the least deprived areas also showed the higher mean dose in comparison with women living in the medium and the most deprived areas. The mean dose received by women living in the least deprived areas increased by 45.7% between 2007–2013 (0.70 mSv) and 2020–2021 (1.02 mSv). In women living in the most deprived area, the mean dose received increased by 28.1% between 2007–2013 (0.64 mSv) and 2020–2021 (0.82 mSv) ($p < 0.001$).

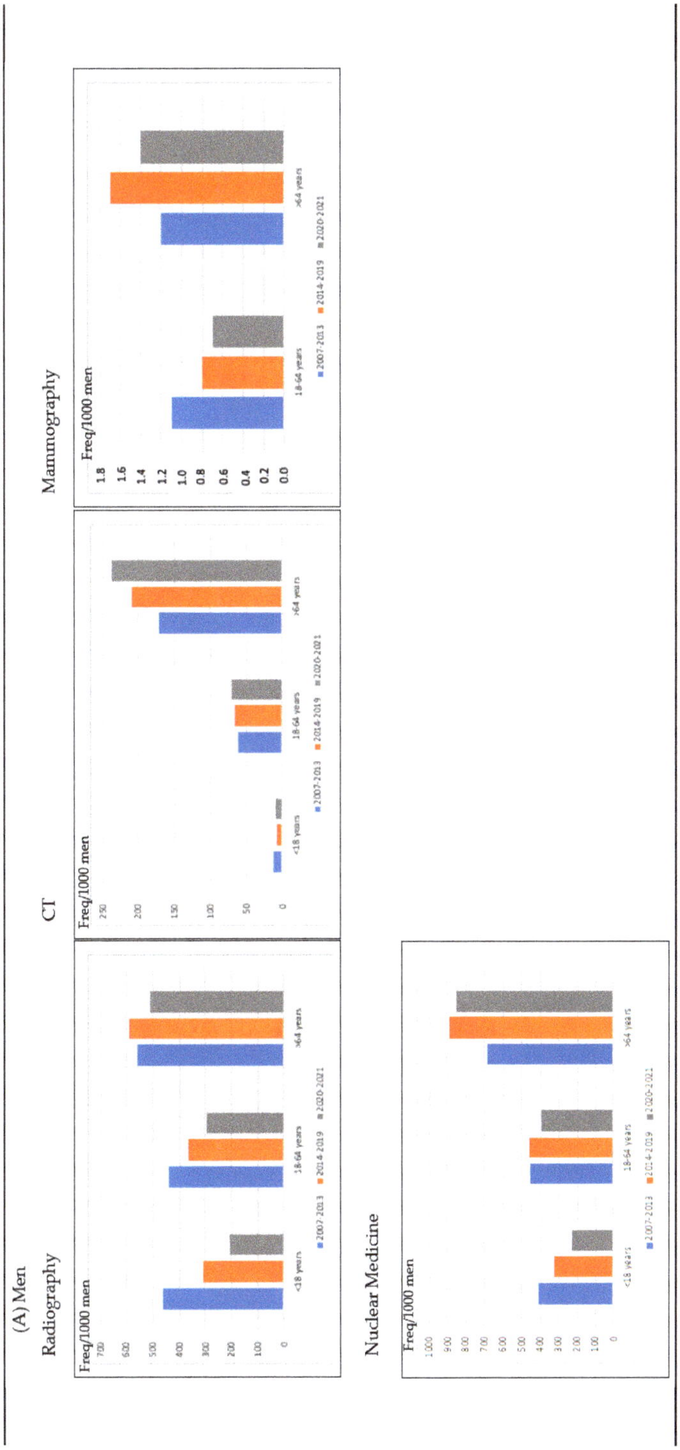

Figure 1. *Cont.*

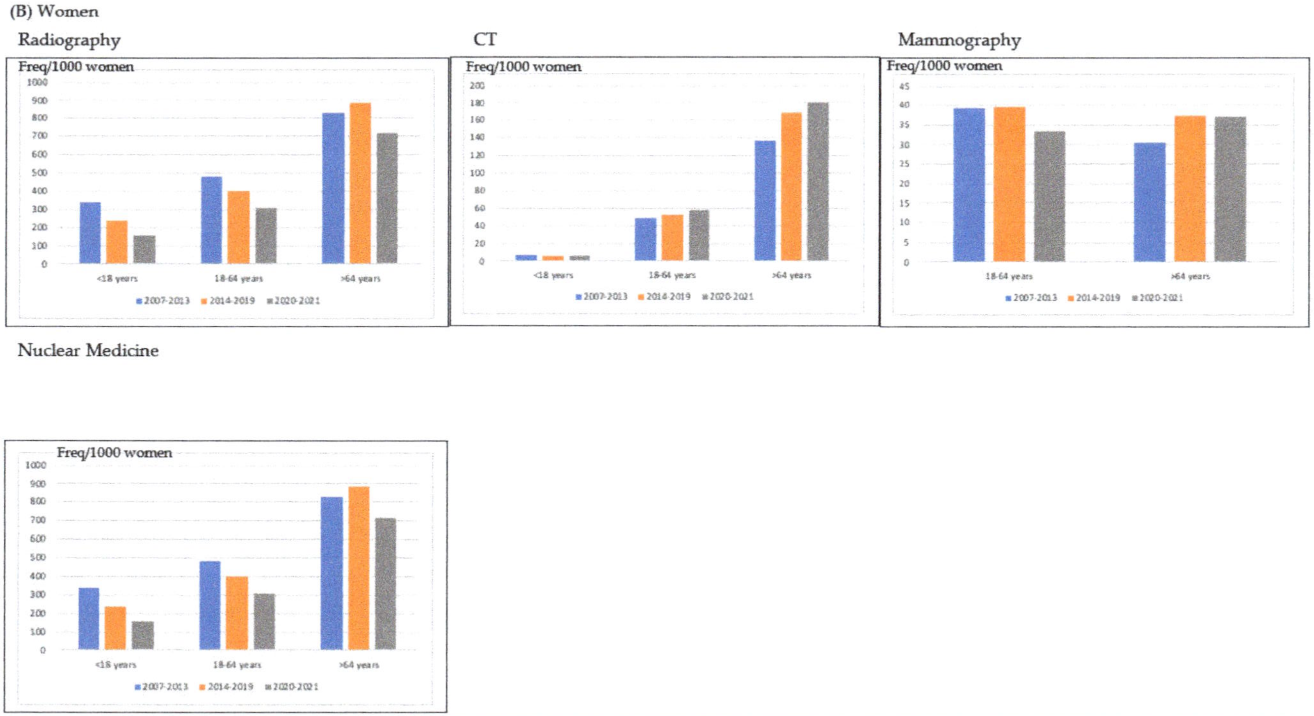

Figure 1. Distribution of the frequency of imaging test per 1000 habitants per men and women according to the patients' age and calendar time.

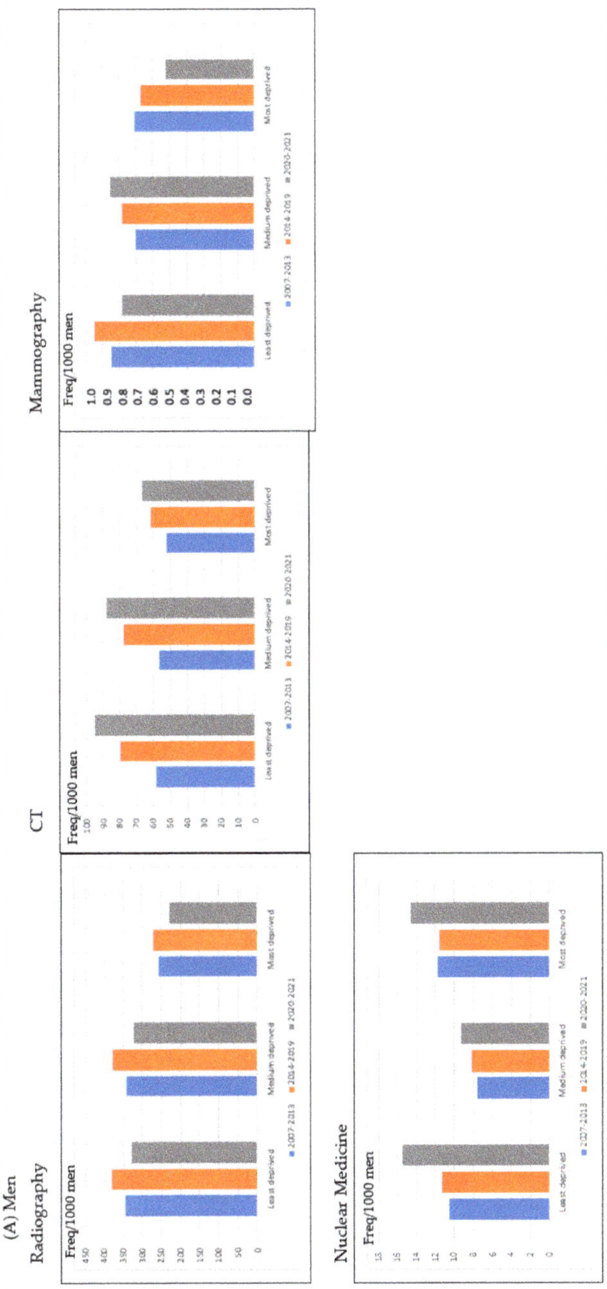

Figure 2. *Cont.*

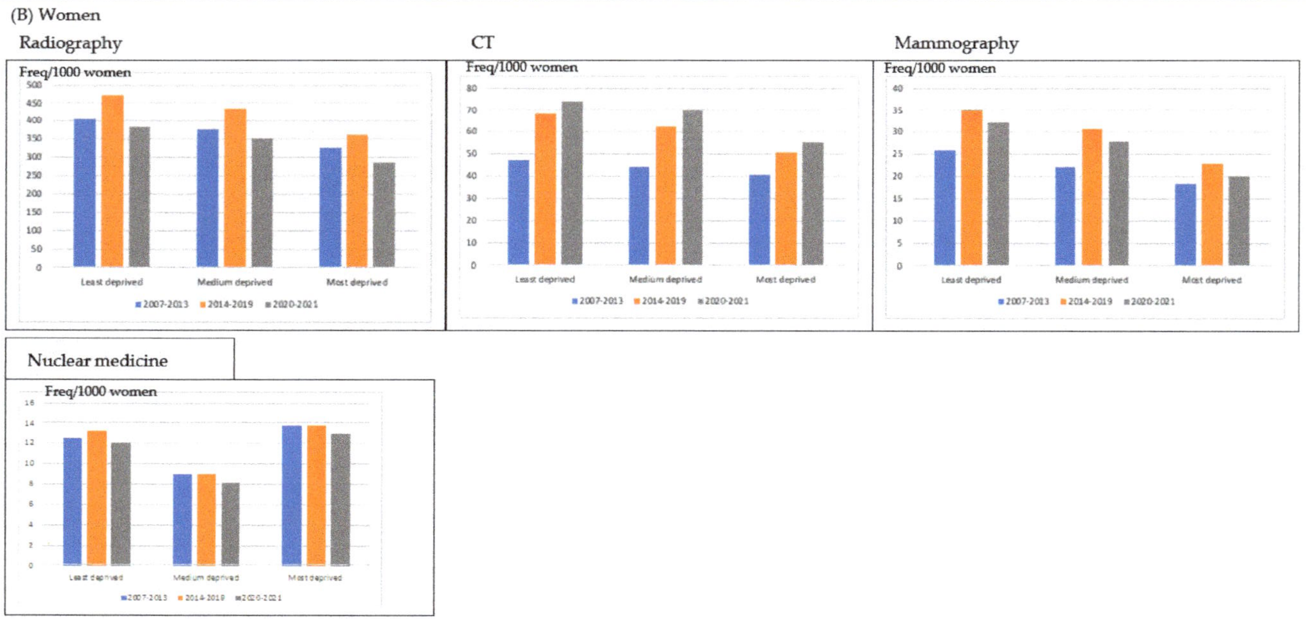

Figure 2. Distribution of the frequency of imaging test per 1000 habitants per men and women according to the patients' deprivation index and calendar time.

4. Discussion

The results provide important information on the imaging diagnostic test trend and the collective effective dose, according to the available recommendations in a university hospital for men and women. We found an increase in the number of imaging tests received by men and women after the publication of the recommendations in 2013, and this increase was higher in women than in men. The frequency of imaging tests, including all imaging modalities, decreased during the pandemic period (2020–2021), but the frequency of CT increased even during these years in men and women, and the frequency of nuclear medicine tests in men and thus, the overall effective mean dose. The population aged > 64 years showed the highest frequency of imaging tests and the frequency increased between 2007–2013 and 2014–2019 for both women and men. Moreover, women and men living in the least deprived areas had a higher frequency of imaging test (and the higher mean dose received) than those living in the most deprived areas.

The publication of several recommendations to decrease the population's exposure to radiation did not have an impact on the imaging tests that were carried out. A previous study on clinicians' awareness of the European recommendations showed that nearly 80% of the clinicians surveyed had never heard of the European recommendations [36]. In addition, they did not take into account the need to consider radiation exposure when ordering imaging tests and the requirement to inform the patient about the risks associated with medical radiation exposure [36]. Women showed a greater increase in both the number of CT scans and conventional radiographies received than men after the publication of the recommendations, especially among patients older than 64 years.

The frequency of nuclear medicine diagnostic procedures increased in 2014–2019 in comparison with 2007–2013 and it also increased in 2020–2021 in men. Nuclear medicine examinations contributed approximately 6% to the total frequency of all diagnostic medical procedures included in this study, leading to a high exposure to radiation. Similar data have been shown in other countries which followed DoseDataMed2 methodology [37], and therefore, it is essential to raise awareness about the potential dose optimization. Although previous studies found a remarkable increase in the number of CTs performed annually [38–40], they also found that CT utilization plateaued due to several reasons such as the acknowledgement of the harms of excess imaging, an increase in the use of other modes of imaging, or the cost [41]. However, we found that the number of CTs increased during the period of study regardless of the publication of recommendations and the pandemic period. During the pandemic in 2020–2021, the lower number of imaging tests reflected the much smaller number of patients treated that year. However, as previous research showed [13], the COVID-19 lung imaging recommendations which covered all radiological modalities, in particular the CT, could reflect the further increase in the number of CTs in 2020–2021. Moreover, the reason for the increase in CTs during the period of study could be that modern helical CT enables faster examination; thus, more examinations can be performed in one day [42]. The decreasing number of conventional radiographies is also probably due to the tendency of replacing it with CT [28]. CTs only represent 13% of all radiological procedures in our setting, but they are the major source of exposure to the population and recent reports have shed light on the increasing frequency of CT. For example, a recent study comprising 2.5 million patients found that patients underwent a median of six CT exams in a year and that some patients received up to 109 exams over five years [43]. Thus, more efforts are needed to increase optimisation and justification according to established recommendations.

The number of CTs carried out was higher in men than in women. In contrast, the number of conventional radiographies was higher in women than in men, mainly due to breast cancer examinations. Nevertheless, the percentage of change between 2007–2013 and 2014–2019 for both conventional radiographies and CTs was higher in women than in men. In addition, the number of nuclear medicine diagnostic procedures was higher in women than in men in 2007–2013 and 2014–2019; however, in the pandemic years the frequency increased in men and decreased in women. This could reflect different clinical management strategies in practice for women and men. A previous study on the clinical management of solitary pulmonary nodule found that women were more likely than men to have a follow-up rather than have an immediate intervention [44]. As a result, accumulative radiation was higher in women than in men. It is necessary to include sex as a variable in future lines of research, as well as in protocols for action in relation to imaging tests and their indications in clinical guidelines since, despite the evidence on the differences between men and women, the incorporation of sex as a variable is not very common at present. Reviews of clinical guidelines have shown that only 35% of them take sex into account as a specific factor for the detection, diagnosis, or management of different diseases in clinical settings [45].

The healthcare system in Spain is public; that is, it is free of charge to anyone living and working in Spain, and the general taxation funds the Spanish state healthcare system. However, patients living in the least deprived areas had a higher frequency of imaging tests and received a higher effective dose than those living in the most deprived areas. On the one hand, if we consider the risk of excessive radiation, this could be interpreted with an equity lens in a positive manner (i.e., lower radiation among the most vulnerable population). However, it would be important to explore the pertinence of the imaging tests solicited and ensure that this reduction in imaging among the most deprived population is not due to a reduction of the pertinent tests due to problems with access, as this could lead to diagnostic delays.

We included a general hospital and its catchment area (with a total population of over 200,000 people). Even though our results could have some limited generalizability in other settings, the population included in this study is similar to the general Spanish population. Moreover, analysing this population provides important insights, showing as far as we know, the first evaluation of the impact of the available recommendations on the frequency of imaging tests carried out for men and women. Limitations of the study need to be included. This study did not consider images carried out in private health and it may hide differences by socioeconomical status. In addition, the deprivation index is a population level type indicator. Its advantage is that it represents a summary measure of the socio-economic characteristics of the population residents in each census section, which allows the study of socio-economic inequalities in health. It can be considered a measure of socio-economic deprivation of the census section, combining information related to individuals (compositional) and to the context. Although this method has the advantage of a similar population size dimension, it is sensitive to variation and it could change over time and given that, it includes 21 different areas. We obtained the data from the Medical Image Bank of the Valencian Community from the Department of Universal Health and Public Health Service and did not distinguish between the slice-spiral CT and the multi-slice CT. However, according to a previous study [46], the average effective dose to patients was only slightly changed from 7.4 mSv at single-slice to 5.5 mSv and 8.1 mSv at dual- and quad-slice scanners, respectively. In our study, we applied an estimation of the average values of the effective dose of imaging tests, and thus, we do not consider that this variation according to the type of CT could have influenced on the results.

5. Conclusions

In conclusion, this is the first study to provide insight into the impact of available recommendations on population exposure due to radiological medical procedures for men and women during a long period of study. By far, the largest increase in the number of imaging tests is due to CTs and nuclear medicine tests, which account for the higher amount of the effective dose. The difference in the increase of imaging tests carried out in men and women and according to the socioeconomic status could reflect different management strategies in clinical practice. Given the low impact of the available recommendations on the population's exposure to radiation, the performance of high-dose procedures, such as CT and nuclear medicine tests, deserve special attention when it comes to justification and optimization, especially in women.

Author Contributions: Conceptualization, B.L. and L.A.P.; methodology, B.L., L.A.P., A.M.-G., E.C.-R. and F.L.; software, B.L., L.A.P. and A.M.-G.; validation, B.L., L.A.P., A.M.-G., E.C.-R. and F.L.; formal analysis, B.L., L.A.P. and A.M.-G.; investigation, B.L., L.A.P., A.M.-G., E.C.-R. and F.L.; resources, B.L.; data curation, B.L. and A.M.-G. writing—original draft preparation, B.L. and L.A.P.; writing—review and editing, B.L., L.A.P., A.M.-G., E.C.-R. and F.L.; visualization, B.L., L.A.P., A.M.-G., E.C.-R. and F.L.; supervision, B.L.; project administration, B.L. All authors have read and agreed to the published version of the manuscript.

Funding: This research did not received funding.

Institutional Review Board Statement: The study was conducted according to the guidelines of the Declaration of Helsinki and approved by the Ethics Committee of Clinical Research Ethics Committee (CEIC) of the Hospital Sant Joan d'Alacant (protocol code 22/037 and date of approval 27 April 2022).

Informed Consent Statement: Patient consent was waived since the data collection was only obtained from the clinical history of patients (with a large sample size), and we considered it unfeasible to obtain the informed consent of patients without large losses of cases and significant selection biases. For this reason, and due to the absence of any significant risk to the patients from their records being accessed, the CEIC approved a waiver of the informed consent requirement. In the research database, patients will be anonymized using dissociated codes, unidentifiable and meaningless to any other information system and will not allow the identification of individual patients or their crossing over with other databases. Since the project database will not contain any data that would allow the identification of patients, no declaration to the Data Protection Agency is required.

Data Availability Statement: The data presented in this study are available on request from the corresponding author.

Acknowledgments: We acknowledge Jessica Gorlin for the English editing of the manuscript.

Conflicts of Interest: The authors declare no conflict of interest.

References

1. Mettler, F.A., Jr.; Bhargavan, M.; Faulkner, K.; Gilley, D.B.; Gray, J.E.; Ibbott, G.S.; Lipoti, J.A.; Mahesh, M.; McCrohan, J.L.; Stabin, M.G.; et al. Radiologic and nuclear medicine studies in the United States and worldwide: Frequency, radiation dose, and comparison with other radiation sources—1950–2007. *Radiology* **2009**, *253*, 520–531. [CrossRef] [PubMed]
2. Skrk, D.; Zontar, D. Estimated collective effective dose to the population from nuclear medicine examinations in Slovenia. *Radiol. Oncol.* **2013**, *47*, 304–310.
3. Huppmann, M.V.; Johnson, W.B.; Javitt, M.C. Radiation risks from exposure to chest computed tomography. *Semin. Ultrasound CT MR* **2010**, *31*, 14–28. [CrossRef]
4. Sodickson, A.; Baeyens, P.F.; Andriole, K.P.; Prevedello, L.M.; Nawfel, R.D.; Hanson, R.; Khorasani, R.; Recurrent, C.T. Cumulative radiation exposure, and associated radiation-induced cancer risks from CT of adults. *Radiology* **2009**, *251*, 175–184. [CrossRef]
5. Lumbreras, B.; González-Alvarez, I.; Gómez-Sáez, N.; Lorente, M.F.; Hernández-Aguado, I. Management of patients with incidental findings in imaging tests: A large prospective single-center study. *Clin. Imaging* **2014**, *38*, 249–254. [CrossRef] [PubMed]
6. IAEA Smart Card/SmartRadTrack Project. Available online: https://rpop.iaea.org/RPOP/RPoP/Content/News/smart-card-project.htm (accessed on 30 October 2022).
7. Council Directive 97/43/Euratom of 30 June 1997 on Health Protection against Dangers Arising from Ionising Radiation in Medical Exposures. Available online: https://www.eumonitor.eu/9353000/1/j9vvik7m1c3gyxp/vitgbgi170zi (accessed on 30 October 2022).
8. Medical Radiation Protection Education and Training (Medrapet Project). Available online: http://www.medrapet.eu/ (accessed on 30 October 2022).
9. European Medical ALARA Network (EMAN). Available online: http://www.eman-network.eu/ (accessed on 30 October 2022).
10. European Commission Project: Guidelines on Medical Physics Expert. Available online: http://ec.europa.eu/energy/nuclear/events/2011_05_09_mpe_workshop_en.htm.2012;366:780-1 (accessed on 30 October 2022).
11. Council Directive 2013/59/Euratom: A revised Basic Safety Standards Directive Was Adopted by the European Union. Available online: https://eur-lex.europa.eu/legal-content/ES/TXT/PDF/?uri=CELEX:32013L0059&from=EN (accessed on 30 October 2022).
12. Real Decreto 601/2019, De 18 De Octubre, Sobre Justificación y Optimización Del Uso De Las Radiaciones Ionizantes Para La Protección Radiológica De Las Personas Con Ocasión De Exposiciones Médicas. Available online: https://www.boe.es/diario_boe/txt.php?id=BOE-A-2019-15604 (accessed on 30 October 2022).
13. Winder, M.; Owczarek, A.J.; Chudek, J.; Pilch-Kowalczyk, J.; Baron, J. Are We Overdoing It? Changes in Diagnostic Imaging Workload during the Years 2010–2020 including the Impact of the SARS-CoV-2 Pandemic. *Healthcare* **2021**, *16*, 1557. [CrossRef] [PubMed]

14. Redondo-Sendino, A.; Guallar-Castillón, P.; Banegas, J.R.; Rodríguez-Artalejo, F. Gender differences in the utilization of health-care services among the older adult population of Spain. *BMC Public Health* **2006**, *6*, 155. [CrossRef]
15. Ruiz-Cantero, M.T.; Blasco-Blasco, M.; Chilet-Rosell, E.; Peiró, A.M. Sesgos de género en el esfuerzo terapéutico: De la investigación a la atención sanitaria. *Farm. Hosp.* **2020**, *44*, 109–113.
16. Vilar-Palop, J.; Hernandez-Aguado, I.; Pastor-Valero, M.; Vilar, J.; González-Alvarez, I.; Lumbreras, B. Appropriate use of medical imaging in two Spanish public hospitals: A cross-sectional analysis. *BMJ Open* **2018**, *8*, e019535. [CrossRef]
17. Gupta, A.; Tsiaras, S.V.; Dunsiger, S.I.; Tilkemeier, P.L. Gender disparity and the appropriateness of myocardial perfusion imaging. *J. Nucl. Cardiol.* **2011**, *18*, 588–594. [CrossRef]
18. Berrington de Gonzalez, A.; Darby, S. Risk of cancer from diagnostic X-rays: Estimates for the UK and 14 other countries. *Lancet* **2004**, *363*, 345–351. [CrossRef] [PubMed]
19. Berrington de González, A.; Mahesh, M.; Kim, K.P.; Bhargavan, M.; Lewis, R.; Mettler, F.; Land, C. Projected cancer risks from computed tomographic scans performed in the United States in 2007. *Arch. Intern. Med.* **2009**, *169*, 2071–2077. [CrossRef] [PubMed]
20. Smith-Bindman, R.; Lipson, J.; Marcus, R.; Kim, K.P.; Mahesh, M.; Gould, R.; Berrington de González, A.; Miglioretti, D.L. Radiation dose associated with common computed tomography examinations and the associated lifetime attributable risk of cancer. *Arch. Intern. Med.* **2009**, *169*, 2078–2086. [CrossRef]
21. Fazel, R.; Krumholz, H.M.; Wang, Y.; Ross, J.S.; Chen, J.; Ting, H.H.; Shah, N.D.; Nasir, K.; Einstein, A.J.; Nallamothu, B.K. Exposure to low-dose ionizing radiation from medical imaging procedures. *N. Engl. J. Med.* **2009**, *361*, 849–857. [CrossRef] [PubMed]
22. Smith-Bindman, R. Environmental causes of breast cancer and radiation from medical imaging: Findings from the Institute of Medicine report. *Arch. Intern. Med.* **2012**, *172*, 1023–1027. [CrossRef] [PubMed]
23. National Research Council of the National Academies. BEIR VII Phase 2. In *Health Risks from Exposure to Low Levels of Ionizing Radiation*; The National Academies Press: Washington, DC, USA, 2006.
24. Einstein, A.J.; Pascual, T.N.; Mercuri, M.; Karthikeyan, G.; Vitola, J.V.; Mahmarian, J.J.; Better, N.; Bouyoucef, S.E.; Hee-Seung Bom, H.; Lele, V.; et al. Current worldwide nuclear cardiology practices and radiation exposure: Results from the 65 country IAEA Nuclear Cardiology Protocols Cross- Sectional Study (INCAPS). *Eur. Heart J.* **2015**, *36*, 1689–1696. [CrossRef] [PubMed]
25. Droomers, M.; Westert, G.P. Do lower socioeconomic groups use more health services, because they suffer from more illnesses? *Eur. J. Pub. Health* **2004**, *14*, 311–313. [CrossRef]
26. Agborsangaya, C.B.; Lau, D.; Lahtinen, M.; Cooke, T.; Johnson, J.A. Multimorbidity prevalence and patterns across socioeconomic determinants: A cross- sectional survey. *BMC Public Health* **2012**, *12*, 201. [CrossRef]
27. Tadiri, C.P.; Gisinger, T.; Kautzky-Willer, A.; Kublickiene, K.; Herrero, M.T.; Norris, C.M.; Raparelli, V.; Pilote, L.; GOING-FWD Consortium. Determinants of perceived health and unmet healthcare needs in universal healthcare systems with high gender equality. *BMC Public Health* **2021**, *31*, 1488. [CrossRef]
28. Jahnen, A.; Järvinen, H.; Olerud, H.; Vassilieva, J.; Vogiatzi, S.; Shannoun, F.; Bly, R. Analysis of factors correlating with medical radiological examination frequencies. *Radiat. Prot. Dosim.* **2015**, *165*, 133–136. [CrossRef]
29. Lumbreras, B.; Salinas, J.M.; Gonzalez-Alvarez, I. Cumulative exposure to ionising radiation from diagnostic imaging tests: A 12-year follow-up population-based analysis in Spain. *BMJ Open* **2019**, *18*, e030905. [CrossRef] [PubMed]
30. Marant-Micallef, C.; Shield, K.D.; Vignat, J.; Cléro, E.; Kesminiene, A.; Hill, C.; Rogel, A.; Vacquier, B.; Bray, F.; Laurier, D.; et al. The risk of cancer attributable to diagnostic medical radiation: Estimation for France in 2015. *Int. J. Cancer* **2019**, *144*, 2954–2963. [CrossRef] [PubMed]
31. Vilar-Palop, J.; Vilar, J.; Hernández-Aguado, I.; González-Álvarez, I.; Lumbreras, B. Updated effective doses in radiology. *J. Radiol. Prot.* **2016**, *36*, 975–990. [CrossRef] [PubMed]
32. Shrimpton, P.C.; Hillier, M.C.; Lewis, M.A.; Dunn, M. National survey of doses from CT in the UK: 2003. *Br. J. Radiol.* **2006**, *79*, 968–980. [CrossRef]
33. Mettler, F.A., Jr.; Huda, W.; Yoshizumi, T.T.; Mahesh, M. Effective doses in radiology and diagnostic nuclear medicine: A catalog. *Radiology* **2008**, *248*, 254–263. [CrossRef]
34. Cohnen, M.; Poll, L.J.; Puettmann, C.; Ewen, K.; Saleh, A.; Mödder, U. Effective doses in standard protocols for multi-slice CT scanning. *Eur. Radiol.* **2003**, *13*, 1148–1153. [CrossRef]
35. Duque, I.; Domínguez-Berjón, M.F.; Cebrecos, A.; Prieto-Salceda, M.D.; Esnaola, S.; Calvo Sánchez, M.; Marí-Dell'Olmo, M. en nombre del Grupo de Determinantes Sociales de la Salud, iniciativa contexto de la Sociedad Española de Epidemiología. Índice de privación en España por sección censal en 2011 [Deprivation index by enumeration district in Spain, 2011]. *Gac. Sanit.* **2021**, *35*, 113–122. [CrossRef]
36. Lumbreras, B.; Vilar, J.; González-Álvarez, I.; Guilabert, M.; Parker, L.A.; Pastor-Valero, M.; Domingo, M.L.; Fernández-Lorente, M.F.; Hernández-Aguado, I. Evaluation of clinicians' knowledge and practices regarding medical radiological exposure: Findings from a mixed-methods investigation (survey and qualitative study). *BMJ Open* **2016**, *6*, e012361. [CrossRef]
37. Kralik, I.; Štefanić, M.; Brkić, H.; Šarić, G.; Težak, S.; Grbac Ivanković, S.; Griotto, N.; Štimac, D.; Rubin, O.; Salha, T.; et al. Estimated collective effective dose to the population from nuclear medicine diagnostic procedures in Croatia: A comparison of 2010 and 2015. *PLoS ONE* **2017**, *12*, e0180057. [CrossRef]

38. Brenner, D.J.; Hall, E.J. Computed tomography—An increasing source of radiation exposure. *N. Engl. J. Med.* **2007**, *357*, 2277–2284. [CrossRef]
39. Stein, E.G.; Haramati, L.B.; Bellin, E.; Ashton, L.; Mitsopoulos, G.; Schoenfeld, A.; Amis, E.S., Jr. Radiation exposure from medical imaging in patients with chronic and recurrent conditions. *J. Am. Coll. Radiol.* **2010**, *7*, 351–359. [CrossRef]
40. Smith-Bindman, R.; Miglioretti, D.L.; Johnson, E.; Lee, C.; Feigelson, H.S.; Flynn, M.; Greenlee, R.T.; Kruger, R.L.; Hornbrook, M.C.; Roblin, D.; et al. Use of diagnostic imaging studies and associated radiation exposure for patients enrolled in large integrated health care systems, 1996–2010. *JAMA* **2012**, *307*, 2400–2409. [CrossRef]
41. Lee, D.W.; Levy, F. The sharp slowdown in growth of medical imaging: An early analysis suggests combination of policies was the cause. *Health Aff.* **2012**, *31*, 1876–1884. [CrossRef] [PubMed]
42. Herts, B.R.; Perl J 2nd Seney, C.; Lieber, M.L.; Davros, W.J.; Baker, M.E. Comparison of examination times between CT scanners: Are the newer scanners faster? *AJR Am. J. Roentgenol.* **1998**, *170*, 13–18. [CrossRef] [PubMed]
43. Rehani, M.M.; Yang, K.; Melick, E.R.; Heil, J.; Šalát, D.; Sensakovic, W.F.; Liu, B. Patients undergoing recurrent CT scans: Assessing the magnitude. *Eur. Radiol.* **2020**, *30*, 1828–1836. [CrossRef] [PubMed]
44. Chilet-Rosell, E.; Parker, L.A.; Hernández-Aguado, I.; Pastor-Valero, M.; Vilar, J.; González-Álvarez, I.; Salinas-Serrano, J.M.; Lorente-Fernández, F.; Domingo, M.L.; Lumbreras, B. Differences in the clinical management of women and men after detection of a solitary pulmonary nodule in clinical practice. *Eur. Radiol.* **2020**, *30*, 4390–4397. [CrossRef] [PubMed]
45. Tannenbaum, C.; Clow, B.; Haworth-Brockman, M.; Voss, P. Sex and gender considerations in Canadian clinical practice guidelines: A systematic review. *CMAJ Open* **2017**, *5*, E66–E73. [CrossRef]
46. Brix, G.; Nagel, H.D.; Stamm, G.; Veit, R.; Lechel, U.; Griebel, J.; Galanski, M. Radiation exposure in multi-slice versus single-slice spiral CT: Results of a nationwide survey. *Eur. Radiol.* **2003**, *8*, 1979–1991. [CrossRef]

life

MDPI

Article

Health Effects of Natural Environmental Radiation during Burning Season in Chiang Mai, Thailand

Narongchai Autsavapromporn [1,*], Chutima Krandrod [2], Pitchayaponne Klunklin [1], Rawiwan Kritsananuwat [3], Churdsak Jaikang [4], Kittikun Kittidachanan [1], Imjai Chitapanarux [1], Somchart Fugkeaw [5], Masahiro Hosoda [2,6] and Shinji Tokonami [2]

[1] Division of Radiation Oncology, Department of Radiology, Faculty of Medicine, Chiang Mai University, Chiang Mai 50200, Thailand; pitchayaponne.kl@cmu.ac.th (P.K.); kittikun.k@cmu.ac.th (K.K.); imjai.chitapanarux@cmu.ac.th (I.C.)
[2] Institute of Radiation Emergency Medicine, Hirosaki University, Hirosaki 036-8564, Japan; kranrodc@hirosaki-u.ac.jp (C.K.); m_hosoda@hirosaki-u.ac.jp (M.H.); tokonami@hirosaki-u.ac.jp (S.T.)
[3] Natural Radiation Survey and Analysis Research Unit, Department of Nuclear Engineering, Faculty of Engineering, Chulalongkorn University, Bangkok 10330, Thailand; rawiwan.kr@chula.ac.th
[4] Toxicology Section, Department of Forensic Medicine, Faculty of Medicine, Chiang Mai University, Chiang Mai 50200, Thailand; churdsak.j@cmu.ac.th
[5] School of Information, Computer, and Communication Technology, Sirindhorn International Institute of Technology, Thammasat University, Pathum Thani 12120, Thailand; somchart@siit.tu.ac.th
[6] Graduate School of Health Science, Hirosaki University, Hirosaki 036-8564, Japan
* Correspondence: narongchai.a@cmu.ac.th

Citation: Autsavapromporn, N.; Krandrod, C.; Klunklin, P.; Kritsananuwat, R.; Jaikang, C.; Kittidachanan, K.; Chitapanarux, I.; Fugkeaw, S.; Hosoda, M.; Tokonami, S. Health Effects of Natural Environmental Radiation during Burning Season in Chiang Mai, Thailand. *Life* **2022**, *12*, 853. https://doi.org/10.3390/life12060853

Academic Editors: Supitcha Chanyotha and Fabrizio Ambrosino

Received: 20 May 2022
Accepted: 7 June 2022
Published: 8 June 2022

Abstract: This paper presents the first measurement of the investigation of the health impacts of indoor radon exposure and external dose from terrestrial radiation in Chiang Mai province during the dry season burning between 2018 and 2020. Indoor radon activity concentrations were carried out using a total of 220 RADUET detectors in 45 dwellings of Chiang Mai (7 districts) during burning and non-burning seasons. Results show that indoor radon activity concentration during the burning season (63 ± 33 Bq/m^3) was significantly higher ($p < 0.001$) compared to the non-burning season (46 ± 19 Bq/m^3), with an average annual value of 55 ± 28 Bq/m^3. All values of indoor radon activity concentration were greater than the national (16 Bq/m^3) and worldwide (39 Bq/m^3) average values. In addition, the external dose from terrestrial radiation was measured using a car-borne survey during the burning season in 2018. The average absorbed rate in the air was 66 nGy/h, which is higher than the worldwide average value of 59 nGy/h. This might be due to the high activity concentrations of ^{238}U and ^{323}Th in the study area. With regards to the health risk assessment, the effective dose due to indoor radon exposure, external (outdoor) effective dose, and total annual effective dose were 1.6, 0.08, and 1.68 mSv/y, respectively. The total annual effective dose is higher than the worldwide average of 1.15 mSv/y. The excess lifetime cancer risk and radon-induced lung cancer risk during the burning season were 0.67% and 28.44 per million persons per year, respectively. Our results substantiate that indoor radon and natural radioactive elements in the air during the burning season are important contributors to the development of lung cancer.

Keywords: lung cancer; natural environmental radiation; indoor radon; external dose; burning season

1. Introduction

According to International Agency for Research on Cancer (IARC), lung cancer (LC) is one of the leading causes of cancer mortality among both men and women worldwide [1,2]. In Thailand, LC is the second cause of incidence and mortality particularly in Upper Northern Thailand (UNT) [3,4]. Chiang Mai is the largest city in UNT and LC is one of the most common cancers for both genders as reported by World Health Organization (WHO) [5]. Multiple risk factors can cause LC in Chiang Mai, such as cigarette smoking, air pollution, and natural background radiation (e.g., radon and gamma) [3,4,6]. Cigarette smoking is the main cause of LC development, while radon (^{222}Rn), the most stable isotope of radon element is identified as the second leading cause of LC and the major risk factor among non-smokers [7–9].

Radon (and its progeny) is the major contributor (more than 50%) of natural environmental radiation on the surface of the earth reported by WHO [7,9] and have been classified as a human carcinogen (group 1) that can cause LC by IARC [8]. It is a radioactive gas (half-life of 3.82 days), invisible, odorless, and colorless. It naturally occurs as a decay product of radium-226 (^{226}Ra) and is ultimately a member of the uranium-238 (^{238}U) series, found in the soil, rocks, groundwater, and air [7,9]. Approximately, 8–33% of all LC deaths worldwide are likely caused by indoor radon exposure [7,10–12]. Therefore, chronic exposure to radon and its decay products can induce DNA damage through chromosome alterations and double-strand breaks (DSBs), which subsequently increase the risk of LC [11,13]. Moreover, it indicates that radon and their decay products may exist in air pollutants including particulate matter (PM) with a diameter of less than 10 μm (PM_{10}), smoke haze, and small dust particles, and all these elements together lead to LC development [14].

Lately, Chiang Mai has been annually facing adverse health impacts of airborne PM including LC and respiratory diseases during the dry season burning for over 20 years. This is because farmers burn biowaste materials from agricultural land and forest fires [15]. The highest levels of PM are seen between November and April every year and the peak tends to occur around the middle of March. Our previous study [16] indicates that the annual average indoor radon activity concentration (57 Bq/m^3) in Chiang Mai is considered to be higher than the worldwide average (39 Bq/m^3) and national average (16 Bq/m^3) values [7,12,17]. An indoor and outdoor-radon activity concentration during the burning season (Mid-March) were 5.5 and 4-fold higher than the worldwide average, respectively. Therefore, it is important to elucidate the long-term measurements of indoor radon levels, particularly during the dry season burning. This paper provides the first attempt that investigates the indoor radon activity concentration and external dose from terrestrial radiation conducted between 2018 to 2020, particularly during the burning season in the Chiang Mai province. Additionally, we assessed the health risk for the potential impact of human health outcomes based on natural environmental radiation.

2. Materials and Methods

2.1. Study Area and Selection of Measurement Locations

Chiang Mai is the second-largest city in Thailand and the largest city in UNT. There are divided into 25 districts with a population of approximately 1.19 million residents which represents 6% of the total population in Thailand (Figure 1a). It is located on the Mae Ping River and surrounded by mountains in particular granitic rock (high background radiation area), such as Daen Lao and Thanon Thong Chai. Chiang Mai has lower humidity and a tropical climate characterized by three seasons: the winter (November–February), summer (March–May), and rainy (June–October).

This research was carried out in seven districts (high radon potential zone) located in different areas in Chiang Mai (Figure 1b). This area is affected by a high number of LC patients from 2009 to 2018 (Figure 1c). Between 2016 to 2018, indoor radon activity concentration measurements were carried out in a total of 172 randomly selected dwellings (1–5 dwellings in each subdistrict randomly depending on the district size). The districts surveyed are Mueang, Hang Dong, Saraphi, and San Pa Tong. In addition, 45 randomly selected dwellings (Mueang, Hang Dong, Saraphi, San Pa Tong, San Sai, San Kamphaeng, and Doi Saket) were selected for the study of indoor radon during the dry season burning in the period between 2018 to 2020. Most of the selected dwellings in the study area were built of cement and wood along with concrete floors.

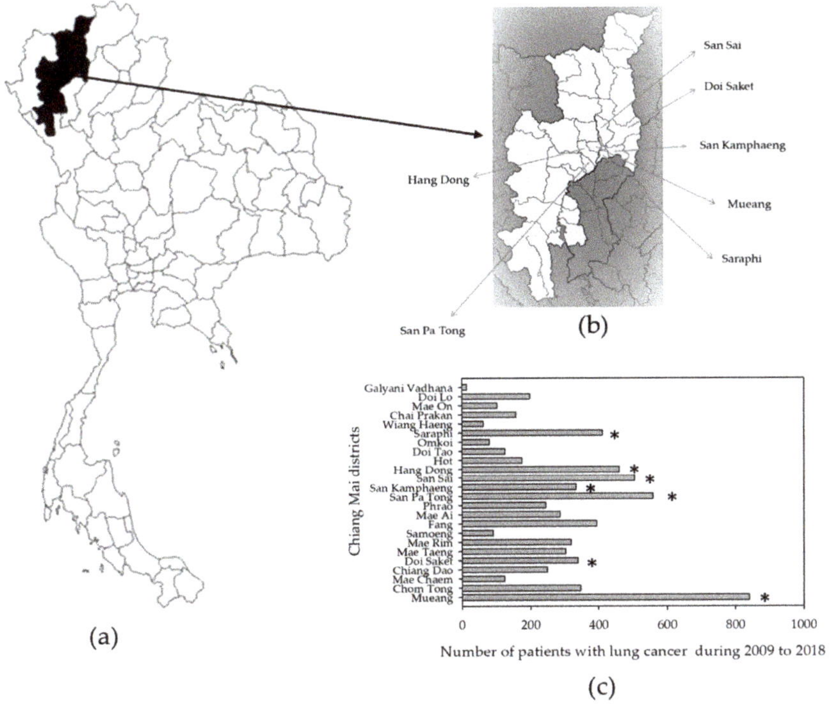

Figure 1. Location of measurement sites in Thailand. (**a**) Map of Chiang Mai, Thailand. (**b**) Map of Chiang Mai showing the study areas. (**c**) Number of lung cancer patients in Chiang Mai (*: study area).

The study encompassed fieldwork and data collection from participants by interviews (questionnaire concerning information about dwelling characteristics, family histories of LC, and lifestyle). All participants were informed of the study information about indoor radon measurements, risks, or benefits that may occur from the study. Informed consent was obtained from all participants prior to the enrollment.

2.2. Radon Activity Concentration Measurement

A passive type of radon-thoron discriminative monitor (RADUET) using an α track type radon detector (CR-39) was used to measure indoor radon in the bedroom (ground floor) of selected dwellings (172 RADUET detectors) for a period of six months between 2016 and 2018 [16]. The RADUET detectors were placed away from sunlight, windows, doors, and electric devices, at a distance of 20 cm from the internal wall and a height of 100 to 200 cm from the floor as representative of human breathing inside the bedroom. At the end of the measurement, all RADUET detectors were collected, wrapped in a plastic bag, shipped, and measured at the Institute of Radiation Emergency Medicine, Hirosaki University. Briefly, CR-39 was chemically etched using a solution of 6.25 M NaOH at 90 °C for 6 h, then washed with distilled water and dried. Afterward, α particles in CR-39 were taken by digital camera and counted with an automatic reading system to evaluate the indoor radon activity concentration.

The radon activity concentration (C) is calculated using Equation (1):

$$C = \frac{\rho}{kt} \tag{1}$$

where ρ is α particles track density corrected by background track density(track/cm^2), k is the conversion factor from α particles track density to indoor radon activity concentration [(tracks/cm^2/h)/(Bq/m^3)], t is exposure time (h). It should be noted that the contribution of thoron and its progeny in this study is relatively small compared to radon and should be negligible (data not shown) [16].

In order to study the effects of biomass burning on indoor radon levels (total of 220 RADUET detectors), the experiments were performed in two sets of 45 random dwellings for periods of 12 months (replaced with a new RADUET detector every 2–3 months to cover the burning season (November–April) and non-burning season (May–October) in Chiang Mai); the first period in 2018–2019 (20 dwellings) and the second period in 2019–2020 (25 dwellings).

2.3. *Car-Borne Survey*

A car-borne survey technique is an effective method to evaluate the external radiation dose from terrestrial gamma radiation [uranium (^{238}U) series, thorium (^{232}Th) series, and potassium (^{40}K)] in the Saraphi district which is subdivided into twelve subdistricts for a short period during the burning season from March 16–17 and 19 in 2018 using a 3-in × 3-in NaI(Tl) scintillation spectrometer (EMF-211, EMF Japan Co., Osaka, Japan) [18]. We selected the Saraphi district as one of the target areas because it is well documented that this area has a higher number of LC patients and is one of the most polluted areas in Chiang Mai [3]. The detector was installed inside a car at 1 m from the ground level and gamma radiation counting was carried out every 30 s in a moving car along the survey route with a global positioning system (the latitude and longitude for each measurement point). During the survey (a total of 821 measurement points), the car was moving at a speed of around 30–40 km/h depending on the road conditions, and the shielding factor of the car body was calculated at 18 measurement points by measurements outside and inside of the car. The methodology of the car-borne survey, the calculation of dose rates in air, and the activity concentration of natural radionuclides (^{40}K, ^{238}U, and ^{232}Th) to absorbed dose rate in the air was followed as previously described by Hosoda et al. [19,20].

2.4. *Health Risk Assessment*

2.4.1. The Annual Effective Dose (H) of Inhalation Dose

H is the total exposure of indoor radon activity concentration (and its progeny) on residents in the study area (in a year), which corresponds to the average indoor radon calculated using Equation (2), based on the United Nations Scientific Committee on the Effects of Atomic Radiation (UNSCEAR) report [21]:

$$\mathrm{H\ (mSv/y) = C \times F \times O \times T \times D} \quad (2)$$

where C is an annual average indoor radon activity concentration in the dwellings (in Bq/m^3), F is the equilibrium factor between indoor radon (and its progeny) (0.4), O is the occupancy factor for the residential population (0.8), T is an average exposure period (24 h × 365 days = 8760 h), and D is an inhalation dose conversion factor (9 × 10^{-6} mSv/h per Bq/m^3) [16].

2.4.2. The Annual Effective Dose to Lungs (H_L)

H_L was calculated using Equation (3):

$$\mathrm{H_L\ (mSv/y) = H \times W_R \times W_T} \quad (3)$$

where W_T (the radiation-weighting factor) is 20 for α particles and W_T (the tissue weighting factor for lungs) is 0.12, as recommended by the Internal Commission on Radiological Protection (ICRP) [22].

2.4.3. The External (Outdoor) Annual Effective Dose (H_e)

H_e was estimated using Equation (4) based on the measured absorbed dose rate in air in the Saraphi district.

$$H_e\ (\text{mSv/y}) = D_a \times DCF \times T \times O \times 10^{-3} \quad (4)$$

where D_a is an average outdoor absorbed dose rate in air (nGy/h), DCF is dose conversion factor received by an adult (0.7 Sv/Gy), T is 8760 h and O is the occupancy factor for the residential population (0.2) [21,23].

2.4.4. Excess Lifetime Cancer Risk (ELCR)

ELCR was estimated using Equation (5):

$$ELCR = H \times DL \times RF \quad (5)$$

where DL is the mean of life estimated to 77 years in Thailand and RF is the risk of fatal cancer per Sievert (0.055 Sv^{-1}), as reported by ICRP [24,25].

2.4.5. The Number of LC Cases per Year per Million (LCC)

LCC was given according to Equation (6):

$$LCC = H \times RFLC \quad (6)$$

where RFLC is the risk factor for LC induction per million per person of 18×10^{-6} mSv^{-1} y as recommended by ICRP [26,27].

2.5. Statistical Analysis

The software Sigma Plot10 (Sigma, St. Louis, MO, USA) and Microsoft Excel were used to conduct all statistical analyses in this study. All data presented were determined based on the mean ± standard deviation (SD), median, and geometric. The Wilcoxon signed rake test was performed to test for the mean difference between two groups of data and a *p*-value of 0.05 between groups was considered to be significant.

3. Results and Discussion

3.1. Indoor Radon Activity Concentration and Health Risk Assessment Due to Indoor Radon Exposure

The indoor radon activity concentration and health risk assessment (H,H_L, ELCR, and LCC) in a total of 172 dwellings in four districts of Chiang Mai province (Mueang; Hang Dong; Saraphi, and San Pa Tong) were measured between 2016 and 2018 as shown in Table 1. The indoor radon activity concentration is found to be varied from 23 (Saraphi) to 209 (San Pa Tong) Bq/m^3 (an average value of 48 ± 20 Bq/m^3) with a geometric mean of 45 Bq/m^3. The highest maximum value of indoor radon in the San Pa Tong district was 209 Bq/m^3 with approximately two times higher than the reference level (100 Bq/m^3) imposed by WHO but this value is below the ICRP reference level of 300 Bq/m^3 [7,26]. This might be due to the difference in radioactive elements in the soil (geological condition), dwelling characteristics, and ventilation condition. The average value of indoor radon in Chiang Mai is higher compared to the worldwide average of 39 Bq/m^3 and national average values of 16 Bq/m^3 [7,12,17]. About 64% of dwellings in the study area exceeded the worldwide average value of indoor radon. Our finding is in agreement with the previous study (Table 2) reported that Chiang Mai has a higher radon activity concentration than in average obtained by national and worldwide.

Table 1. Indoor radon activity concentration and the potential risk of lung cancer to residents of Chiang Mai, Thailand.

	Mueang	Hang Dong	Saraphi	San Pa Tong	Total
No. of dwellings	48	25	83	16	172
Max (Bq/m^3)	99	65	100	209	209
Min (Bq/m^3)	24	33	23	31	23
Mean ± SD (Bq/m^3)	47 ± 17	43 ± 10	49 ± 16	50 ± 43	48 ± 20
Geometric mean (Bq/m^3)	45	42	46	43	45
No. of Dwellings >39 Bq/m^3	28	15	60	7	110 (64%)
No. of Dwellings >100 Bq/m^3	0	0	0	1	1 (0.6%)
H (mSv/y)	1.2	1.1	1.2	1.3	1.2
H_L (mSv/y)	2.9	2.6	3.0	3.0	2.9
ELCR (%)	0.51	0.47	0.52	0.53	0.51
LCC ($\times 10^{-6}$)	21.6	19.8	22.1	22.7	21.6

Abbreviations: SD, standard deviation; H, annual effective dose; H_L, annual effective dose to lungs; ELCR, excess lifetime cancer risk; LCC, the number of LC cases per year per million.

Table 2. Comparison of the average indoor radon activity concentration in Chiang Mai.

Study Areas (Districts)	No. of Detectors/Period	Indoor Radon (Bq/m^3)	Ref.
Saraphi	50/99 days	21	[28]
Mueang, Hang Dong, Saraphi and San Pa Tong	110/1 year	57	[16]
Doi Saket	30/4 months	53	[29]
Not available	46/3 months	110	[10]
Mueang, Hang Dong, Saraphi and San Pa Tong	172/6 months	48	This study

As the risk of individual LC development increases with duration and exposure to indoor radon, it is very pivotal to estimate the effect on human health from long-term exposure to indoor radon exposure [6,11]. Table 1 shows the estimated health risk assessment of indoor radon exposure to residents in the study area. The total values of H were calculated between 0.6 to 5.3 mSv/y (data not shown) with an average of 1.2 mSv/y. The average H is less than the action level limit of 3–10 mSv/y, as recommended by ICRP [30]. However, the average H is found to be higher than the worldwide average of 1.15 mSv/y (inhalation dose), while the average H_L is 2.9 mSv/y [21]. This value is higher than the worldwide average due to the stressful effects of the α particle on the lungs [6]. To consider the risk of LC due to indoor radon exposure, ELCR is used to predict the probability of cancer development by residential radon over a lifetime. The average ELCR for indoor radon exposure in the study area was 0.51%, which is lower than the action level of 1.3%, and due to indoor radon levels of 148 Bq/m^3 as recommended by the United States Environmental Protection Agency (USEPA) [31]. However, this value is higher than the worldwide average of 0.145%, which may be related to the high radiation area in Chiang Mai [10,32]. Therefore, LCC average value in Chiang Mai caused by radon exposure was estimated to be 21.6 per million people per year. This value is lower than the limit range between 170 to 230 per million persons per year as reported by ICRP [30]. Based on the estimated values, our data show that the impact of health risk for LC development received by residents in the study area is related to chronic exposure to indoor radon. To this end, our future work will focus on the investigation of long-term indoor radon measurements with a larger sample size.

3.2. Indoor Radon Activity Concentration and Health Risk Assessment during Burning- and Non-Burning Seasons

Long-term exposure to natural environmental radiation and outdoor air pollution may be associated with an increased risk of LC development [33]. Lately, Chiang Mai is facing the highest air pollution in the world, caused by the open burning of biomass during the harvest season. Despite biomass burning being important; this condition causes Chiang Mai to have a high level of radon in UNT and Thailand [10]. To our knowledge, there is little understanding of the relationship between indoor radon exposure and air pollution during the burning season in Chiang Mai, which can affect LC development and other diseases [7]. In this study, indoor activity concentration and health risk assessment were recorded in 45 dwellings in seven districts in the Chiang Mai province (Mueang (n = 6), Hang Dong (n = 7), Saraphi (n = 21), San Pa Tong (n = 5), San Sai (n = 2), San Kamphaeng (n = 2) and Doi Saket (n = 2)) during burning- and non-burning seasons between May 2018 and October 2020 are reported in Table 3.

Table 3. Indoor radon activity concentration and potential risk of lung cancer during burning and non-burning seasons to Chiang Mai residents during 2018–2020.

	Burning Season	Non-Burning Season	Total
No. of dwellings	45	45	45
Period (months)	6	6	12
Max (Bq/m^3)	230	139	230
Min (Bq/m^3)	28	19	19
Mean $\pm$ SD (Bq/m^3)	63 $\pm$ 33	46 $\pm$ 19	55 $\pm$ 28
Median (IQR) (Bq/m^3)	61 (35)	42 (16)	48 (27)
Geomean (Bq/m^3)	57	44	50
No. of Dwellings >39 Bq/m^3	34	24	19 (42.2%)
No. of Dwellings >100 Bq/m^3	2	1	1 (2.2%)
H (mSv/y)	1.6	1.2	1.4 $\pm$ 0.3
H_L (mSv/y)	3.8	2.8	3.3 $\pm$ 0.7
ELCR (%)	0.67	0.5	0.58 $\pm$ 0.12
LCC ($\times 10^{-6}$)	28.44	21	24.72 $\pm$ 5.26

Abbreviations: SD, standard deviation; IQR, interquartile range; H, annual effective dose; H_L, annual effective dose to lungs; ELCR, excess lifetime cancer risk; LCC, the number of LC cases per year per million.

As can be seen in Table 3, the annual indoor radon activity concentration measurement ranged from 19 to 230 Bq/m^3 with an average value of 55 $\pm$ 28 Bq/m^3 (geometric mean is found to be 50 Bq/m^3). The highest indoor radon activity concentration was observed in the San Pa Tong district (data not shown). Overall, 42.2% of dwellings presented indoor radon activity concentration comparatively higher than the world average value and 2.2% had a value above 100 Bq/m^3. This clearly suggests that the annual average value of indoor radon in Chiang Mai is greater than the national and worldwide average value. Interestingly, a comparison result (Table 3 and Figure 2), shows a significant statistical difference ($p < 0.001$) between indoor radon activity concentration during burning-and non-burning seasons. The average indoor radon level during the burning season (63 $\pm$ 33 Bq/m^3 with a geomean of 57 Bq/m^3) was found to be higher than those measured in the non-burning season (46 $\pm$ 19 Bq/m^3 with a geomean of 44 Bq/m^3). The difference in the radon level during biomass burning season may be due to high levels of natural background radiation in air pollution, high levels of radioactive elements in the soil, climatic parameters (such as high concentrations of radon in the winter burning season), home ventilation and building materials [34–36].

In estimating human health risk due to indoor radon exposure (and its progeny) during burning- and non-burning seasons (Table 3), H values in the study area during burning-and non-burning seasons were found to be 1.6 and 1.2, respectively, (with varies from 0.5 to 5.8 mSv/y, data not shown), with an average of 1.4 mSv/y. The estimated average H during burning- and non-burning seasons and average annual H are higher when compared with a worldwide average value of 1.15 mSv/y [21]. The calculated H_L due to indoor radon exposure during burning-and non-burning seasons were 3.8 and 2.8, respectively, with an average of 3.3 mSv/y. These results are higher than the action level by ICRP [30]. The ECLR attributable to residential radon during burning-and non-burning seasons were 0.67 and 0.5%, respectively, with an average value of 0.58%. All estimated values in the present data were lower than the action level reported by USEPA [31]. However, these values are higher than average worldwide [32]. The radon-induced LC during burning-and non-burning seasons were 28.44 and 21 per million people per year, respectively. While the LCC average of 24.84 per million people per year is lower than the range recommended by ICRP [30]. All together, these findings also show a significant difference ($p < 0.05$, data not shown) in all human health risk assessments on residential radon exposure (H, H_L, ECLR, and LCC) between burning-and non-burning seasons. Therefore, this comparison indicates the potential risk of natural background radiation during the burning season to human health; hence it can be a major public health problem for the UNT of Thailand due to chronic exposures to radon (and its progeny) along their lifetime.

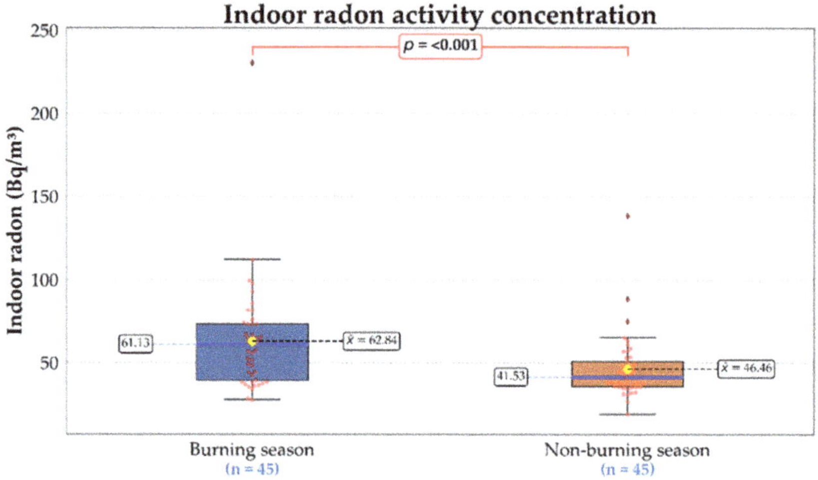

Figure 2. Variation of indoor radon activity concentration during burning and non-burning seasons in seven districts of Chiang Mai during 2018–2020.

To the best of our knowledge, our study is the first attempt in dealing with long-term indoor radon measurements within a human health risk assessment in the Chiang Mai province during the burning season. However, the lack of measurement of natural radioactivity concentrations such as ^{40}K, ^{232}Th, and ^{238}U present in dust particles in the air, and outdoor radon activity concentration in the ambient air during burning- and non-burning seasons is the limitation in this study. Additional research is needed to obtain more detailed results.

3.3. *External Radiation Dose and External Annual Effective Dose Estimation from Natural Environmental Radiation during burning Season*

For a further understanding of the health effects of natural environmental radiation during the burning season in Chiang Mai, an external dose of terrestrial radiation for residents living in the Saraphi area was obtained by car-borne measurement during the peak of the dry season burning in 2018 (16, 17 and 19 March). The survey route consists of twelve subdistricts of Saraphi and variations of external radiation dose in the air (outdoor absorbed gamma dose rates) are shown in Figure 3c. The shielding factor (Figure 3a) and dose conversion factor (Figure 3b) were determined as 2.44 and 0.0018 nGy/h, respectively.

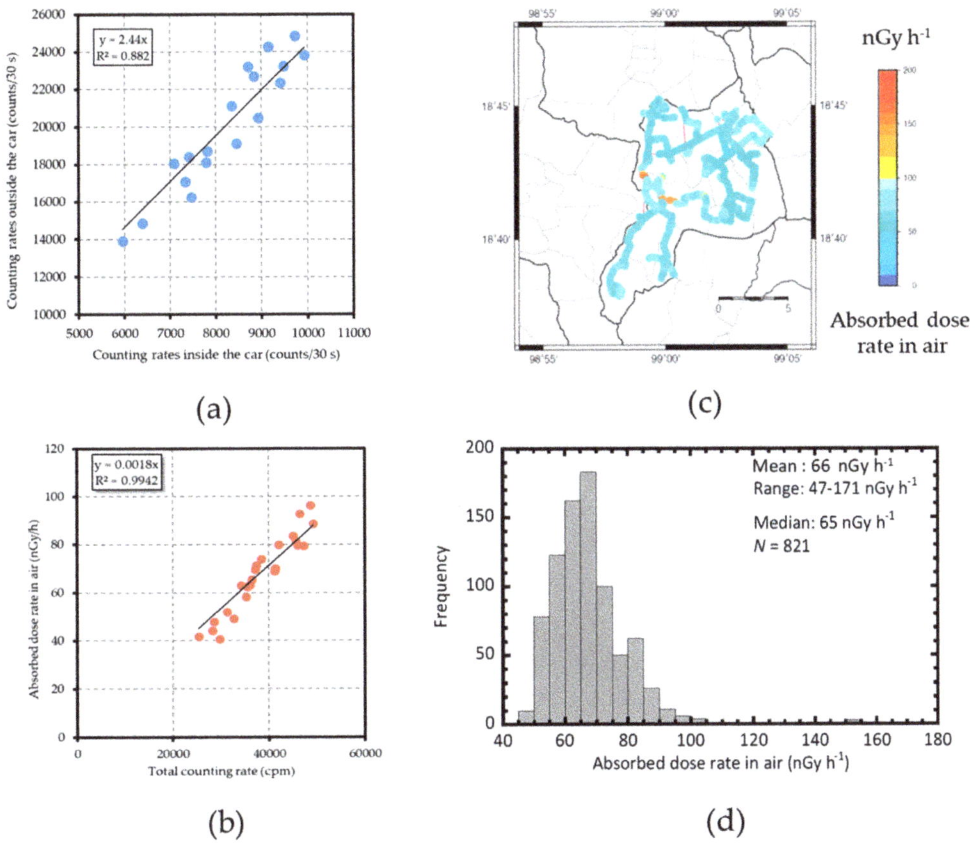

Figure 3. Map of outdoor gamma dose rates in the air in Saraphi district during the dry season burning measured in 2018. (**a**) Correlation between count rates outside and inside the car. (**b**) Correlation between air karma and total count rate outside the car. (**c**) Map showing the survey route and distribution of air kerma rate. (**d**) Absorbed dose rate in air in Saraphi district.

A total of 821 measurement points were collected in the study areas. Figure 3c shows the outdoor gamma dose rates range from 47 to 171 nGy/h with an average value of 66 nGy/h (median value of 65 nGy/h). This average value was found to be higher than the world average value of 59 nGy/h as reported by UNSCEAR [21]. This study shows that a high outdoor gamma dose rate in air is an important contribution to external radiation dose in Chiang Mai during the dry season burning. In addition, the average gamma dose rate in air in Chiang Mai, UNT was higher than in other parts of Thailand Western (44 nGy/h), Eastern (35 nGy/h), and Southern (42 nGy/h), which may be related to high activity radioactivity area on the UNT of Thailand [37].

Furthermore, airborne gamma-ray spectrometry measurement was carried out to determine the radionuclides activity concentration contributing to natural environmental radiation during the burning season. Table 4 represents the activity concentration of ^{40}K, ^{238}U, and ^{232}Th in soil, and the absorbed dose rate in the air 1 m above the ground was measured at 24 points in the Saraphi district. The contribution to absorbed dose rate in the air of ^{40}K, ^{238}U, and ^{232}Th ranged from 13% to 27%, 30% to 41%, and 34% to 48% with an average value of 22%, 36%, and 42%, respectively (data not shown). As displayed in Table 4, the average activity concentration of ^{40}K, ^{238}U, and ^{232}Th were 346 ± 90, 56 ± 16 and 43 ± 14 Bq/kg, respectively. Based on this result, the activity concentration of ^{238}U and ^{232}Th were higher than the worldwide average values of 35 and 30 Bq/kg, respectively [21].

Table 4. The measured activity concentration of ^{40}K, ^{238}U, and ^{232}Th and absorbed dose rate in air in Saraphi district during the dry season burning.

Point No.	Location		Activity Concentration (Bq/kg)			Absorbed Dose Rate in Air (nGy/h)
	Latitude (°)	Longitude (°)	^{40}K	^{238}U	^{232}Th	
1	18.6659	98.9742	234	30	21	42
2	18.633	98.9637	306	43	30	52
3	18.6514	98.9698	372	42	37	59
4	18.686	98.9954	404	44	40	63
5	18.6521	99.0001	298	35	28	48
6	18.6754	99.0043	398	46	46	69
7	18.6969	98.9956	437	69	38	70
8	18.7048	98.9969	416	70	51	80
9	18.6894	99.0134	439	65	51	80
10	18.6932	99.0274	451	76	62	89
11	18.6862	99.0573	330	46	36	63
12	18.6821	99.0387	147	30	26	41
13	18.7054	99.0384	413	77	68	93
14	18.7169	99.0263	173	34	30	44
15	18.72	99.0633	297	88	72	96
16	18.6986	99.0591	294	69	61	81
17	18.745	99.0405	338	59	37	66
18	18.7416	99.0568	396	65	38	74
19	18.7367	99.0661	335	72	54	84
20	18.7203	99.0136	173	49	29	49
21	18.739	99.0214	409	59	44	71
22	18.7288	98.9943	438	66	55	80
23	18.7499	98.9997	427	54	41	70
24	18.7158	99.004	376	47	37	63
	Average		346 ± 90	56 ± 16	43 ± 14	68 ± 16

These findings suggest that ^{238}U and ^{232} Th-series elements are the main sources of external natural radiation exposure during the burning season in the study area. High activity concentrations of ^{238} U and ^{232} Th can be explained by the geometrical environment (such as granites) and the mechanisms of soil information [19,20,38,39]. However, it should be noted that 38% of measurement points (n = 9) have an activity concentration of ^{40}K above the worldwide average value of 400 Bq/kg and were found to be a minor contribution to the total absorbed dose rate in the air [21]. Further investigation of radionuclides activity concentration in this study area is needed to confirm these results.

With regards to the health risk assessment, the mean, minimum and maximum values of external (outdoor) annual effective dose (H_e) were estimated to be 0.08, 0.06, and 0.21 mSv/y, respectively (data not shown). The average H_e value of this study area was higher than the worldwide average value of 0.07 mSv/y, as reported by UNSCEAR [21]. Therefore, the outcome of radiation dose assessment indicates the relevant effects of natural environmental radiation during the dry burning season on human health.

4. Conclusions

We have presented our first study that provides an understanding of the impacts of biomass burning on natural environmental radiation in the Chiang Mai province, particularly indoor radon exposure and external dose from terrestrial radiation. The findings show indoor radon activity concentration (63 ± 33 Bq/m^3) and external dose from terrestrial radiation (66 nGy/h) during the burning season was higher than the national and worldwide average value. The activity concentration of ^{238}U and ^{232}Th found in the soil of area studies during the burning season was higher than the worldwide average value. The estimated value of effective dose due to exposure to indoor radon (and its progeny), external (outdoor) effective dose, and the total annual effective dose received by Chiang Mai residents were 1.6, 0.08, and 1.68 mSv/y, respectively. The total annual effective dose is higher than the worldwide representative value of 1.15 mSv/y. The excess lifetime cancer risk was found to be 0.67%, which is higher than the worldwide average. The radon-induced LC risk during the burning season presents a value of 28.44 per million persons per year. With all results obtained from the fieldwork, indoor radon (and its progeny) and terrestrial radiation represent the major contributions of human exposure to natural radiation during the dry season burning and may increase the possibility of LC developing in their lifetime. Future research related to natural environmental radiation and air pollution during the burning season is required to confirm these findings.

Author Contributions: Conceptualization, N.A.; methodology, N.A.; software, K.K. and S.F.; validation, N.A., C.K. and R.K.; formal analysis, N.A.; investigation, N.A., P.K., R.K., M.H. and S.T.; data curation, N.A., M.H. and S.F.; writing—original draft preparation, N.A.; writing—review and editing, N.A., C.K., P.K., R.K., C.J., K.K., I.C., S.F., M.H. and S.T.; visualization, N.A.; project administration, N.A.; funding acquisition, N.A. and S.T. All authors have read and agreed to the published version of the manuscript.

Funding: This research was funded by the National Research Council of Thailand, and Thailand Science Research and Innovation (grant number RSA6280045), Faculty of Medicine, Chiang Mai University (grant number MT01/2562), the International Atomic Energy Agency (research contract number 21062) and Environmental Research Projects from The Sumitomo Foundation (grant number 173017).

Institutional Review Board Statement: The study was conducted in accordance with the Declaration of Helsinki, and approved by The Research Ethics Committee in the Faculty of Medicine, Chiang Mai University [Research ID: 2559-04011(4 October 2016), 2559-04252 (26 January 2017) and 2562-06213 (28 June 2019)] for studies involving humans.

Informed Consent Statement: Informed consent was obtained from all participants.

Data Availability Statement: Not applicable.

Acknowledgments: We gratefully acknowledge the contribution of Naofumi Akata, Takahito Suzuki, Hiromi Kudo, Chanis Pornnumpa Rattanapongs, Thamaborn Ploykrathok, Tengku Ahbrizal Farizal Tengku Ahmad for all their support. We would also like to thank the residents of the study area.

Conflicts of Interest: The authors declare no conflict of interest.

References

1. IARC Working Group on the Evaluation of Carcinogenic Risks to Humans; World Health Organization; International Agency for Research on Cancer. *Tobacco Smoke and Involuntary Smoking*; IARC: Lyon, France, 2004; Volume 83, pp. 1–1438.
2. Sung, H.; Ferlay, J.; Siegel, R.L.; Laversanne, M.; Soerjomataram, I.; Jemal, A.; Bray, F. Global cancer statistic 2020: GLO-BOCAN estimates of incidence and mortality worldwide for 36 cancers in 185 countries. *CA Cancer J. Clin.* **2021**, *71*, 209–249. [CrossRef] [PubMed]
3. Wiwatanadate, P. Lung cancer related to environmental and occupational hazards and epidemiology in Chiang Mai, Thailand. *Genes Environ.* **2011**, *33*, 120–127. [CrossRef]
4. Rankantha, A.; Chitapanarux, I.; Pongnikorn, D.; Prasitwattanaseree, S.; Bunyatisai, W.; Sripan, P.; Traisathit, P. Risk patterns of lung cancer mortality in Northern Thailand. *BMC Public Health* **2018**, *18*, 1138. [CrossRef] [PubMed]
5. Thailand-Global Cancer Observatory. Available online: https://gco.iarc.fr/today/data/factsheets/populations/764-thailand-fact-sheets.pdf (accessed on 20 September 2021).
6. Autsavapromporn, N.; Klunklin, P.; Chitapanarux, I.; Jaikang, C.; Chewaskulyong, B.; Sripan, P.; Hosoda, M.; Tokonami, S. A potential serum biomarker for screening lung cancer risk in high level environmental radon areas: A pilot study. *Life* **2021**, *11*, 1273. [CrossRef]
7. World Health Organization (WHO). *Handbook on Indoor Radon*; WHO: Geneva, Switzerland, 2009.
8. International Agency for Research on Cancer (IARC); World Health Organization (WHO). *Man-made Mineral Fibres and Radon*; World Health Organization (WHO): Geneva, Switzerland, 1988; Volume 43, pp. 1–300.
9. National Research Council. *Biological Effects of Ionizing Radiation (BEIR) VI Report: The Health Effects of Exposure to Radon*; National Academy Press: Washington, DC, USA, 1999.

10. Somsunun, K.; Prapamontol, T.; Pothirat, C.; Liwsrisakun, C.; Pomhnikorn, D.; Fongmoon, D.; Chantata, S.; Wongpoomchai, R.; Naksen, W.; Autsavapromporn, N.; et al. Estimation of lung cancer deaths attributable to indoor radon exposure in Upper Northern Thailand. *Sci. Rep.* **2022**, *12*, 5169. [CrossRef]
11. Al-Zoughool, M.; Krewski, D. Health effects of radon: A review of the literature. *Int. J. Radiat. Biol.* **2009**, *85*, 57–69. [CrossRef]
12. Gaskin, J.; Coyle, D.; Whyte, J.; Krewksi, D. Global estimate of lung cancer mortality attributable to residential radon. *Environ. Health Perspect.* **2018**, *126*, 057009. [CrossRef]
13. Autsavapromporn, N.; de Toledo, S.M.; Little, J.B.; Jay-Gerin, J.P.; Harris, A.L.; Azzam, E.I. The role of gap junction communication and oxidative stress in the propagation of toxic effects among high-dose-α particle-irradiated human cells. *Radiat. Res.* **2011**, *175*, 347–357. [CrossRef]
14. Tokonami, S. Experimental verification of the attachment theory of radon progeny onto ambient aerosols. *Health Phys.* **2000**, *78*, 74–79. [CrossRef]
15. Phariuang, W.; Suwattiga, P.; Chetiyanukornkul, T.; Hongtieab, S.; Limpaseni, W.; Ikemori, F.; Hata, M.; Furuuchi, M. The influence of the open burning of agricultural biomass and forest fires in Thailand on the carbonaceous components in size-fractionated particles. *Environ. Pollut.* **2019**, *247*, 238–247. [CrossRef]
16. Autsavapromporn, N.; Klunklin, P.; Threeatana, C.; Tuntiwechapikul, W.; Hosada, M.; Tokonami, S. Short telomere length as a biomarker risk of lung cancer development induced by high radon levels: A pilot study. *Int. J. Environ. Res. Public Health* **2018**, *15*, 2152. [CrossRef]
17. International Atomic Energy Agency (IAEA). *National and Regional Surveys of Radon Concentration in Dwelling*; IAEA: Viena, Austria, 2014.
18. Tanigaki, M.; Okumura, R.; Takamiya, K.; Sato, N.; Yoshino, H.; Yamana, H. Development of a car-borne γ-ray survey system, KURAMA. *Nucl. Instrum. Methods Phys. Res. A* **2013**, *726*, 162–168. [CrossRef]
19. Hosoda, M.; Tokonami, S.; Omori, Y.; Sahoo, S.K.; Akiba, S.; Sorimachi, A.; Ishikawa, T.; Nair, R.R.; Jayalekshmi, P.A.; Sebastian, P.; et al. Estimation of external dose by car-borne survey in Kerala, India. *PLoS ONE* **2015**, *10*, e0124433. [CrossRef] [PubMed]
20. Hosoda, M.; Nugraha, E.D.; Akata, N.; Yamada, R.; Tamakuma, Y.; Sasaki, M.; Kelleher, K.; Yoshinaga, S.; Suzuki, T.; Rattanapongs, C.P.; et al. A unique high natural background radiation area—dose assessment and perspectives. *Sci. Total Environ.* **2021**, *750*, 142346. [CrossRef] [PubMed]
21. United Nations Scientific Committee on the Effect of Atomic Radiation (UNSCEAR). *Sources and Effects on Ionizing Radiation*; United Nation: New York, NY, USA, 2000.
22. International Commission on Radiological Protection (ICRP). 1990 Recommendations of the International Commission on Radiological Protection. ICRP 60. *Ann. ICRP* **1991**, *21*, 1–3.
23. United Nations Scientific Committee on the Effect of Atonic Radiation (UNSCEAR). *UNSCEAR 2013 Report to the General Assembly with Scientific Annexes*; United Nation: New York, NY, USA, 2014.
24. International Commission on Radiological Protection (ICRP). *The 2007 Recommendations of the International Commission on Radiological Protection*; ICRP 103; Elsevier: Amsterdam, The Netherlands, 2007.
25. Darwish, D.A.E.; Abul-Nasr, K.T.M.; El-Khayatt, A.M. The assessment of natural radioactivity and its associated radiological hazards and dose parameters in granite samples from South Sinai, Egypt. *J. Radiat. Res. Appl. Sci.* **2015**, *8*, 17–25. [CrossRef]
26. International Commission on Radiological Protection (ICRP). Lung cancer risk from exposure to radon daughters. ICRP 50. *Ann. ICRP* **1987**, *17*, 1–57.
27. International Commission on Radiological Protection (ICRP). Lung cancer risk from radon and progeny and statement on radon. ICRP 115. *Ann. ICRP* **2010**, *40*, 1–64. [CrossRef]
28. Wanabongse, P.; Tokonami, S.; Bovornkitti, S. Current studies on radon gas in Thailand. *Int. Congr. Ser.* **2005**, *1276*, 208–209. [CrossRef]
29. Thumvijit, T.; Chanyotha, S.; Sriburee, S.; Hongsriti, P.; Tapanya, M.; Krandrod, C.; Tokonami, S. Identifying indoor radon sources in Pa Miang, Chiang Mai, Thailand. *Sci. Rep.* **2020**, *10*, 17723. [CrossRef]
30. International Commission on Radiological Protection (ICRP). Protection against radon-222 at home and at work. ICRP 65. *Ann. ICRP* **1993**, *23*, 1–45.
31. United States Environmental Protection Agency (USEPA). *EPA Assessment of Risks from Radon in Homes*; USEPA: Washington, DC, USA, 2003.
32. Qureshi, A.A.; Tariq, S.; Din, K.U.; Manzoor, S.; Calligaris, C.; Waheed, A. Evaluation of excessive lifetime cancer risk due to natural radioactivity in the rivers sediments of Northern Pakistan. *J. Radiat. Res. Appl. Sci.* **2014**, *7*, 438–447. [CrossRef]
33. Nyhan, M.M.; Rice, M.; Blomberg, A.; Coull, B.A.; Garshick, E.; Vokonas, P.; Schwartz, J.; Gold, D.R.; Koutrakis, P. Associations between ambient particle radioactivity and lung function. *Environ. Int.* **2019**, *130*, 104795. [CrossRef] [PubMed]
34. Bochicchio, F.; Campos-Venuti, G.; Piermattei, S.; Nuccetelli, C.; Risica, S.; Tommasino, L.; Torri, G.; Magnoni, M.; Agnesod, G.; Sgorbati, G.; et al. Annual average and seasonal variations of residential radon concentration for all the Italian Regions. *Radiat. Meas.* **2005**, *40*, 686–694. [CrossRef]
35. Stabile, L.; Dell'Isola, M.; Frattolillo, A.; Massimo, A.; Russi, A. Effect of natural ventilation and manual airing on indoor air quality in naturally ventilated Italian classrooms. *Build. Environ.* **2016**, *98*, 180–189. [CrossRef]
36. Tchorz-Trzeciakiewicz, D.E.; Rysiukiewicz, M. Ambient gamma dose rate as an indicator of geogenic radon potential. *Sci. Total Environ.* **2021**, *755*, 142771. [CrossRef] [PubMed]

37. Kranrod, C.; Chanyotha, S.; Pengvanich, P.; Kritsananuwat, R.; Ploykrathok, T.; Sriploy, P.; Hosoda, M.; Tokonami, S. Car-borne survey of natural background gamma radiation in Western, Eastern and Southern Thailand. *Radiat. Prot. Dosim.* **2019**, *188*, 174–180. [CrossRef]
38. Suman, G.; Reddy, V.K.K.; Reddy, M.S.; Reddy, C.G.; Reddy, P.Y. Radon and thoron levels in the dwellings of buddonithanda: A village in the environs of proposed uranium mining site, Nalgonda district, Telangana state, India. *Sci. Rep.* **2021**, *11*, 6199. [CrossRef] [PubMed]
39. Ambrosino, F.; Stellato, L.; Sabbarese, C. A case study on possible radiological contamination in the Lo Uttaro landfill site (Caserta, Italy). *J. Phys. Conf. Ser.* **2020**, *1548*, 012001. [CrossRef]

life

MDPI

Article

Ingestional Toxicity of Radiation-Dependent Metabolites of the Host Plant for the Pale Grass Blue Butterfly: A Mechanism of Field Effects of Radioactive Pollution in Fukushima

Akari Morita [1], Ko Sakauchi [1], Wataru Taira [2] and Joji M. Otaki [1,*]

[1] The BCPH Unit of Molecular Physiology, Department of Chemistry, Biology and Marine Science, Faculty of Science, University of the Ryukyus, Okinawa 903-0213, Japan; e183422@eve.u-ryukyu.ac.jp (A.M.); yamatoshijimi@sm1044.skr.u-ryukyu.ac.jp (K.S.)

[2] Research Planning Office, University of the Ryukyus, Okinawa 903-0213, Japan; wataira@lab.u-ryukyu.ac.jp

* Correspondence: otaki@sci.u-ryukyu.ac.jp; Tel.: +81-98-895-8557

Abstract: Biological effects of the Fukushima nuclear accident have been reported in various organisms, including the pale grass blue butterfly *Zizeeria maha* and its host plant *Oxalis corniculata*. This plant upregulates various secondary metabolites in response to low-dose radiation exposure, which may contribute to the high mortality and abnormality rates of the butterfly in Fukushima. However, this field effect hypothesis has not been experimentally tested. Here, using an artificial diet for larvae, we examined the ingestional toxicity of three radiation-dependent plant metabolites annotated in a previous metabolomic study: lauric acid (a saturated fatty acid), alfuzosin (an adrenergic receptor antagonist), and ikarugamycin (an antibiotic likely from endophytic bacteria). Ingestion of lauric acid or alfuzosin caused a significant decrease in the pupation, eclosion (survival), and normality rates, indicating toxicity of these compounds. Lauric acid made the egg-larval days significantly longer, indicating larval growth retardation. In contrast, ikarugamycin caused a significant increase in the pupation and eclosion rates, probably due to the protection of the diet from fungi and bacteria. These results suggest that at least some of the radiation-dependent plant metabolites, such as lauric acid, contribute to the deleterious effects of radioactive pollution on the butterfly in Fukushima, providing experimental evidence for the field effect hypothesis.

Keywords: radioactive pollution; Fukushima nuclear accident; lauric acid; alfuzosin; ikarugamycin; plant secondary metabolite; artificial diet; *Zizeeria maha*; *Oxalis corniculata*; low-dose exposure

Citation: Morita, A.; Sakauchi, K.; Taira, W.; Otaki, J.M. Ingestional Toxicity of Radiation-Dependent Metabolites of the Host Plant for the Pale Grass Blue Butterfly: A Mechanism of Field Effects of Radioactive Pollution in Fukushima. *Life* **2022**, *12*, 615. https://doi.org/10.3390/life12050615

Academic Editors: Fabrizio Ambrosino and Supitcha Chanyotha

Received: 30 March 2022
Accepted: 19 April 2022
Published: 20 April 2022

1. Introduction

Anthropogenic impacts on wild organisms have been an important scientific and political issue worldwide in this century. Human activities often involve local and global scale pollution of air, water, soil, and ocean, leading to climate changes and human health disorders [1,2]. For example, anthropogenic radionuclides from nuclear bomb tests and nuclear power plant accidents can be found worldwide [3–7]. It is thus important to understand precisely how severely wild organisms are affected by human activities and in what ways. To this end, butterflies have often been used as ecological indicators because of their advantages over other organisms [8,9]. For example, (1) butterflies are often conspicuous and abundant in the field and easy to identify at the species level, (2) a wealth of information on life history is available, (3) rich museum specimens are often available, and (4) many amateur lepidopterists may join field studies covering a wide geographic range. These advantages of using butterflies are invaluable for field studies. Not surprisingly, changes in butterfly species in abundance, range, phenology, and diversity have been used as key factors to understand recent environmental influences in many studies [10–16]. Occasionally, studies have focused on a single or a few indicator species [17–23]. An advantage of a single-species approach is to couple field surveys and laboratory experiments to understand what occurs in the field.

The pale grass blue butterfly, *Zizeeria maha*, has been used as a field indicator and laboratory model species to understand evolutionary and developmental plasticity in response to environmental changes [17,24–28]. In this butterfly, an environmentally induced plastic phenotype was genetically assimilated in the laboratory experiment and in the field, which was probably one of the best pieces of evidence for genetic assimilation [29,30]. Just after the establishment of the pale grass blue butterfly as a laboratory model species that can also be used as a field indicator species, the Fukushima nuclear accident occurred in March 2011. Anthropogenic radioactive materials from the Fukushima Dai-ichi Nuclear Power Plant (FDNPP) heavily polluted the east side of Tohoku district in Japan. The northern range margin of the pale grass blue butterfly was located 380 km away from the FDNPP [17], and the polluted area in Fukushima is completely covered by the distribution range of the pale grass blue butterfly. Without question, the pale grass blue butterfly was the logical choice for studying the biological effects of the Fukushima nuclear accident.

The Fukushima nuclear accident was reported to have caused various biological and ecological effects on animals, such as birds [31–33], insects [34–39], Japanese monkeys [40–42], and intertidal invertebrates [43], plants such as rice [44,45], fir trees [46], red pine trees [47], and the creeping wood sorrel *Oxalis corniculata* [48–50], and soil microbes [51]. A series of our studies [31–34,52–63] demonstrated that the pale grass blue butterfly has been severely affected by the Fukushima nuclear accident. One of the pieces of important evidence was provided by the internal exposure experiment, in which the contaminated host plant leaves collected from Fukushima were given to larvae from Okinawa (where radioactive contamination is minimal), resulting in high mortality and abnormality rates. However, when an artificial diet containing pure radioactive cesium (^{137}Cs) was given to larvae, no change in the survival rate was observed [64]. A similar discrepancy between field and laboratory results has been observed in the case of the Chernobyl nuclear accident [65,66]. This field-laboratory paradox was explained by the field effect hypothesis: the host plant in the field responded to low-level radiation stress by upregulating metabolites that were toxic to larvae as a part of plant defense mechanisms [59,67]. Subsequent studies have reported upregulated and downregulated metabolites and nutrients in plant leaves [48–50], supporting this field effect hypothesis.

In a previous metabolomic study [49], the creeping wood sorrel in Okinawa was irradiated by contaminated soil collected from Fukushima, and the leaf samples (the edible part for larvae) were subjected to GC–MS (gas chromatography–mass spectrometry) and LC–MS (liquid chromatography–mass spectrometry) analyses. Under the acute low-dose radiation conditions, 5.7 mGy (34 μGy/h for seven days), many peaks were significantly upregulated, although most of them were annotated as multiple compounds or not annotated at all. One of the upregulated peaks was singularly annotated as lauric acid by targeted GC–MS analysis, and two of the upregulated peaks were singularly annotated as alfuzosin and ikarugamycin by LC–MS analysis. Therefore, the potential toxic effects of these three compounds are of great interest.

Lauric acid is a saturated fatty acid, also called dodecanoic acid, that can be found widely in plants. Lauric acid shows a wide variety of bioactivities as a plant defense, volatile against *Staphylococcus* [68,69], *Mycobacterium tuberculosis* [70], fungus [71], and *Phytophthora sojae*, an agriculturally important plant pathogen that belongs to Protista [72]. Notably, extracts from *Vitex* species containing lauric acid have larvicidal activity against a mosquito species, *Culex quinquefasciatus* [73]. Lauric acid is likely sensed at least by an insect, *Holotrichia parallela* [74], as an odorant. Accordingly, it is reasonable to hypothesize a larvicidal activity of lauric acid in *O. corniculata* against larvae of the pale grass blue butterfly.

Alfuzosin is a synthetic α_1-adrenergic receptor antagonist used widely for the treatment of benign prostatic hyperplasia [75–78]. Because it is synthetic, it is unlikely to be present naturally in the plant. However, because the LC–MS peak annotated as alfuzosin has very similar (virtually identical) elution time and exact mass with alfuzosin, examination of MS/MS (mass spectrometry/mass spectrometry, i.e., tandem mass spectrometry) spectrograms was required to differentiate alfuzosin and this unknown plant metabolite (called the alfuzosin-related compound hereafter) [49]. Faced with the fact that the exact identity of the alfuzosin-related compound cannot be determined easily, we tested the toxicity of alfuzosin itself in this study, assuming that alfuzosin and its related metabolite may have similar biological effects on larvae of the pale grass blue butterfly.

Ikarugamycin is an antiprotozoal agent isolated originally from the soil bacterium *Streptomyces phaeochromogenes* var. *ikaruganensis* [79]. Importantly, ikarugamycin has also been detected from an endophytic actinomycete, *Streptomyces harbinensis*, from soybean root [80]. Endophytic bacteria have been widely observed in plants [81–86], including *O. corniculata* [87,88]. Accordingly, ikarugamycin detected from leaves of *O. corniculata* is likely from an endophytic *Streptomyces* sp. in leaves or roots, which responded to low-level radiation [49,50]. Ikarugamycin and its derivatives are antifungal [89] and antibacterial [89,90] agents and inhibit clathrin-mediated endocytosis in eukaryotic cell lines [91].

These three upregulated metabolites have been hypothesized to function as toxicants for larvae of the pale grass blue butterfly under low-level radiation stress. In other words, these compounds are candidate causal substances for the ecological field effects of low-level radiation pollution in Fukushima. In this study, we tested the above hypothesis by investigating the ingestional toxicity of these compounds using a novel artificial diet that has a reduced leaf content of *O. corniculata*.

To detect their potential toxicity, we examined three aspects of development: metamorphosis rates, developmental periods, and adult wing size. The metamorphosis rates were used to detect the number of surviving individuals after metamorphosis and included the following: the pupation rate, the eclosion (survival) rate, and the normality rate. The developmental periods were used to detect developmental retardation or acceleration and included the following: the egg-larval period, the pupal period, and the immature period. Adult wing size included both male and female forewing sizes. In this way, we examined high mortality and abnormalities, growth retardation, and smaller forewing size, which have been detected as biological effects of the Fukushima nuclear accident in previous studies [34–37].

2. Materials and Methods

2.1. Egg Collection and Larval Rearing

Egg collection and rearing were performed according to Hiyama et al. (2010) [24] and other related publications [34,52,56] with some minor modifications, as described briefly below. Adults of the pale grass blue butterfly *Z. maha* and its host plant, the creeping wood sorrel *O. corniculata*, were collected at the University of the Ryukyus and its vicinity. The whole plant was placed in a pot and set in a glass tank (300 mm × 300 mm × 300 mm) in which approximately three female butterflies and a few male butterflies were confined at a time. A single trial of egg collection was performed for a period of four days. All rearing processes were executed under the conditions where light was automatically turned on from 6:00 a.m. to 10:00 p.m. (L16:D8) and room temperature was set at 27 °C.

After eggs were deposited on the leaves of the host plant, the plant pot was removed from the glass tank and covered with a plastic bag. When the leaves were eaten enough by newly hatched larvae, they were transferred to a transparent plastic container (150 mm × 150 mm × 55 mm), to which a new bunch of the host plant leaves was supplied every day. Larvae were reared with fresh leaves for 14 days from the beginning of egg collection. Larvae were then randomly divided into different treatment groups: a fresh leaf group, an artificial diet group with no test additive (0 mg/g), and a few artificial diet groups with different concentrations of a test additive (0.01 mg/g, 0.1 mg/g, and 1 mg/g). One group was reared in one container that housed 15–25 larvae, depending on the total number of larvae that were obtained simultaneously from a single egg collection trial. The larvae obtained from a single trial were all siblings; they constitute a sibling group. In this way, genetic bias was minimized. The artificial diet was given as four small square lumps (10 mm × 8 mm × 3 mm per lump) at four corners in a container.

We cleaned containers and changed old lumps of the artificial diet for fresh ones every day, but unexpected deaths that were apparently unrelated to the toxicity of the test additives could not be entirely avoided. This is probably partly because we use fresh leaves collected from the field without sterilization because the sterilization process may breakdown or vaporize some ingredients in fresh leaves important for larvae to initiate eating behavior (such as oxalic acid as an eating initiator [92]). Pupation and eclosion were checked every morning, roughly from 8:30 am to 1:00 pm, so that the data on the developmental periods (days) could be obtained later. We set a criterion that the eclosion (survival) rate of the sibling group of the artificial diet without a test additive should be more than 45% to be considered a successful rearing trial. Sibling groups with eclosion rates below this criterion were considered technical failures and excluded from subsequent analyses. This threshold was set as the gap based on our rearing experience. After pupation, pupae inside the container were transferred to small petri dishes individually. Soon after eclosion, adult butterflies were frozen until subsequent analyses.

2.2. *Artificial Diet Preparation*

Larvae require leaves (or some plant chemicals) as a component of an artificial diet to eat, but the leaf content in an artificial diet should be minimized to test toxicological effects of a metabolite in leaves themselves. Furthermore, the process of collecting fresh leaves is the most laborious and time-consuming process for preparing an artificial diet. To meet these demands, we developed a novel artificial diet for the pale grass blue butterfly for the feeding experiments in this study. We used a commercially available diet, Silk Mate L4M (Nosan Corporation, Yokohama, Kanagawa, Japan), for rearing silkworm. This diet was supplied as powder containing defatted soybean, starch, sugar, cellulose, formative agent, citric acid, mulberry leaf powder, vitamins, minerals, preservative, and antibiotics (diet additives), according to the manufacturer's specification. For the current study, fresh leaves of the creeping wood sorrel, Silk Mate L4M, and deionized water were mixed at a weight ratio of 1:3:5. Thus, this new diet was named AD-FSW-135.

A possibility that Silk Mate L4M contains lauric acid, the alfuzosin-related compound, and ikarugamycin from mulberry leaves was not considered in this study based on the following reasons, although it cannot be excluded completely. First, ikarugamycin and alfuzosin (and thus the alfuzosin-related compound) have never been reported from mulberry leaves to the best of our knowledge. Besides, their concentrations in the creeping wood sorrel were low (see below). Second, lauric acid is known to be contained in mulberry leaves [93,94]. However, because lauric acid is volatile, it may be minimized during an autoclave sterilization process of Silk Mate L4M. The basal levels of the three compounds that were carried from the fresh leaves in AD-FSW-135 (prepared in this study) were shown to be low (see Section 4).

2.3. *Lauric Acid, Alfuzosin, and Ikarugamycin*

Lauric acid (catalog No. 042-23281, Wako Special Grade; FUJIFILM Wako Chemicals, Tokyo, Japan), alfuzosin (catalog No. PHR1638, Pharmaceutical Secondary Standard, Certified Reference Material; Sigma–Aldrich, St. Louis, MO, USA), and ikarugamycin (catalog No. 15386; Cayman Chemical, Ann Arbor, MI, USA) were purchased. They were in powder form and were added to the diet preparation directly. They were then mixed well manually or using an electric mixer until the diet preparation was visually judged to be homogeneous. In this way, we assumed that the test additives were incorporated evenly into the diet and became solubilized. However, they might not have been solubilized completely at higher concentrations (see Section 4). Concentrations of these test additives were expressed in milligrams per gram of artificial diet (mg/g) throughout this paper. For lauric acid, we tested 0 mg/g (control), 0.01 mg/g, 0.1 mg/g, and 1 mg/g. This range covered the estimated concentration of lauric acid in leaves (see below). For alfuzosin, we tested 0 mg/g (control), 0.01 mg/g, and 0.1 mg/g. For ikarugamycin, we tested 0 mg/g (control) and 0.01 mg/g. These ranges of alfuzosin and ikarugamycin were much greater than the concentrations of the alfuzosin-related compound and ikarugamycin in leaves (see below). Nevertheless, we tested these concentrations because excessive doses are often necessary to obtain the median toxic dose TD_{50} and the median lethal dose LD_{50} and because comparison among the three compounds at the same doses may provide us with valuable information on diverse effects of leaf compounds. Basal levels of the compounds from leaves in AD-FSW-135 were not taken into account for analyses due to uncertainty of the estimated leaf concentrations.

2.4. Concentration of Lauric Acid in Leaves

The concentration of lauric acid in leaves was roughly estimated as follows. Lauric acid in leaves was discovered by the targeted method of GC–MS [49] (Appendix A). Because this metabolite was targeted based on previously known chemical information, the identification and peak area data were more credible than the nontargeted method. Although each compound has a different detection efficiency in GC–MS, it is possible to roughly compare peak area values among targeted metabolites detected simultaneously from the same samples. One of the targeted metabolites was oxalic acid. The concentration of oxalic acid in leaves of *O. corniculata* has been reported to be 16.9 mg/g (leaf) [95]. On the other hand, the mean peak area value of oxalic acid (No. 15) in nonirradiated samples in the targeted GC–MS analysis was 6,409,017 (n = 3) (Figure A1) [49]. Similarly, the mean peak area value of lauric acid (No. 175) in nonirradiated samples in the targeted GC–MS analysis was 18,899 (n = 3) (Figure A1) [49]. Therefore, the concentration of lauric acid in nonirradiated leaves was calculated to be 0.050 mg/g (leaf). When irradiated, the mean peak area value of lauric acid was 23,977 (n = 3) [49], and it increased approximately 1.27 times to 0.063 mg/g (leaf) under the previous experimental conditions [49].

2.5. Concentration of the Alfuzosin-Related Compound in Leaves

The concentration of the unknown alfuzosin-related compound in *Oxalis* leaves was estimated by HPLC spectrograms of LC–MS newly performed in this study (Figure A2). Leaf samples for the previous study (Figure A1) [49] and the current study (Figure A2) were identical. Sample preparation procedures followed a previous LC–MS study [49]. Washed fresh leaves of *O. corniculata* (100 mg) were frozen, ground, and thoroughly homogenized with methanol (300 μL). After a brief centrifugation, 200 μL was recovered, from which 10 μL was subjected to LC–MS analysis. This extraction process can be considered a total volume increase to 400 μL, assuming that leaf density is close to that of water (1.0 g/mL). The experimental conditions for the LC–MS analysis in the present study are described in Appendix A.

The alfuzosin-related compound in the leaf extract had a mean peak area value of 201.81 in triplicate of a sample (Figure A2a). This peak area value was similar to that of the alfuzosin standard (Sigma–Aldrich), 199.36, when 0.10 ng/mL methanol extract was analyzed (Figure A2b). Although the alfuzosin standard spectrum showed an additional peak, this peak was found to be an impurity peak in methanol (Figure A2c). Considering the volume conversion factor, ×4.0, the concentration of the alfuzosin-related compound in nonirradiated leaves was estimated to be 0.40 ng/g (leaf). In a previous metabolomic study, the mean peak area values of the alfuzosin-related compound (No. 4746) were 133,169 (n = 3) (without irradiation) and 534,069 (n = 3) (irradiated) (Figure A1) [49]. Thus, when irradiated, the area value increased 4.01 times to 1.6 ng/mg (leaf) under the previous experimental conditions.

2.6. Concentration of Ikarugamycin in Leaves

The concentration of ikarugamycin in leaves was estimated based on the previous peak area values of LC–MS (Figure A1) [49]. Alfuzosin-related compound (No. 4746) had a mean peak area value of 133,169 (n = 3) in nonirradiated samples, whereas ikarugamycin had a peak area value of 76,713 (n = 3) from the same samples [49]. Assuming that it is possible to roughly compare peak area values among metabolites detected simultaneously from the same samples, the concentration of ikarugamycin was calculated to be 0.20 ng/g (leaf) based on the estimated concentration of the alfuzosin-related compound in leaves. When irradiated, the mean peak area value of ikarugamycin was 99,905. Thus, the area value increased 1.30 times to 0.26 ng/g (leaf) under the previous experimental conditions.

2.7. Toxicological Output Data

To understand the toxicity of the three metabolites, we recorded three metamorphosis-related data as the number of individuals as follows: the number of individuals that successfully pupated (the number of pupae), the number of individuals that successfully eclosed (the number of adults), and the number of individuals that successfully eclosed without wrinkled wings (the number of normal adults), as shown in Appendix B (Tables A1–A4). These three numbers were used for calculating the three metamorphosis rates: the pupation rate, the eclosion rate, and the normality rate. For calculations, these numbers were divided by the starting number of larvae, and the results were expressed as a percentage. The eclosion rate was also called the "survival rate".

We also recorded three developmental period data as the number of days: the number of days from the time point when egg collection started to pupation (the egg-larval days), the number of days from pupation to eclosion (the papal days), and the number of days from the time point when egg collection started to eclosion (the immature days). The immature days are simple summation of the egg-larval days and the pupal days. The egg-larval days included prepupal days.

Additionally, we measured adult forewing size from the wing base to the apex using a desktop digital microscope SKM-2000 with its associated software SK-measure (Saito Kogaku, Yokohama, Kanagawa, Japan). Because male and female forewing sizes are known to be different in this species [24,64], forewing size data were compiled sex-dependently. Individuals with wrinkled wings were not subjected to size measurements. The developmental period data and the forewing size data were compiled in Table S1. The numbers of individuals in repeated biological trials were also shown in Table S1.

2.8. Statistical Analysis

A treatment group (0.01 mg/g, 0.1 mg/g, or 1 mg/g of a test additive of interest) was statistically compared to a corresponding no treatment (control) group (0 mg/g). We performed the χ^2 test for the data on the number of individuals that produced the pupation rate, the eclosion rate, and the normality rate. The χ^2 test was also performed for evaluating the performance of the artificial diets and for comparing the normalized eclosion and normality rates between lauric acid and alfuzosin. Yates' correction was not performed. We performed either Student's *t*-test (equal variance) or Welch's *t*-test (unequal variance) for the egg-larval days, the pupal days, the immature days, and the forewing size, assuming that they were normally distributed. Bonferroni or other correction was not performed. Statistical analyses were performed using Microsoft Excel (Office 365), JSTAT (Yokohama, Kanagawa, Japan), and MetaboAnalyst [96]. MetaboAnalyst was also used to produce box plots of metabolites obtained in a previous study [49].

3. Results

3.1. Performance of the Artificial Diet AD-FSW-135

The new artificial diet AD-FSW-135 developed for this study was compared with the previous diets, AD-F (artificial diet with fresh leaves) [24] and AD-FSI-112 (artificial diet with fresh leaves, soy powder, and Insecta F-II (Nosan Corporation)) [64], in terms of ingredients (Figure 1a). The most important difference among the three artificial diets was the leaf content. *Oxalis* leaves occupied 58.7% of the diet in AD-F [24] and 32.2% in AF-FSI-112 [64]. In contrast, *Oxalis* leaves occupied only 11.1% in the new diet AD-FSW-135. In other words, AD-FSW-135 (this study) contained approximately one-fifth and one-third of fresh leaves of the previous diets AD-F [24] and AD-FSI-112 [64], respectively. The reduced leaf content in the new diet AD-FSW-135 was considered important for toxicological tests (see Section 2). This leaf-content reduction was achieved by the introduction of Silk Mate L4M (Nosan Corporation), a commercially available artificial diet for silkworms. In a previous diet, AD-FSI-112, Insecta F-II from the same manufacturer was used [64]. Simplification of the contents for quick and easy preparation with just three ingredients was also an important advantage of the new artificial diet AD-FSW-135 developed for this study.

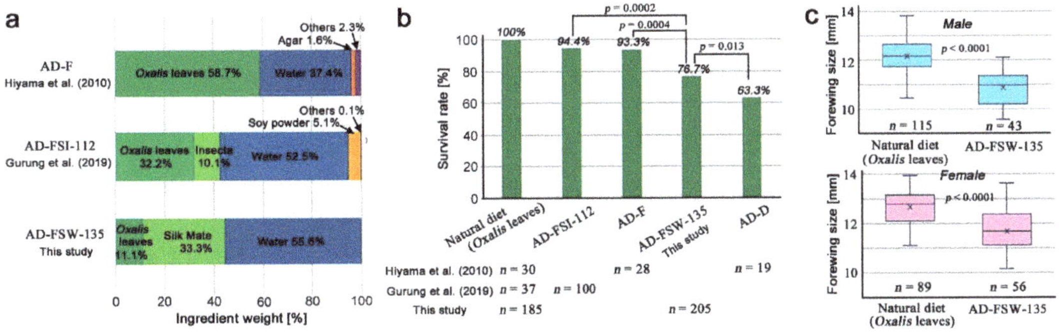

Figure 1. Ingredients and performance of artificial diets. (**a**) Ingredients and their weight percentages. AD-FSW-135 has a reduced leaf content. It also has simplified contents with just three ingredients. (**b**) Survival rate (eclosion rate). AD-FSW-135 shows a higher survival rate than AD-D [24] and a lower rate than AD-F [24] and AD-FSI-112 [64]. The *p*-values obtained from the χ^2 test are shown. The AD-FSW-135 results were obtained from ten biological repeats (see Supplementary Table S1). (**c**) Male (top) and female (bottom) forewing size. The *p*-values obtained from the *t*-test between the natural diet and AD-FSW-135 are indicated. Both natural diet and AD-FSW-135 results were obtained from ten biological repeats (see Supplementary Table S1).

Throughout the rearing experiments with the new artificial diet AD-FSW-135 containing a compound of interest (called a test additive), we always simultaneously reared a group of larvae with natural fresh leaves of the creeping wood sorrel and another group of larvae with AD-FSW-135 without a test additive (0 mg/g) (Appendix B; Table A1). To evaluate the performance of the new diet AD-FSW-135, the survival (eclosion) rates of larvae without an additive were compared among the previous and present diets (Figure 1b). AD-FSW-135 (this study) had a significantly higher survival rate than AD-D (artificial diet with dried leaves) [24] and a significantly lower rate than AD-FSI-112 [64] and AD-F [24], indicating that the survival rate of AD-FSW-135 (this study) was not very high but was not very low. AD-FSW-135 (this study) was thus considered acceptable for toxicological tests, as long as the majority of larvae could eat AD-FSW-135 and grow.

The forewing size of adult individuals reared with the new diet AD-FSW-135 was reduced in comparison to that of fresh plant leaves (Figure 1c). Males from the natural diet and AD-FSW-135 groups showed forewing sizes of 12.16 ± 0.71 mm (mean ± standard deviation) and 10.88 ± 0.65 mm, respectively. Females from the natural diet and AD-FSW-135 groups showed forewing sizes of 12.67 ± 0.68 mm and 11.68 ± 0.92 mm, respectively. In both sexes, the forewing size was reduced significantly. However, these results were essentially similar to those of the previous diets in terms of size distributions [24,64]; this level of size reduction seems to be inherent in rearing butterflies in artificial diets. Therefore, the forewing size reduction in the AD-FSW-135 results was considered acceptable for toxicological tests in this study.

3.2. Lauric Acid

We prepared three concentrations of lauric acid in the artificial diet: 0.01 mg/g, 0.1 mg/g, and 1 mg/g, in addition to the diet without it (0 mg/g). These concentrations covered an estimated concentration of lauric acid in the irradiated leaves of *O. corniculata* of

0.063 mg/g. The number of pupae, eclosion, and normal adults were recorded (Appendix B; Table A2). We examined the toxicity of lauric acid from three different viewpoints: metamorphosis rates (the pupation rate, eclosion rate, and normality rate), developmental periods (egg-larval days, pupal days, and immature days), and adult forewing size.

The normality rate linearly decreased in response to the concentration of lauric acid (Figure 2a). The pupation rate and the eclosion rate were largely similar to the normality rate except at 0.1 mg/g, which showed an increase (Figure 2a). In comparison to the diet without lauric acid (0 mg/g), the diet with 1 mg/g showed a significant decrease in the pupation rate, eclosion rate, and normality rate, indicating the toxicity of lauric acid. At 0.1 mg/g and 1 mg/g, the egg-larval days appeared to be significantly longer than the control (0 mg/g) (Figure 2b). The immature days of the 0.01 mg/g and 0.1 mg/g treatments were significantly longer than those of the control (0 mg/g). These results indicate that developmental retardation occurred in the larval periods at all three concentrations of lauric acid. The forewing size of females at 0.1 mg/g was reduced significantly in comparison to the control (0 mg/g), although such a reduction was not observed at other concentrations (Figure 2c).

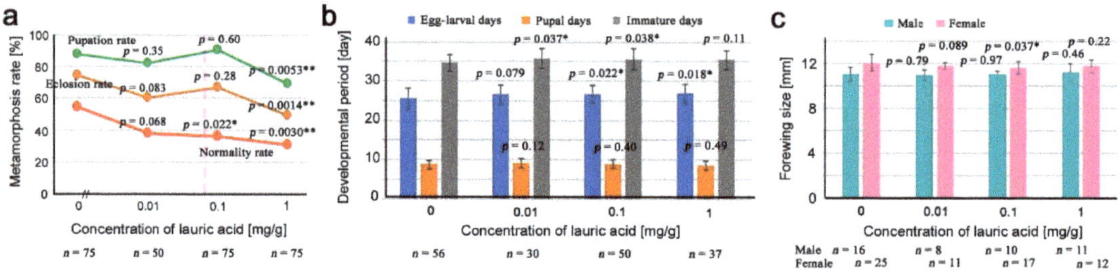

Figure 2. Results of the toxicity test for lauric acid. Asterisks indicate levels of statistical significance in comparison to the control (0 mg/g); *, $p < 0.05$; **, $p < 0.01$. These results were obtained from four biological repeats (see Supplementary Table S1). (**a**) Pupation rate (green), eclosion rate (brown), and normality rate (red). The *p*-values obtained from the χ^2 test are indicated. The pink vertical broken line indicates a rough position of the estimated concentration of lauric acid in irradiated leaves, 0.063 mg/g. (**b**) Egg-larval days (blue), pupal days (brown), and immature days (gray). The mean values (±standard deviation) are shown as bar height. The *p*-values obtained from the *t*-test are indicated. (**c**) Male (blue green) and female (pink) forewing size. The mean values (±standard deviation) are shown as bar height. The *p*-values obtained from the *t*-test are indicated.

3.3. *Alfuzosin*

We prepared two concentrations of alfuzosin in the artificial diet, 0.01 mg/g and 0.1 mg/g, in addition to the diet without it (0 mg/g), to compare the results with those of lauric acid, although the lowest concentration used in the present study, 0.01 mg/g, was 6.3×10^3 times higher than an estimated concentration of the alfuzosin-related compound in irradiated leaves, 1.6 ng/g. As in the case of lauric acid, we examined the metamorphosis rates, developmental periods, and adult forewing size (Appendix B; Table A3).

The pupation rate, eclosion rate, and normality rate all decreased significantly in response to alfuzosin, but not linearly (Figure 3a). Reasons for lower rates at 0.01 mg/g than those at 0.1 mg/g were uncertain but may be technical (see Section 4). The egg-larval days and the immature days were significantly different from those without alfuzosin (0 mg/g) (Figure 3b). Somewhat surprisingly, these differences showed developmental acceleration instead of retardation. The forewing size did not differ from that of the control (0 mg/g), but at 0.1 mg/g in females, the forewing size tended to increase, although the increase was not statistically significant (Figure 3c).

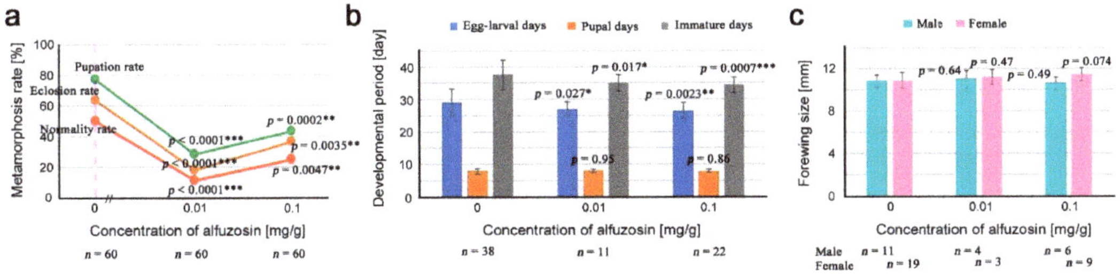

Figure 3. Results of the toxicity test for alfuzosin. Asterisks indicate levels of statistical significance in comparison to the control (0 mg/g); *, $p < 0.05$; **, $p < 0.01$; ***, $p < 0.001$. These results were obtained from three biological repeats (see Supplementary Table S1). (**a**) Pupation rate (green), eclosion rate (brown), and normality rate (red). The p-values obtained from the χ^2 test are indicated. The pink vertical broken line indicates a rough position of the estimated concentration of the alfuzosin-related compound in irradiated leaves, 1.6 ng/g. (**b**) Egg-larval days (blue), pupal days (brown), and immature days (gray). The mean values (±standard deviation) are shown as bar height. The p-values obtained from the t-test are indicated. (**c**) Male (blue green) and female (pink) forewing size. The mean values (±standard deviation) are shown as bar height. The p-values obtained from the t-test are indicated.

3.4. *Ikarugamycin*

We prepared one concentration of ikarugamycin in the artificial diet, 0.01 mg/g, in addition to the diet without it (0 mg/g) to compare the results with those of lauric acid, although the lowest concentration used in the present study, 0.01 mg/g, was 3.8×10^4 times higher than an estimated concentration of ikarugamycin in irradiated leaves, 0.26 ng/g. As in the cases of lauric acid and alfuzosin, we examined the metamorphosis rates, developmental periods, and adult forewing size (Appendix B; Table A4).

Surprisingly, the pupation rate and the eclosion rate increased significantly in response to ikarugamycin, although an increase in the normality rate was not significant (Figure 4a). These results indicate mild drug efficacy of ikarugamycin instead of toxicity. In contrast to lauric acid and alfuzosin, the egg-larval days, pupal days, and immature days at 0.01 mg/g were not different from those without ikarugamycin (0 mg/g) (Figure 4b). The forewing size at 0.01 mg/g did not differ from those without ikarugamycin (0 mg/g) (Figure 4c).

3.5. *Comparison of Three Compounds*

Here, we compared the results of the three compounds tested above. The eclosion (survival) rates and the normality rates were normalized so that they became 100% when no compound was added to the diet (0 mg/g) (Appendix B; Tables A5 and A6) as shown in Figure 5. The eclosion rates (Figure 5a) and the normality rates (Figure 5b) were not very different, but lauric acid exhibited a smooth and gradual dose-dependent decrease in the normality rate curve, although not in the eclosion rate curve, as seen previously (Figure 2a). It is remarkable that the normality rate curves of the three compounds showed different behaviors; in response to concentration, the lauric acid curves decreased dose-dependently, the alfuzosin curves decreased more sharply and not linearly, and the ikarugamycin curves increased (Figure 5b). At the concentration of 0.01 mg/g, where three compounds were able to be compared, the normality rates of lauric acid, alfuzosin, and ikarugamycin were 69.5%, 23.4%, and 125.0%, respectively (Appendix B; Table A6). The differences between lauric acid and alfuzosin appeared to be more significant at the concentration of 0.01 mg/g than 0.1 mg/g in both the eclosion and normality rates, but this may be because of a low solubility of alfuzosin at 0.1 mg/g (see Sections 2 and 4).

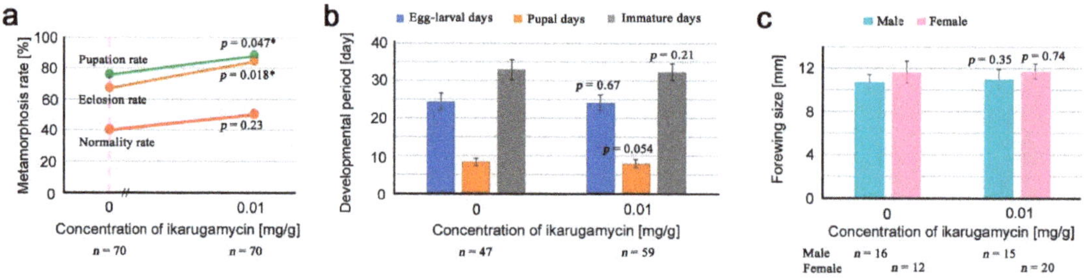

Figure 4. Results of the toxicity test for ikarugamycin. Asterisks indicate levels of statistical significance in comparison to the control (0 mg/g); *, $p < 0.05$. These results were obtained from three biological repeats (see Supplementary Table S1). (**a**) Pupation rate (green), eclosion rate (brown), and normality rate (red). The *p*-values obtained from the χ^2 test are indicated. The pink vertical broken line indicates a rough position of the estimated concentration of ikarugamycin in irradiated leaves, 0.26 ng/g. (**b**) Egg-larval days (blue), pupal days (brown), and immature days (gray). The mean values (±standard deviation) are shown as bar height. The *p*-values obtained from the *t*-test are indicated. (**c**) Male (blue green) and female (pink) forewing size. The mean values (±standard deviation) are shown as bar height. The *p*-values obtained from the *t*-test are indicated.

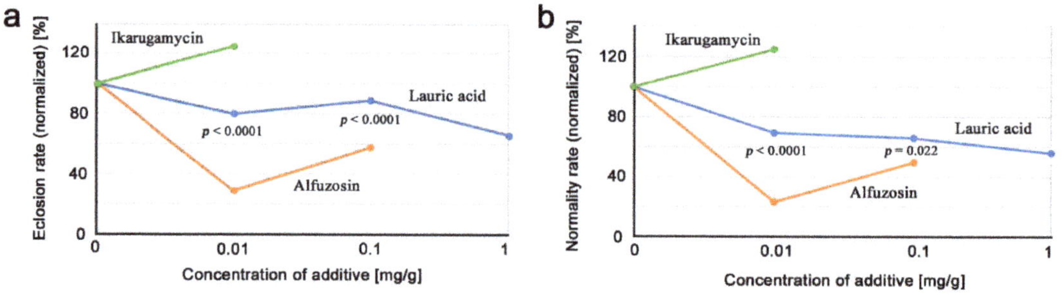

Figure 5. Comparison of the results of the three compounds. The *p*-values obtained from the χ^2 test between lauric acid and alfuzosin are shown. (**a**) Eclosion (survival) rate (normalized). (**b**) Normality rate (normalized).

As a convention of toxicological analysis, a regression line was determined using the normality rates as $y = -24.9x + 79.8$ ($R^2 = 0.40$) for lauric acid, and at the normality rate of 50% ($y = 50$), lauric acid concentration x was determined as 1.2 mg/g. This is considered equivalent to the median toxic dose, TD_{50}. Similarly, a regression line was determined as $y = -185x + 64.6$ ($R^2 = 0.069$) for alfuzosin, and at the normality rate of 50% ($y = 50$), the alfuzosin concentration x was determined to be 0.079 mg/g, which is 15 times smaller than that of lauric acid. Just to be sure, if the value at 0.1 mg/g in alfuzosin was erroneously high due to technical reasons, such as low solubility (see Section 4), the LD_{50} value of alfuzosin should be much lower.

Likewise, using the eclosion (survival) rates, a regression line was determined as $y = -24.8x + 90.8$ ($R^2 = 0.69$) for lauric acid, and at the normality rate of 50% ($y = 50$), lauric acid concentration x was determined as 1.6 mg/g. This is considered equivalent to the median lethal dose, LD_{50}. Similarly, a regression line was determined as $y = -126x + 66.9$ ($R^2 = 0.038$) for alfuzosin, and at the normality rate of 50% ($y = 50$), the alfuzosin concentration x was determined as 0.13 mg/g, which is 12 times smaller than that of lauric acid. As in the case of TD_{50}, the LD_{50} value of alfuzosin should be much lower if the value at 0.1 mg/g was technically erroneous.

4. Discussion

We tested the ingestional toxicity of three compounds, namely, lauric acid, alfuzosin, and ikarugamycin, which were significantly upregulated in *O. corniculata* and annotated by a previous metabolomic study [49]. For convenience, concentration data are compiled in Table 1. In this study, we employed a new artificial diet, AD-FSW-135, which contained a relatively small amount of host plant leaves, occupying just 11% of the entire diet. It is important to keep the leaf content as low as possible in the artificial diet due to an experimental addition of a testing compound. Indeed, the basal levels of the three compounds in the new diet AD-FSW-135 (Table 1) were considered low enough for the current study. This new diet showed acceptable performance based on the survival rate and forewing size, two indexes to evaluate artificial diets [64]. Thus, we believe that the use of AD-FSW-135 in the present study is justifiable but that there is still much room for further improvement of artificial diets.

Table 1. Concentrations of the three metabolites of interest.

Metabolite	Leaf (w/o Radiation)	Leaf (with Radiation) *1	Basal Level in AD-FSW-135	Toxicity Test in AD-FSW-135 *2	Coverage *3	TD_{50}	LD_{50}
Lauric acid	0.050 mg/g	0.063 mg/g	0.0055 mg/g	0, 0.01, 0.1, 1 mg/g	Yes	1.2 mg/g	1.6 mg/g
Alfuzosin *4	0.40 ng/g	1.6 ng/g	0.044 ng/g	0, 0.01, 0.1 mg/g	No	0.079 mg/g	0.13 mg/g
Ikarugamycin	0.20 ng/g	0.26 ng/g	0.022 ng/g	0, 0.01 mg/g	No	NA	NA

*1: For irradiation conditions, see Sakauchi et al. (2021) [49]. *2: Concentrations in toxicity tests ignore the basal levels of these metabolites from leaves in AD-FSW-135. *3: Coverage indicates if leaf concentration was covered by the tested concentration range. *4: Alfuzosin was tested, but the leaf concentrations shown here are those of the alfuzosin-related compound. NA: Not applicable.

With this new diet AD-FSW-135, we demonstrated that lauric acid was toxic to larvae dose-dependently in terms of metamorphosis rates, although the larval response was mild. Lauric acid is present in leaves without radiation, and larvae are certainly tolerant to lauric acid at the leaf level of 0.050 mg/g, explaining the gradual dose–response curves. The mild toxicity of lauric acid is expressed in its TD_{50}, 1.2 mg/g, in contrast to the TD_{50} of alfuzosin, 0.079 mg/g. LD_{50} values also indicated such a relationship. We observed some toxicity even at the level of 0.01 mg/g, but this may be because larvae were exposed to a sudden rise in lauric acid concentration when the artificial diet was first given. In addition to the changes in the metamorphosis rates, growth retardation was detected at the egg-larval period in response to lauric acid. Furthermore, the forewing size reduction was observed, although only at 0.1 mg/g in females. These results indicate the toxicity of lauric acid on the developmental physiology of the butterfly and appear to be biologically significant in the field because the estimated concentration of lauric acid in irradiated leaves, 0.063 mg/g, was covered by the current study.

In a previous study, the fold change values in the upregulation of lauric acid was 1.27 at low-level radiation exposure; the cumulative dose to the plant was 5.7 mGy (34 μSv/h in a period of seven days) [49]. It is somewhat surprising that the plant significantly responded to this low-level exposure, and we expect that the fold change value may increase further in response to higher levels of radiation exposure. According to Nohara et al. (2014) [52], the ground radiation dose rate in Iitate was 18.9 μSv/h, which is indeed lower than the experimental dose rate used in our study, 34 μSv/h. However, experimental irradiation in the present study was only by external exposure during a very limited period of time (seven days), but in the wild, both external and internal exposures are expected for much longer periods of time throughout the entire life span of the butterfly. Importantly, the present results are reminiscent of those found in previous exposure experiments [34–36,52–56] and may also explain the spatiotemporal dynamics of the abnormality rates and collection efficiency (an indicator of population density) in 2011–2013 in Fukushima [37]. Therefore, we conclude that lauric acid acts as a potent toxicant (larvicide) for the pale grass blue butterfly not only in the laboratory but also in wide polluted areas in Fukushima in the field.

This conclusion is consistent with previous studies on lauric acid as a plant defense chemical [68–74]. More tolerance may evolve in larvae in the field, and this scenario may explain the adaptive evolution of the butterfly shown in the polluted areas in Fukushima [56]. Interestingly, lauric acid has been reported to be a feeding stimulant for the silkworm at the concentration of 0.013% in an artificial diet [94]. This percentage corresponds to the 0.1 mg/g level in the present study. In lepidopteran insects, a feeding stimulant for a given species is often toxic to other organisms [97]. Thus, it is reasonable that a feeding stimulant for the silkworm moth, lauric acid, is toxic to the pale grass blue butterfly. Conversely, a feeding stimulant for the pale grass blue butterfly, oxalic acid [92], is probably toxic to other insects including the silkworm moth.

Alfuzosin was also demonstrated to be toxic, but its toxicity was not linearly dose dependent in the metamorphosis rates. We do not understand this nonlinearity, but it might have originated from a technical reason regarding low solubility; alfuzosin might not have been dissolved well in the diet at the relatively high concentrations. Surprisingly, in addition to the reduced metamorphosis rates, alfuzosin appeared to act on the egg-larval period to accelerate growth and tended to increase the forewing size. These results are in sharp contrast to those of lauric acid, indicating different toxic pathways in these two compounds. Because alfuzosin is an antagonist of the α_1-adrenergic receptor [75–78], it may act on insect receptors for biogenic amines, such as octopamine and tyramine [98]. Nonetheless, both alfuzosin and lauric acid seem to affect the larval period but not the pupal period.

Because the alfuzosin concentrations tested in AD-FSW-135 were much higher than those in leaves and because the biological effects of alfuzosin and its related compound are not necessarily similar, direct extrapolations of the alfuzosin results to the alfuzosin-related compound were difficult. However, there may be a possibility that the alfuzosin-related compound was as toxic as alfuzosin due to their structural similarities. If so, the alfuzosin-related compound is 15 times as toxic as lauric acid (based on the TD_{50} values) and 12 times as lethal as lauric acid (based on the LD_{50} values), but the concentration of the alfuzosin-related compound in leaves was much lower than that of lauric acid. Therefore, the presence of the alfuzosin-related compound in leaves would not affect larvae in the field.

In contrast, ikarugamycin showed mild drug efficacy instead of toxicity. This may be simply because it is an antibiotic that inhibits bacterial or fungal growth in the artificial diet, although some antibiotics were contained in Silk Mate L4M, a commercially available ingredient of AD-FSW-135. In that case, ikarugamycin may protect the plant in the wild from fungal and bacterial infection. However, this drug efficacy of ikarugamycin for larvae may not be evident in the field because of the low concentration of ikarugamycin in leaves. Therefore, ikarugamycin would not nullify the toxicity of plant larvicides, such as lauric acid, in the field. Importantly, the present results of ikarugamycin suggest a possible contribution of metabolites from endophytic bacteria to plant and larval immunity under radiation stress. Practically, further addition of ikarugamycin or other antibiotics into AD-FSW-135 may improve its performance in the future.

In reality, in the wild, lauric acid and other upregulated unknown metabolites probably function together to ward off insects. Indeed, in response to radiation exposure, 24 upregulated peaks ($p < 0.05$) were obtained in LC–MS, among which only two of them (alfuzosin and ikarugamycin) were annotated singularly [49]. Only one upregulated peak ($p < 0.05$) was obtained in targeted GC–MS, which was lauric acid [49]. Additionally, 10 upregulated peaks ($p < 0.05$) were obtained in nontargeted GC–MS [49].

It is not possible, at least at this point, to demonstrate collective effects of many upregulated compounds with an artificial diet containing them. On the other hand, the "collective effects" have already been known by internal exposure experiments using contaminated leaves collected from Fukushima, resulting in lower survivorship and growth retardation [34–36,52–56]. We also have evidence that external exposure resulted in similar outcomes [34]. Therefore, the present finding that at least one upregulated metabolite, lauric acid, is larvicidal, is important. It is reasonable to conclude that the intricate balance between the plant and the larva through chemical interactions was affected by the Fukushima nuclear accident.

In addition to revealing the importance of plant-insect interactions in evaluating the biological effects of the Fukushima nuclear accident, this study opened new perspectives. Because ikarugamycin is likely produced by an endophytic bacterium, bacterial, fungal, or other microbial communities in plants and soil may play a role in amplifying the biological effects of low-dose radiation pollution.

5. Conclusions

We demonstrated within a reasonable concentration range (0.01 mg/g to 1 mg/g) that lauric acid is able to function as a toxicant for the pale grass blue butterfly at the leaf concentration (0.063 mg/g with radiation) by lowering metamorphosis rates and by causing growth retardation. Based on its TD_{50} and LD_{50} values (1.2 mg/g and 1.6 mg/g, respectively), lauric acid may be considered a mild larvicide. In the field, lauric acid probably acts as one of the larvicides in leaves in response to radiation exposure. Interpretations of alfuzosin results are not straightforward, but its relatively low TD_{50} and LD_{50} values (0.079 mg/g and 0.13 mg/g, respectively) imply that the alfuzosin-related compound may also be toxic, although it may be irrelevant in the field because of its low leaf concentration (1.6 ng/g with radiation). Because ikarugamycin is an antibiotic likely from endophytic bacteria, its drug efficacy on increasing the metamorphosis rates of larvae may be secondary; it may function to prevent the artificial diet from fungal and bacterial growth. As an extrapolation, ikarugamycin may function to protect leaves from fungi and bacteria under radiation stress. The case of ikarugamycin suggests a contribution of endophytic bacteria to the process of radiation-stress management in the plant.

In conclusion, the present results provide experimental evidence for the field effect hypothesis that concentration changes in radiation-induced metabolites, such as lauric acid, in the host plant leaves of the pale grass blue butterfly caused deterioration of the butterfly at the individual and population levels in radioactively polluted areas in Fukushima.

Supplementary Materials: The following file is available online at https://www.mdpi.com/article/10.3390/life12050615/s1, Table S1: Toxicological Output Data.

Author Contributions: Conceptualization, J.M.O.; methodology, J.M.O.; W.T., K.S. and A.M.; software, A.M. and J.M.O.; validation, J.M.O.; formal analysis and investigation, A.M.; resources, A.M., K.S. and W.T.; data curation, J.M.O.; writing—original draft preparation, J.M.O.; writing—review and editing, J.M.O., A.M., K.S. and W.T.; visualization, J.M.O. and A.M.; supervision, project administration, and funding acquisition, J.M.O. All authors have read and agreed to the published version of the manuscript.

Funding: This research was funded by the Asahi Glass Foundation (Tokyo) and the basic fund from the University of the Ryukyus. The APC was funded by the basic fund from the University of the Ryukyus.

Institutional Review Board Statement: Not applicable.

Informed Consent Statement: Not applicable.

Data Availability Statement: The data presented in this study and the source data are available in this article and in the Supplementary Materials.

Acknowledgments: We thank S. Gima (Center for Research Advancement and Collaboration, University of the Ryukyus) for the LC–MS analysis; Y. Yona, Y. Iraha, A. Hiyama, and R.D. Gurung for their technical assistance for preparing artificial diets; and other members of the BCPH Unit of Molecular Physiology for suggestions and technical help.

Conflicts of Interest: The authors declare no conflict of interest. The funders had no role in the design of the study; in the collection, analyses, or interpretation of data; in the writing of the manuscript; or in the decision to publish the results.

Appendix A

To estimate the concentrations of the three metabolites of interest, lauric acid, alfuzosin-related compound, and ikarugamycin in leaves, the peak area values for these metabolites and oxalic acid obtained in the previous GC–MS and LC–MS analyses were compared (Figure A1). Exact values are mentioned in the Section 2. For the alfuzosin-related compound, LC–MS analyses were newly performed in the present study in triplicate to estimate its concentration in leaves using an alfuzosin standard (Sigma–Aldrich) as a reference material (Figure A2). Leaf samples used for Figures A1 and A2 were identical.

We used a Shimadzu Prominence UFLC XR System (Kyoto, Japan) equipped with a Shimadzu solvent delivery unit LC-20ADXR and a Shimadzu autosampler SIL-20ACXR using a reverse-phase column Inertsil ODS-4 (2.1 mm × 150 mm) (GL Sciences, Tokyo, Japan). Mobile phase A was a 0.1% aqueous solution of formic acid, and mobile phase B was acetonitrile with a time program of its concentrations as follows: 20% (0 min) → 40% (10 min) → 98% (10.01–15 min) → 20% (15.01–23 min). The injection volume was 10 μL, and the flow rate was 0.2 mL/min.

A peak of the alfuzosin-related compound was obtained at 6.7 min from the leaf extract (Figure A2a). The alfuzosin standard also showed a peak at 6.7 min but with an additional peak at 2.1 min (Figure A2b). The latter peak was an impurity from methanol (Figure A2c). MS/MS analyses were also performed simultaneously to confirm the identities of these compounds (not shown).

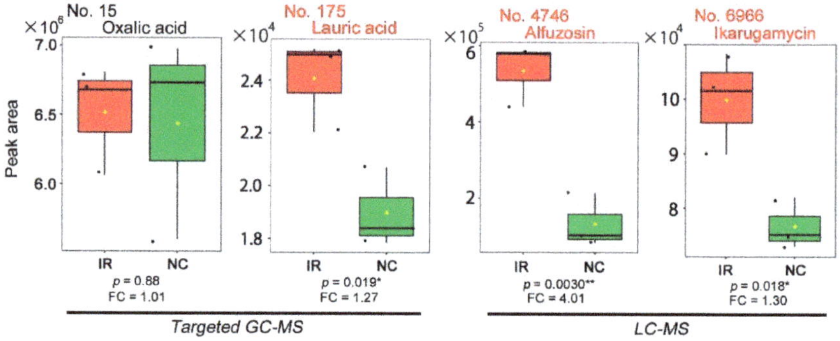

Figure A1. Peak area values of oxalic acid and lauric acid in the targeted GC–MS analysis [49] (left) and peak area values of alfuzosin-related compound and ikarugamycin in the LC–MS analysis [49] (right). Box plots for lauric acid, alfuzosin, and ikarugamycin are also shown in Sakauchi et al. (2021) [49]. IR (shown in red) and NC (shown in green) indicate irradiated samples ($n = 3$) and nonirradiated control samples ($n = 3$), respectively. A single black dot represents the mean value of triplicate of a sample. FC indicates fold change of mean values from nonirradiated to irradiated samples. These plots were produced using MetaboAnalyst [96]. Asterisks indicate levels of statistical significance; *, $p < 0.05$; **, $p < 0.01$ (t-test).

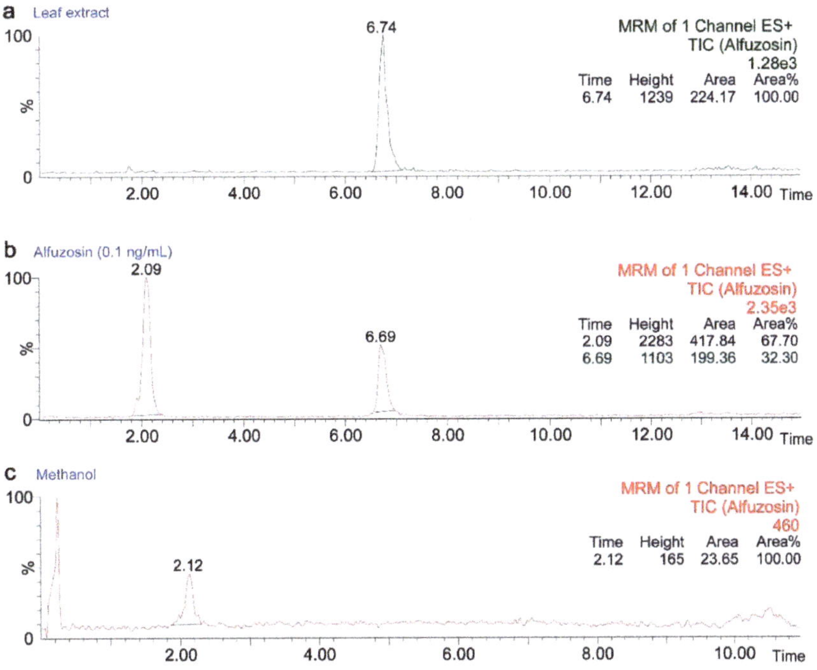

Figure A2. Identification of the alfuzosin-related metabolite by HPLC. (**a**) Leaf extract. A peak at 6.74 min was observed. This is one of the triplicate results. The peak area value here is 224.17. (**b**) Alfuzosin standard (Sigma–Aldrich) 0.1 ng/mL in methanol. The peak at 6.69 min is attributed to alfuzosin, and the peak at 2.09 min is attributed to an impurity in methanol. The peak area value of alfuzosin, 199.36, is similar to that of the alfuzosin-related compound. (**c**) Methanol only. A peak at 2.12 min was observed, demonstrating impurity.

Appendix B

The exact numbers and percentages of individuals obtained after the feeding experiments are shown below for leaf and AD-FSW-135 (without any test additive) controls (Table A1), lauric acid (Table A2), alfuzosin (Table A3), and ikarugamycin (Table A4). Normalized eclosion rates (survival rates) (Table A5) and normalized normality rates (Table A6), used for Figure 5, are also shown. Further information can be found in Table S1.

Table A1. Number of individuals after rearing with live leaves or AD-FSW-135.

Number	Leaf	AD-FSW-135
Number of starting individuals	185 (100%)	205 (100%)
Number of pupae (Pupation rate)	168 (90.8%)	165 (80.5%)
Number of eclosion (Eclosion rate)	166 (89.7%)	141 (68.7%)
Number of normal adults (Normality rate)	156 (84.4%)	99 (48.3%)

Table A2. Number of individuals after oral administration of lauric acid.

Number	0 mg/g	0.01 mg/g	0.1 mg/g	1 mg/g
Number of starting individuals	75 (100%)	50 (100%)	75 (100%)	75 (100%)
Number of pupae (Pupation rate)	66 (88.0%)	41 (82.0%)	68 (90.7%)	52 (69.3%)
Number of eclosion (Eclosion rate)	56 (74.7%)	30 (60.0%)	50 (66.7%)	37 (49.3%)
Number of normal adults (Normality rate)	41 (54.7%)	19 (38.0%)	27 (36%)	23 (30.7%)

Table A3. Number of individuals after oral administration of alfuzosin.

Number	0 mg/g	0.01 mg/g	0.1 mg/g
Number of starting larvae	60 (100%)	60 (100%)	60 (100%)
Number of pupae (Pupation rate)	46 (76.7%)	17 (28.3%)	26 (43.3%)
Number of eclosion (Eclosion rate)	38 (63.3%)	11 (18.3%)	22 (36.7%)
Number of normal adults (Normality rate)	30 (50.0%)	7 (11.7%)	15 (25.0%)

Table A4. Number of individuals after oral administration of ikarugamycin.

Number	0 mg/g	0.01 mg/g
Number of starting individuals	70 (100%)	70 (100%)
Number of pupae (Pupation rate)	53 (75.7%)	62 (88.6%)
Number of eclosion (Eclosion rate)	47 (67.1%)	59 (84.3%)
Number of normal adults (Normality rate)	28 (40.0%)	35 (50.0%)

Table A5. Normalized eclosion rates (survival rates).

Metabolite	0 mg/g	0.01 mg/g	0.1 mg/g	1 mg/g
Lauric acid	100%	80.3%	89.3%	66.0%
Alfuzosin	100%	28.9%	58.0%	NA
Ikarugamycin	100%	125.6%	NA	NA

NA: not applicable.

Table A6. Normalized normality rates.

Metabolite	0 mg/g	0.01 mg/g	0.1 mg/g	1 mg/g
Lauric acid	100%	69.5%	65.8%	56.1%
Alfuzosin	100%	23.4%	50.0%	NA
Ikarugamycin	100%	125.0%	NA	NA

NA: not applicable.

References

1. D'Mello, J.P.F. Preface. In *A Handbook of Environmental Toxicology: Human Disorders and Ecotoxicology*; D'Mello, J.P.F., Ed.; CAB International: Wallingford, UK, 2020; pp. xxv–xxxvi.
2. D'Mello, J.P.F. Unequivocal evidence associating environmental contaminants and pollutants with human morbidity and ecological degradation. In *A Handbook of Environmental Toxicology: Human Disorders and Ecotoxicology*; D'Mello, J.P.F., Ed.; CAB International: Wallingford, UK, 2020; pp. 587–610.
3. Lokobauer, N.; Franić, Z.; Bauman, A.; Maračić, M.; Cesar, D.; Senčar, J. Radiation contamination after the Chernobyl nuclear accident and the effective dose received by the population of Croatia. *J. Environ. Radioact.* **1998**, *41*, 137–146. [CrossRef]
4. Arapis, G.D.; Karandinos, M.G. Migration of ^{137}Cs in the soil of sloping semi-natural ecosystems in Northern Greece. *J. Environ. Radioact.* **2004**, *77*, 133–142. [CrossRef] [PubMed]

5. Tahir, S.N.A.; Jamil, K.; Zaidi, J.H.; Arif, M.; Ahmed, N. Activity concentration of ^{137}Cs in soil samples from Punjab province (Pakistan) and estimation of gamma-ray dose rate for external exposure. *Radiat. Prot. Dosim.* **2006**, *118*, 345–351. [CrossRef] [PubMed]
6. Ambrosino, F.; Stellato, L.; Sabbarese, C. A case study on possible radiological contamination in the Lo Uttara landfill site (Caserta, Italy). *J. Phys. Conf. Ser.* **2020**, *1548*, 012001. [CrossRef]
7. Endo, S.; Kimura, S.; Takatsuji, T.; Nanasawa, K.; Imanaka, T.; Shizuma, K. Measurement of soil contamination by radionuclides due to the Fukushima Dai-ichi Nuclear Ppower Plant accident and associated estimated cumulative external dose estimation. *J. Environ. Radioact.* **2021**, *111*, 18–27. [CrossRef]
8. Thomas, J.A. Monitoring changes in the abundance and distribution of insects using butterflies and other indicator groups. *Philos. Trans. R. Soc. Lond. B Biol. Sci.* **2005**, *360*, 339–357. [CrossRef]
9. Vickery, M. Butterflies as indicators of climate change. *Sci. Prog.* **2008**, *91*, 193–201. [CrossRef]
10. Warren, M.S.; Hill, J.K.; Thomas, J.A.; Asher, J.; Fox, R.; Huntley, B.; Roy, D.B.; Telfer, M.G.; Jeffcoate, S.; Harding, P.; et al. Rapid responses of British butterflies to opposing forces of climate and habitat change. *Nature* **2001**, *414*, 65–69. [CrossRef]
11. MacGregor, C.J.; Thomas, C.D.; Roy, D.B.; Beaumont, M.; Bell, J.; Brereton, T.; Bridle, J.; Dytham, C.; Fox, R.; Gotthard, K.; et al. Climate-induced phenology shifts linked to range expansions in species with multiple reproductive cycles per year. *Nat. Commun.* **2019**, *10*, 4455. [CrossRef]
12. Halsch, C.A.; Shapiro, A.M.; Fordyce, J.A.; Nice, C.C.; Thorne, J.H.; Waetjen, D.P.; Forister, M.L. Insects and recent climate change. *Proc. Natl. Acad. Sci. USA* **2021**, *118*, e2002543117. [CrossRef]
13. Forister, M.L.; Halsch, C.A.; Nice, C.C.; Fordyce, J.A.; Dilts, T.E.; Oliver, J.C.; Prudic, K.L.; Shapiro, A.M.; Wilson, J.K.; Glassberg, J. Fewer butterflies seen by community scientists across the warming and drying landscapes of the American West. *Science* **2021**, *371*, 1042–1045. [CrossRef] [PubMed]
14. Forister, M.L.; McCall, A.C.; Sanders, N.J.; Fordyce, J.A.; Thorne, J.H.; O'Brien, J.; Waetjen, D.P.; Shapiro, A.M. Compounded effects of climate change and habitat alteration shift patterns of butterfly diversity. *Proc. Natl. Acad. Sci. USA* **2010**, *107*, 2088–2092. [CrossRef] [PubMed]
15. Rödder, D.; Schmitt, T.; Gros, P.; Ulrich, W.; Habel, J.C. Climate change drives mountain butterflies towards the summits. *Sci. Rep.* **2021**, *11*, 14382. [CrossRef] [PubMed]
16. Zografou, K.; Swartz, M.T.; Adamidis, G.; Tilden, V.P.; McKinney, E.N.; Sewall, B.J. Species traits affect phenological responses to climate change in a butterfly community. *Sci. Rep.* **2021**, *11*, 3283. [CrossRef] [PubMed]
17. Otaki, J.M.; Hiyama, A.; Iwata, M.; Kudo, T. Phenotypic plasticity in the range-margin population of the lycaenid butterfly *Zizeeria maha*. *BMC Evol. Biol.* **2010**, *10*, 252. [CrossRef] [PubMed]
18. Buckley, J.; Butlin, R.; Bridle, J.R. Evidence for evolutionary change associated with the recent range expansion of the British butterfly, *Aricia agestis*, in response to climate change. *Mol. Ecol.* **2012**, *21*, 267–280. [CrossRef] [PubMed]
19. Kingsolver, J.; Buckley, L. Evolution of plasticity and adaptive responses to climate change along climate gradients. *Proc. Biol. Sci.* **2017**, *284*, 20170386. [CrossRef]
20. Au, T.F.; Bonebrake, T. Increased suitability of poleward climate for a tropical butterfly (*Euripus nyctelius*) (Lepidoptera: Nymphalidae) accompanies its successful range expansion. *J. Insect Sci.* **2019**, *19*, 2. [CrossRef]
21. Carnicer, J.; Stefanescu, C.; Vives-Ingla, M.; López, C.; Cortizas, S.; Wheat, C.; Vila, R.; Llusià, J.; Peñuelas, J. Phenotypic biomarkers of climatic impacts on declining insect populations: A key role for decadal drought, thermal buffering and amplication effects and host plant dynamics. *J. Anim. Ecol.* **2019**, *88*, 376–391. [CrossRef]
22. Herremans, M.; Gielen, K.; Van Kerckhoven, J.; Vanormelingen, P.; Veraghtert, W.; Swinnen, K.R.R.; Maes, D. Abundant citizen science data reveal that the peacock butterfly *Aglais io* recently became bivoltine in Belgium. *Insects* **2021**, *12*, 683. [CrossRef]
23. Wendt, M.; Senftleben, N.; Gros, P.; Schmitt, T. Coping with environmental extremes: Population ecology and behavioral adaptation of *Erebia pronoe*, an alpine butterfly species. *Insects* **2021**, *12*, 896. [CrossRef]
24. Hiyama, A.; Iwata, M.; Otaki, J.M. Rearing the pale grass blue *Zizeeria maha* (Lepidoptera, Lycaenidae): Toward the establishment of a lycaenid model system for butterfly physiology and genetics. *Entomol. Sci.* **2010**, *13*, 293–302. [CrossRef]
25. Hiyama, A.; Taira, W.; Otaki, J.M. Color-pattern evolution in response to environmental stress in butterflies. *Front. Genet.* **2012**, *3*, 15. [CrossRef]
26. Taira, W.; Iwasaki, M.; Otaki, J.M. Body size distributions of the pale grass blue butterfly in Japan: Size rules and the status of the Fukushima population. *Sci. Rep.* **2015**, *5*, 12351. [CrossRef]
27. Hiyama, A.; Taira, W.; Sakauchi, K.; Otaki, J.M. Sampling efficiency of the pale grass blue butterfly *Zizeeria maha* (Lepidoptera: Lycaenidae): A versatile indicator species for environmental risk assessment in Japan. *J. Asia-Pac. Entomol.* **2018**, *21*, 609–615. [CrossRef]
28. Hiyama, A.; Otaki, J.M. Dispersibility of the pale grass blue butterfly *Zizeeria maha* (Lepidoptera: Lycaenidae) revealed by one-individual tracking in the field: Quantitative comparisons between subspecies and between sexes. *Insects* **2020**, *11*, 122. [CrossRef]
29. Buckley, J.; Bridle, J.R.; Pomiankowski, A. Novel variation associated with species range expansion. *BMC Evol. Biol.* **2010**, *10*, 382. [CrossRef]
30. Gilbert, S.; Epel, D. *Ecological Developmental Biology*; Sinauer Associates: Sunderland, MA, USA, 2016.

31. Møller, A.P.; Hagiwara, A.; Matsui, S.; Kasahara, S.; Kawatsu, K.; Nishiumi, I.; Suzuki, H.; Mousseau, T.A. Abundance of birds in Fukushima as judges from Chernobyl. *Environ. Pollut.* **2012**, *164*, 36–39. [CrossRef]
32. Bonisoli-Alquati, A.; Koyama, K.; Tedeschi, D.J.; Kitamura, W.; Sukuzi, H.; Ostermiller, S.; Arai, E.; Møller, A.P.; Mousseau, T.A. Abundance and genetic damage of barn swallows from Fukushima. *Sci. Rep.* **2015**, *5*, 9432. [CrossRef]
33. Murase, K.; Murase, J.; Horie, R.; Endo, K. Effects of the Fukushima Daiichi nuclear accident on goshawk reproduction. *Sci. Rep.* **2015**, *5*, 9405. [CrossRef]
34. Hiyama, A.; Nohara, C.; Kinjo, S.; Taira, W.; Gima, S.; Tanahara, A.; Otaki, J.M. The biological impacts of the Fukushima nuclear accident on the pale grass blue butterfly. *Sci. Rep.* **2012**, *2*, 570. [CrossRef]
35. Hiyama, A.; Nohara, C.; Taira, W.; Kinjo, S.; Iwata, M.; Otaki, J.M. The Fukushima nuclear accident and the pale grass blue butterfly: Evaluating biological effects of long-term low-dose exposures. *BMC Evol. Biol.* **2013**, *13*, 168. [CrossRef]
36. Nohara, C.; Hiyama, A.; Taira, W.; Tanahara, A.; Otaki, J.M. The biological impacts of ingested radioactive materials on the pale grass blue butterfly. *Sci. Rep.* **2014**, *4*, 4946. [CrossRef]
37. Hiyama, A.; Taira, W.; Nohara, C.; Iwasaki, M.; Kinjo, S.; Iwata, M.; Otaki, J.M. Spatiotemporal abnormality dynamics of the pale grass blue butterfly: Three years of monitoring (2011–2013) after the Fukushima nuclear accident. *BMC Evol. Biol.* **2015**, *15*, 15. [CrossRef]
38. Akimoto, S. Morphological abnormalities in gall-forming aphids in a radiation-contaminated area near Fukushima Daiichi: Selective impact of fallout? *Ecol. Evol.* **2014**, *4*, 355–369. [CrossRef]
39. Akimoto, S.I.; Li, Y.; Imanaka, T.; Sato, H.; Ishida, K. Effects of radiation from contaminated soil and moss in Fukushima on embryogenesis and egg hatching of the aphid *Prociphilus oriens*. *J. Hered.* **2018**, *109*, 199–205. [CrossRef]
40. Ochiai, K.; Hayama, S.; Nakiri, S.; Nakanishi, S.; Ishii, N.; Uno, T.; Kato, T.; Konno, F.; Kawamoto, Y.; Tsuchida, S.; et al. Low blood cell counts in wild Japanese monkeys after the Fukushima Daiichi nuclear disaster. *Sci. Rep.* **2014**, *4*, 5793. [CrossRef]
41. Hayama, S.; Tsuchiya, M.; Ochiai, K.; Nakiri, S.; Nakanishi, S.; Ishii, N.; Kato, T.; Tanaka, A.; Konno, F.; Kawamoto, Y.; et al. Small head size and delayed body weight growth in wild Japanese monkey fetuses after the Fukushima Daiichi nuclear disaster. *Sci. Rep.* **2017**, *7*, 3528. [CrossRef]
42. Urushihara, Y.; Suzuki, T.; Shimizu, Y.; Ohtaki, M.; Kuwahara, Y.; Suzuki, M.; Uno, T.; Fujita, S.; Saito, A.; Yamashiro, H.; et al. Haematological analysis of Japanese macaques (*Macaca fuscata*) in the area affected by the Fukushima Daiichi Nuclear Power Plant accident. *Sci. Rep.* **2018**, *8*, 16748. [CrossRef]
43. Horiguchi, T.; Yoshii, H.; Mizuno, S.; Shiraishi, H. Decline in intertidal biota after the 2011 Great East Japan Earthquake and Tsunami and the Fukushima nuclear disaster: Field observations. *Sci. Rep.* **2016**, *6*, 20416. [CrossRef]
44. Hayashi, G.; Shibato, J.; Imanaka, T.; Cho, K.; Kubo, A.; Kikuchi, S.; Satoh, K.; Kimura, S.; Ozawa, S.; Fukutani, S.; et al. Unraveling low-level gamma radiation-responsive changes in expression of early and late genes in leaves of rice seedlings at Iitate Village, Fukushima. *J. Hered.* **2014**, *105*, 723–738. [CrossRef] [PubMed]
45. Rakwal, R.; Hayashi, G.; Shibato, J.; Deepak, S.A.; Gundimeda, S.; Simha, U.; Padmanaban, A.; Gupta, R.; Han, S.; Kim, S.T.; et al. Progress toward rice seed OMICS in low-level gamma radiation environment in Iitate Village, Fukushima. *J. Hered.* **2018**, *109*, 2089–2211. [CrossRef] [PubMed]
46. Watanabe, Y.; Ichikawa, S.; Kubota, M.; Hoshino, J.; Kubota, Y.; Maruyama, K.; Fuma, S.; Kawaguchi, I.; Yoschenko, V.I.; Yoshida, S. Morphological defects in native Japanese fir trees around the Fukushima Daiichi Nuclear Power Plant. *Sci. Rep.* **2015**, *5*, 13232. [CrossRef] [PubMed]
47. Yoschenko, V.; Nanba, K.; Yoshida, S.; Watanabe, Y.; Takase, T.; Sato, N.; Keitoku, K. Morphological abnormalities in Japanese red pine (*Pinus densiflora*) at the territories contaminated as a result of the accident at Fukushima Dai-ichi Nuclear Power Plant. *J. Environ. Radioact.* **2016**, *165*, 60–67. [CrossRef] [PubMed]
48. Sakauchi, K.; Taira, W.; Toki, M.; Tsuhako, M.; Umetsu, K.; Otaki, J.M. Nutrient imbalance of the host plant for larvae of the pale grass blue butterfly may mediate the field effect of low-dose radiation exposure in Fukushima: Dose-dependent changes in the sodium content. *Insects* **2021**, *12*, 149. [CrossRef] [PubMed]
49. Sakauchi, K.; Taira, W.; Otaki, J.M. Metabolomic response of the creeping wood sorrel *Oxalis corniculata* to low-dose radiation exposure from Fukushima's contaminated soil. *Life* **2021**, *11*, 990. [CrossRef]
50. Sakauchi, K.; Taira, W.; Otaki, J.M. Metabolomic profiles of the creeping wood sorrel *Oxalis corniculata* in radioactivelt contaminated fields in Fukushima: Dose-dependent changes in key metabolites. *Life* **2022**, *12*, 115. [CrossRef]
51. Yamanouchi, K.; Tsujiguchi, T.; Shiroma, Y.; Suzuki, T.; Tamakuma, Y.; Sakamoto, Y.; Hegedüs, K.; Iwaoka, K.; Hosoda, M.; Kashiwakura, I.; et al. Comparison of bacterial flora in river sediments from Fukushima and Aomori prefectures by 16S rDNA sequence analysis. *Radiat. Prot. Dosim.* **2019**, *184*, 504–509. [CrossRef]
52. Nohara, C.; Taira, W.; Hiyama, A.; Tanahara, A.; Takatsuji, T.; Otaki, J.M. Ingestion of radioactively contaminated diets for two generations in the pale grass blue butterfly. *BMC Evol. Biol.* **2014**, *14*, 193. [CrossRef]
53. Taira, W.; Hiyama, A.; Nohara, C.; Sakauchi, K.; Otaki, J.M. Ingestional and transgenerational effects of the Fukushima nuclear accident on the pale grass blue butterfly. *J. Radiat. Res.* **2015**, *56*, i2–i18. [CrossRef]
54. Taira, W.; Nohara, C.; Hiyama, A.; Otaki, J.M. Fukushima's biological impacts: The case of the pale grass blue butterfly. *J. Hered.* **2014**, *105*, 710–722. [CrossRef] [PubMed]
55. Taira, W.; Toki, M.; Kakinohana, K.; Sakauchi, K.; Otaki, J.M. Developmental and hemocytological effects of ingesting Fukushima's radiocesium on the cabbage white butterfly *Pieris rapae*. *Sci. Rep.* **2019**, *9*, 2625. [CrossRef] [PubMed]

56. Nohara, C.; Hiyama, A.; Taira, W.; Otaki, J.M. Robustness and radiation resistance of the pale grass blue butterfly from radioactively contaminated areas: A possible case of adaptive evolution. *J. Hered.* **2018**, *109*, 188–198. [CrossRef] [PubMed]
57. Hancock, S.; Vo, N.T.K.; Omar-Nazir, L.; Batlle, J.V.I.; Otaki, J.M.; Hiyama, A.; Byun, S.H.; Seymour, C.B.; Mothersill, C. Transgenerational effects of historic radiation dose in pale grass blue butterflies around Fukushima following the Fukushima Dai-ichi Nuclear Power Plant meltdown accident. *Environ. Res.* **2019**, *168*, 230–240. [CrossRef]
58. Sakauchi, K.; Taira, W.; Hiyama, A.; Imanaka, T.; Otaki, J.M. The pale grass blue butterfly in ex-evacuation zones 5.5 years after the Fukushima nuclear accident: Contributions of initial high-dose exposure to transgenerational effects. *J. Asia-Pac. Entomol.* **2020**, *23*, 242–252. [CrossRef]
59. Otaki, J.M. Fukushima's lessons from the blue butterfly: A risk assessment of the human living environment in the post-Fukushima era. *Integr. Environ. Assess. Manag.* **2016**, *12*, 667–672. [CrossRef]
60. Otaki, J.M.; Taira, W. Current status of the blue butterfly in Fukushima research. *J. Hered.* **2018**, *109*, 178–187. [CrossRef]
61. Otaki, J.M. The pale grass blue butterfly as an indicator for the biological effect of the Fukushima Daiichi Nuclear Power Plant accident. In *Low-Dose Radiation Effects on Animals and Ecosystems*; Fukumoto, M., Ed.; Springer: Singapore, 2020; pp. 239–247. [CrossRef]
62. Hiyama, A.; Taira, W.; Iwasaki, M.; Sakauchi, K.; Iwata, M.; Otaki, J.M. Morphological abnormality rate of the pale grass blue butterfly *Zizeeria maha* (Lepidoptera: Lycaenidae) in southwestern Japan: A reference data set for environmental monitoring. *J. Asia-Pac. Entomol.* **2017**, *20*, 1333–1339. [CrossRef]
63. Hiyama, A.; Taira, W.; Iwasaki, M.; Sakauchi, K.; Gurung, R.; Otaki, J.M. Geographical distribution of morphological abnormalities and wing color pattern modifications of the pale grass blue butterfly in northeastern Japan. *Entomol. Sci.* **2017**, *20*, 100–110. [CrossRef]
64. Gurung, R.D.; Taira, W.; Sakauchi, K.; Iwata, M.; Hiyama, A.; Otaki, J.M. Tolerance of high oral doses of nonradioactive and radioactive caesium chloride in the pale grass blue butterfly *Zizeeria maha*. *Insects* **2019**, *10*, 290. [CrossRef]
65. Garnier-Laplace, J.; Geras'kin, S.; Della-Vedova, C.; Beaugelin-Seiller, K.; Hinton, T.G.; Real, A.; Oudalova, A. Are radiosensitivity data derived from natural field conditions consistent with data from controlled exposures? A case study of Chernobyl wildlife chronically exposed to low dose rates. *J. Environ. Radioact.* **2013**, *121*, 12–21. [CrossRef] [PubMed]
66. Beaugelin-Seiller, K.; Della-Vedova, C.; Garnier-Laplace, J. Is non-human species radiosensitivity in the lab a good indicator of that in the field? Making the comparison more robust. *J. Environ. Radioact.* **2020**, *211*, 105870. [CrossRef]
67. Otaki, J.M. Understanding low-dose exposure and field effects to resolve the field-laboratory paradox: Multifaceted biological effects from the Fukushima nuclear accident. In *New Trends in Nuclear Science*; Awwad, N.S., AlFaify, S.A., Eds.; IntechOpen: London, UK, 2018; pp. 49–71. [CrossRef]
68. Tangwatcharin, P.; Khopaibool, P. Activity of virgin coconut oil, lauric acid or monolaurin in combination with lactic acid against *Staphylococcus aureus*. *Southeast Asian J. Trop. Med. Public Health* **2012**, *43*, 969–985.
69. Herdiyati, Y.; Astrid, Y.; Shadrina, A.A.N.; Wiani, I.; Satari, M.H.; Kurnia, D. Potential fatty acid as antibacterial agent against oral bacteria of *Streptococcus mutans* and *Streptococcus sanguinis* from basil (*Ocimum americanum*): *In vitro* and *in silico* studies. *Curr. Drug Discov. Technol.* **2021**, *18*, 532–541. [CrossRef] [PubMed]
70. Muniyan, R.; Gurunathan, J. Lauric acid and myristic acid from *Allium sativum* inhibit the growth of *Mycobacterium tuberculosis* H37Ra: *In silico* analysis reveals possible binding to protein kinase B *Pharm. Biol.* **2016**, *54*, 2814–2821. [CrossRef] [PubMed]
71. Walters, D.R.; Walker, R.L.; Walker, K.C. Lauric acid exhibits antifungal activity against plant pathogenic fungi. *J. Phytopathol.* **2003**, *151*, 228–230. [CrossRef]
72. Liang, C.; Gao, W.; Ge, T.; Tan, X.; Wang, J.; Liu, H.; Wang, Y.; Han, C.; Xu, Q.; Wang, Q. Lauric acid is a potent biological control agent that damages the cell membrane of *Phytophthora sojae*. *Front. Microbiol.* **2021**, *12*, 666761. [CrossRef]
73. Kannathasan, K.; Senthilkumar, A.; Venkatesalu, V.; Chandrasekaran, M. Larvicidal activity of fatty acid methyl esters of *Vitex* species against *Culex quinquefasciatus*. *Parasitol. Res.* **2008**, *103*, 999–1001. [CrossRef]
74. Wang, X.; Wang, S.; Yi, J.; Li, Y.; Liu, J.; Wang, J.; Xi, J. Three host plant volatiles, hexanal, lauric acid, and tetradecane, are detected by an antenna-biased expressed odorant receptor 27 in the dark black chafer *Holotrichia parallela*. *J. Agric. Food Chem.* **2020**, *68*, 7316–7323. [CrossRef]
75. Langer, S.Z. History and nomenclature of α_1-adrenoceptors. *Eur. Urol.* **1999**, *36*, 2–6. [CrossRef]
76. MacDonald, R.; Wilt, T.J. Alfuzosin for treatment of lower urinary tract symptoms compatible with benign prostatic hyperplasia: A systematic review of efficacy and adverse effects. *Urology* **2005**, *66*, 780–788. [CrossRef] [PubMed]
77. McKeage, K.; PLoSker, G.L. Alfuzosin: A review of the therapeutic use of the prolonged-release formulation given once daily in the management of benign prostatic hyperplasia. *Drugs* **2002**, *62*, 633–653. [CrossRef] [PubMed]
78. Wilde, M.I.; Fitton, A.; McTavish, D. Alfuzosin. A review of its pharmacodynamic and pharmacokinetic properties, and therapeutic potential in benign prostatic hyperplasia. *Drugs* **1993**, *45*, 410–429. [CrossRef] [PubMed]
79. Jomon, K.; Kuroda, Y.; Ajisaka, M.; Sakai, H. A new antibiotic, ikarugamycin. *J. Antibiot.* **1972**, *25*, 271–280. [CrossRef] [PubMed]
80. Liu, C.; Wang, X.; Zhao, J.; Liu, Q.; Wang, L.; Guan, X.; He, H.; Xiang, W. *Streptomyces harbinensis* sp. nov., an endophytic, ikarugamycin-producing actinomycete isolated from soybean root [*Glycine max* (L.) Merr]. *Int. J. Syst. Evol. Microbiol.* **2013**, *63*, 3579–3584. [CrossRef]
81. Qin, S.; Xing, K.; Jiang, J.-H.; Xu, L.-H.; Li, W.-J. Biodiversity, bioactive natural products and biotechnological potential of plant-associated endophytic actinobacteria. *Appl. Microbiol. Biotechnol.* **2011**, *89*, 457–473. [CrossRef]

82. Shimizu, M. Endophytic actinomycetes: Biocontrol agents and growth promoters. In *Bacteria in Agrobiology: Plant Growth Responses*; Maheshwari, D., Ed.; Springer: Berlin/Heidelberg, Germany, 2011; pp. 201–220. [CrossRef]
83. Golinska, P.; Wypij, M.; Agarkar, G.; Rathod, D.; Dahm, H.; Rai, M. Endophytic actinobacteria of medicinal plants: Diversity and bioactivity. *Antonie van Leeuwenhoek* **2015**, *108*, 267–289. [CrossRef]
84. Grover, M.; Bodhankar, S.; Maheswari, M.; Srinivasarao, C. Actinomycetes as mitigators of climate change and abiotic stress. In *Plant Growth Promoting Actinobacteria*; Subramaniam, G., Arumugam, S., Rajendran, V., Eds.; Springer: Singapore, 2016; pp. 203–212. [CrossRef]
85. Kuldau, G.; Bacon, C. Clavicipitaceous endophytes: Their ability to enhance resistance of grasses to multiple stresses. *Biol. Control* **2008**, *46*, 57–71. [CrossRef]
86. Rodriguez, R.J.; Henson, J.; Volkenburgh, E.V.; Hoy, M.; Wright, L.; Beckwith, F.; Kim, Y.-O.; Redman, R.S. Stress tolerance in plants via habitat-adapted symbiosis. *ISME J.* **2008**, *2*, 404–416. [CrossRef]
87. Peng, A.; Liu, J.; Gao, Y.; Chen, Z. Distribution of endophytic bacteria in *Alopecurus aequalis* Sobol and *Oxalis corniculata* L. from soils contaminated by polycyclic aromatic hydrocarbons. *PLoS ONE* **2013**, *8*, e83054. [CrossRef]
88. Mufti, R.; Amna; Rafique, M.; Haq, F.; Munis, M.F.H.; Masood, S.; Mumtaz, A.S.; Chaudhary, H.J. Genetic diversity and metal resistance assessment of endophytes isolated from *Oxalis corniculata*. *Soil Environ.* **2015**, *34*, 89–99.
89. Lacret, R.; Oves-Costales, D.; Gómez, C.; Diaz, C.; de la Cruz, M.; Pérez-Victoria, I.; Vicente, F.; Genilloud, O.; Reyes, F. New ikarugamycin derivatives with antifungal and antibacterial properties from *Streptomyces zhaozhouensis*. *Mar. Drugs* **2014**, *13*, 128–140. [CrossRef] [PubMed]
90. Saeed, S.I.; Aklilu, E.; Mohammedsalih, K.M.; Adekola, A.A.; Mergani, A.E.; Mohamad, M.; Kamaruzzaman, N.F. Antibacterial activity of ikarugamycin against intracellular *Staphylococcus aureus* in bovine mammary epithelial cells in vitro infection model. *Biology* **2021**, *10*, 985. [CrossRef] [PubMed]
91. Elkin, S.R.; Oswald, N.W.; Reed, D.K.; Mettlen, M.; MacMillan, J.B.; Schmid, S. Ikarugamycin: A natural product inhibitor of clathrin-mediated endocytosis. *Traffic* **2016**, *17*, 1139–1149. [CrossRef] [PubMed]
92. Yamaguchi, M.; Matsuyama, S.; Yamaji, K. Oxalic acid as a larval feeding stimulant for the pale grass blue butterfly *Zizeeria maha* (Lepidoptera: Lycaenidae). *Appl. Entomol. Zool.* **2016**, *51*, 91–98. [CrossRef]
93. Ito, T. Nutritional requirements and artificial diets for the silkworm, *Bombyx mori* L. *J. Sericult. Sci. Jpn.* **1967**, *36*, 315–319. [CrossRef]
94. Lin, K.; Yamada, H.; Kato, M. Free fatty acids promote feeding behavior of the silk-worm, *Bombyx mori* L. *Memories Faculty Sci. Kyoto Univ. Ser. Biol.* **1971**, *4*, 108–115. Available online: http://hdl.handle.net/2433/258804 (accessed on 29 March 2022).
95. Furukawa, Y.; Sakiyama, T.; Wakamiya, K.; Ichikou, J.; Matsumura, T.; Mizokami, H. Comparison of oviposition and growth in response to the host plant differences in the pale grass blue butterfly. *Chem. Biol.* **2020**, *58*, 640–643. (In Japanese) [CrossRef]
96. Pang, Z.; Chong, J.; Zhou, G.; de Lima Morais, D.A.; Chang, L.; Barrette, M.; Gauthier, C.; Jacques, P.É.; Li, S.; Xia, J. MetaboAnalyst 5.0: Narrowing the gap between raw spectra and functional insights. *Nucl. Acids Res.* **2021**, *49*, W388–W396. [CrossRef]
97. Nishida, R. Sequestration of plant secondary compounds by butterflies and moths. *Chemoecology* **1994**, *5*, 127–138. [CrossRef]
98. Roeder, T. The control of metabolic traits by octopamine and tyramine in invertebrates. *J. Exp. Biol.* **2020**, *223*, jeb194282. [CrossRef] [PubMed]

life

MDPI

Article

Metabolomic Profiles of the Creeping Wood Sorrel *Oxalis corniculata* in Radioactively Contaminated Fields in Fukushima: Dose-Dependent Changes in Key Metabolites

Ko Sakauchi [1], Wataru Taira [1,2] and Joji M. Otaki [1,*]

[1] The BCPH Unit of Molecular Physiology, Department of Chemistry, Biology and Marine Science, Faculty of Science, University of the Ryukyus, Okinawa 903-0213, Japan; yamatoshijimi@sm1044.skr.u-ryukyu.ac.jp (K.S.); wataira@lab.u-ryukyu.ac.jp (W.T.)

[2] Research Planning Office, University of the Ryukyus, Okinawa 903-0213, Japan

* Correspondence: otaki@sci.u-ryukyu.ac.jp; Tel.: +81-98-895-8557

Abstract: The biological impacts of the Fukushima nuclear accident, in 2011, on wildlife have been studied in many organisms, including the pale grass blue butterfly and its host plant, the creeping wood sorrel *Oxalis corniculata*. Here, we performed an LC–MS-based metabolomic analysis on leaves of this plant collected in 2018 from radioactively contaminated and control localities in Fukushima, Miyagi, and Niigata prefectures, Japan. Using 7967 peaks detected by LC–MS analysis, clustering analyses showed that nine Fukushima samples and one Miyagi sample were clustered together, irrespective of radiation dose, while two Fukushima (Iitate) and two Niigata samples were not in this cluster. However, 93 peaks were significantly different (FDR < 0.05) among the three dose-dependent groups based on background, low, and high radiation dose rates. Among them, seven upregulated and 15 downregulated peaks had single annotations, and their peak intensity values were positively and negatively correlated with ground radiation dose rates, respectively. Upregulated peaks were annotated as kudinoside D (saponin), andrachcinidine (alkaloid), pyridoxal phosphate (stress-related activated vitamin B6), and four microbe-related bioactive compounds, including antibiotics. Additionally, two peaks were singularly annotated and significantly upregulated ($K_1R_1H_1$; peptide) or downregulated (DHAP(10:0); decanoyl dihydroxyacetone phosphate) most at the low dose rates. Therefore, this plant likely responded to radioactive pollution in Fukushima by upregulating and downregulating key metabolites. Furthermore, plant-associated endophytic microbes may also have responded to pollution, suggesting their contributions to the stress response of the plant.

Keywords: metabolome; LC–MS; Fukushima nuclear accident; plant physiology; radioactive pollution; *Oxalis corniculata*; creeping wood sorrel; endophytic microbe; stress response

Citation: Sakauchi, K.; Taira, W.; Otaki, J.M. Metabolomic Profiles of the Creeping Wood Sorrel *Oxalis corniculata* in Radioactively Contaminated Fields in Fukushima: Dose-Dependent Changes in Key Metabolites. *Life* **2022**, *12*, 115. https://doi.org/10.3390/life12010115

Academic Editors: Fabrizio Ambrosino and Supitcha Chanyotha

Received: 23 December 2021
Accepted: 11 January 2022
Published: 13 January 2022

1. Introduction

Environmental pollution caused by human activities is widespread around the globe in the 21st century. Major incidents of pollution after World War II include the Great Smog in London, UK (1952) caused by particulates and gaseous mixtures, the Minamata disease outbreak in Japan (1956) caused by methylmercury, Agent Orange used during the Vietnam War (1961–1971), and the *Deepwater Horizon* oil spill accident (2010) in the Gulf of Mexico [1]. Additionally, recent human history has seen a series of pollution incidents by anthropogenic radionuclides: atomic bombs used in Hiroshima and Nagasaki, Japan (1945); atomic and hydrogen bomb experiments in Bikini Atoll (1946–1958); the Three Mile Island accident in the USA (1979); the Chernobyl nuclear accident in the Ukraine (1986); and the Fukushima nuclear accident, Japan (2011) [1]. The Fukushima nuclear accident in 2011 was the second largest nuclear accident next to the Chernobyl nuclear accident in 1986. Without question, one of the most serious environmental pollutants in this century is a group of radioactive materials released from nuclear bombs and the collapse of nuclear power plants. Today, anthropogenic ^{137}Cs is detected from soil worldwide [2–5].

In the case of the Chernobyl nuclear accident, there have been inconsistencies in the biological impacts of relatively low-level radiation exposure on organisms in the surrounding environments [6–10]. There seem to be many reasons for these inconsistencies, but one reason may be political; the Chernobyl nuclear accident occurred in the former Soviet Union, and access to the polluted areas was limited. Another important reason may be technical. At the time of the Chernobyl nuclear accident, none of the currently available analysis technologies based on genomics, proteomics, and metabolomics had been developed. In the case of the Fukushima nuclear accident, some scientists began investigating the biological effects soon after the accident using various wild animals and plants because access to the polluted areas was not difficult from a political standpoint. There is now accumulating field-based evidence that the Fukushima nuclear accident impacted animals and plants, including birds [11–13], butterflies [14–17], aphids [18,19], Japanese monkeys [20–22], intertidal invertebrates [23], and plants [24–29], even at relatively low levels of anthropogenic radiation. However, the application of advanced technologies such as metabolomics in studies on Fukushima has not yet been sufficient.

In this study, we focused on a weed plant, the creeping wood sorrel *Oxalis corniculata*, in a contaminated field in Fukushima. This plant is the host plant of the pale grass blue butterfly, which has been used as an indicator species in Fukushima-based studies. Larvae of this butterfly eat only this plant. It has been demonstrated that the pale grass blue butterfly was impacted both genetically and physiologically by the Fukushima nuclear accident. More precisely, in view of genetic damage, the inheritance of mutation-related phenotypes over generations has been demonstrated [14,15,30,31]. In terms of physiological damage, it has been demonstrated that the ingestion of contaminated plants by butterfly larvae caused internal radiation exposure and resulted in abnormal and fatal phenotypes [14,15,32–35], although adaptive evolution to tolerate radioactive pollutants may occur over generations [36]. However, the ingestion of a ^{137}Cs-containing artificial diet by larvae did not decrease survival rate, pupation rate, and eclosion rate [37]. Therefore, a positive involvement of the plant itself has been suggested to cause abnormal or fatal phenotypes in butterflies based on internal exposure experiments [38]. The plant may have experienced biochemical changes in leaves in response to radiation exposure, which has led to harmful consequences in butterflies. This field effect hypothesis is reasonable, considering that at least some plants responded to Fukushima pollution at the levels of gene expression and phenotype [24–29]. Physiological damage to butterflies is likely mediated by multiple pathways, but one of them includes biochemical plant changes in response to radiation exposure, such as changes in nutritional contents [28] and changes in secondary metabolites [29].

Plants produce secondary metabolites and proteins that are toxic to herbivorous animals such as insects. These phytotoxins include a wide variety of chemical compounds, such as cyanogens, glycoalkaloids, glucosinolates, saponins, flavones, nonprotein amino acids, furanocoumarins, condensed tannins, gossypol, protease inhibitors, lectins, and threonine dehydratase [39]. It is generally believed that herbivorous insects have evolved to cope with phytotoxins; the larvae of many species of butterflies feed on leaves containing phytotoxins such as cyanogenic glucosides and have the ability to sequester them [40]. These butterflies use these chemicals for their own defense, although many other lepidopteran insects can de novo synthesize cyanogenic glucosides [41]. The field effect hypothesis above, thus, posits that a delicate balance between phytotoxins in plants and the tolerance of phytotoxins in insects in ecosystems may have been affected by radioactive pollution.

In the present study, to examine changes in the metabolites found in *O. corniculata* in response to anthropogenic environmental radiation, we performed an LC–MS-based metabolomic analysis using plant leaf samples collected from 14 localities with various levels of ^{137}Cs contamination, including Fukushima, Miyagi, and Niigata prefectures, and examined whether there were any LC–MS peaks that changed based on the ground radiation dose rate. In this way, we found candidate compounds that were upregulated or downregulated in response to different levels of radiation exposure in the plant.

2. Materials and Methods

2.1. Field Sampling

We visited 14 localities in the period from 29 July 2018 to 17 September 2018 (Figure 1a–e), and two people collected leaves of the creeping wood sorrel *O. corniculata*. These localities were not affected by the tsunamis from the Great East Japan Earthquake on 11 March 2011, excluding its potential effects on the plant. Information on sampling sites and dates is listed in Table 1. Leaf samples were named OC01 to OC16 for each locality. OC05 (Minamisoma-1) and OC12 (Iwaki) were collected but were not analyzed for financial reasons.

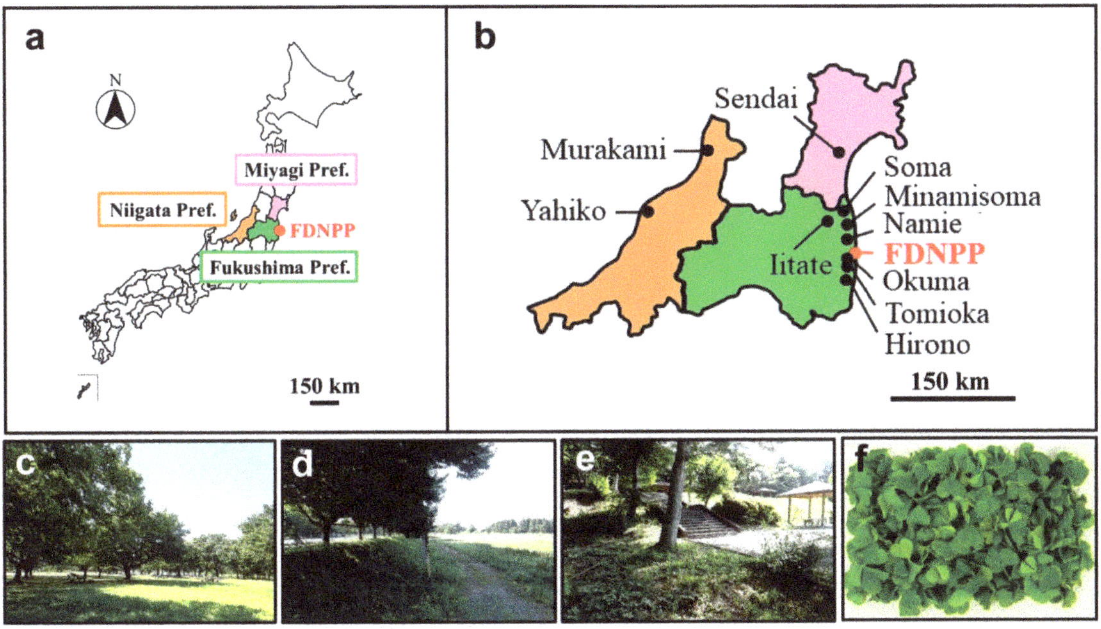

Figure 1. Leaf sampling: (**a**) Prefectures that include sampling localities in this study. The Fukushima Daiichi Nuclear Power Plant (FDNPP) is indicated in red; (**b**) municipalities that include 14 sampling localities in Fukushima, Miyagi, and Niigata prefectures. Minamisoma, Namie, and Iitate each have 2 or 3 independent sampling localities (see Table 1); (**c**–**e**) landscapes of sampling sites in Murakami (**c**), Minamisoma-2 (**d**), and Hirono (**e**); (**f**) Stem-separated leaves of *Oxalis corniculata* collected in Tomioka (photographed at the Kazusa DNA Institute upon sample receipt). Leaf samples from all other localities were similarly healthy when they arrived at the Institute.

Table 1. Sampling information for the leaf samples, ground radiation dose rates, and radioactivity concentrations.

Sample Name	Sampling Locality	Date (2018)	Ground Dose Rate [μSv/h] (*1)	^{137}Cs Radioactivity Concentration [Bq/kg]
OC01	Murakami City, Niigata Pref.	29 Jul	0.07 (B)	0
OC02	Yahiko Village, Niigata Pref.	30 Jul	0.04 (B)	0
OC03	Sendai City, Miyagi Pref.	31 Jul	0.04 (B)	4.54
OC04	Soma City, Fukushima Pref.	31 Jul	0.10 (L)	74.45
OC06	Minamisoma City, Fukushima Pref. (Minamisoma-2)	31 Jul	0.42 (L)	84.27
OC07	Hirono Town, Fukushima Pref.	1 Aug	0.11 (L)	7.96
OC08	Namie Town, Fukushima Pref. (Namie-1)	1 Aug	0.97 (L)	64.10
OC09	Namie Town, Fukushima Pref. (Namie-2)	1 Aug	2.45 (H)	551.16
OC10	Okuma Town, Fukushima Pref.	1 Aug	0.60 (L)	424.45
OC11	Tomioka Town, Fukushima Pref.	1 Aug	0.27 (L)	135.98
OC13	Iitate Village, Fukushima Pref. (Iitate-1)	17 Sep	3.50 (H)	213.72
OC14	Iitate Village, Fukushima Pref. (Iitate-2)	17 Sep	2.94 (H)	494.74
OC15	Namie Town, Fukushima Pref. (Namie-3)	17 Sep	4.55 (H)	717.65
OC16	Minamisoma City, Fukushima Pref. (Minamisoma-3)	17 Sep	1.46 (H)	175.23

(*1) Three groups were set depending on the relative levels of ground radiation dose rate (R): H (high, $R \geq 1.00$); L (low, $0.10 \leq R < 1.00$); and B (background, $R < 0.10$). Samples OC05 (Minamisoma-1) and OC12 (Iwaki) were collected but not analyzed.

Leaf sample collection procedures followed those described in a previous study [28]. Briefly, the plant leaves and stems were handpicked with disposable gloves so as not to damage the leaves. Leaves were further isolated from the stem. We collected leaves that were healthy and showed no signs of leaf necrosis, chlorosis, or other abnormalities. In other words, we observed no phenotypic changes under radiation stress. Leaves with damage (by insect bites, handpicking, or other unknown reasons), dead or dying leaves, leaves of different species, and other objects were eliminated manually. A minimum of 40 g of leaf samples per site was collected. Leaf samples were washed with Evian bottled natural mineral water (Evian les Bains, France).

The leaf samples (minimum of 10 g per site) were sent to the Kazusa DNA Research Institute, Kisarazu, Chiba, Japan, under refrigeration (unfrozen) conditions (0−10 °C) for LC–MS analyses. The samples arrived at the Institute within a day. At the time of arrival, leaf quality was visually checked again at the Institute; the leaves were reasonably fresh and green (Figure 1f). A portion of the leaf samples (approximately 30 g per site) was saved for an analysis of radioactivity concentration at the University of the Ryukyus.

2.2. *Measurements of Ground Radiation Dose Rates and Radioactivity Concentrations*

At the sampling sites, we measured the ground radiation dose rate (often simply called the ground dose) using a Hitachi Aloka Medical TCS-172B scintillation survey meter (Tokyo, Japan) for 90 s at 3 points in the area of leaf collection with the probe at 0 cm from the ground surface. The ground dose was measured similarly in two localities of Iitate Village (Iitate-1 and Iitate-2), one locality of Namie Town (Namie-3), and one locality of Minamisoma (Minamisoma-3) using a Polimaster handheld gamma monitor PM1710A (Minsk, Republic of Belarus). The measured values were averaged, and they are shown in Table 1.

Procedures for measuring radioactivity concentrations were described elsewhere [28]. Briefly, the radioactivity concentration of a dried leaf sample was measured using a Canberra GCW-4023 germanium semiconductor radiation detector (Meriden, CT, USA). Measurements were conducted to obtain ^{137}Cs signals until the error rate became less than 5% of the measured value within 14 days of the measurement period. Otherwise, the measurements were terminated at the end of the 14th day. In that case, a measurement value was not obtained, and it was considered zero. The results were listed in Table 1.

Ground dose and radioactivity concentration were not perfectly correlated (Supplementary Results and Supplementary Discussion, Figure S1). Based on the following considerations, we decided to preferentially use ground dose values. The leaves were more likely to be subjected to external irradiation from the ground than to internal irradiation from absorbed ^{137}Cs because the plant was small, the leaves were very close to the ground, and the ground radiation included complete radiation doses of various radionuclides.

2.3. LC–MS: Analysis, Peak Detection, Alignment, and Annotation

The procedures for LC–MS, including analysis, peak detection, alignment, and annotation, were described elsewhere [29]. Briefly, leaf samples were prepared using methanol and MonoSpin M18 columns (GL Sciences, Tokyo, Japan). Samples were analyzed using a SHIMADZU Nexera X2 high-performance liquid chromatography (HPLC) instrument (Kyoto, Japan) with an InertSustain AQ-C18 column (2.1 × 150 mm, 3 μm particle size) (GL Sciences) connected to a Thermo Fisher Scientific Q Exactive Plus high-resolution mass analyzer (Waltham, MA, USA).

The LC–MS data obtained above were converted to mzXML format using ProteoWizard (Palo Alto, CA, USA). Peak detection, determination of ionizing states, and peak alignments were performed automatically using the data analysis software PowerGet-Batch developed by the Kazusa DNA Research Institute [42,43]. The exact mass values of the nonionized compounds calculated from the adducts were used to search candidate compounds against the UC2 chemical mass databases [44] (i.e., a combination of two databases, KNApSAcK [45] and the Human Metabolome Database [46,47]) with the search program MFSearcher [48]. The LC–MS results were compiled in the Microsoft Excel file "LCMS_Result Field Data KDRI" (Supplementary File S1).

2.4. Statistical Analysis of the Peak Area Data

The output peak area (intensity) data from LC–MS were compiled in the Microsoft Excel file "LCMS Peak Data" (Supplementary File S2). These data were subjected to statistical analyses using MetaboAnalyst 5.0 [49–51], as described elsewhere [29]. We performed one-way ANOVA (analysis of variance) and used FDR (false discovery rate) < 0.05 as the criterion to consider statistical significance, and the peaks that met this criterion were examined independently, after which a Student's t-test was performed as necessary, using Microsoft Excel. A principal component analysis (PCA) and heatmap analysis were performed to obtain possible relationships among the samples. In the latter, the Euclidean distance and the Ward linkage method were employed for clustering. Scatter plots were made, and mathematical model fits for linear ($y = ax + b$) and logarithmic curves ($y = a \times ln(x) + b$) were performed using Microsoft Excel. Correlation coefficients for linear and logarithmic curves were obtained using JSTAT (Yokohama, Japan).

3. Results

3.1. Clustering Analyses: PCA and Heatmaps

In the LC–MS analysis, 9554 peaks were detected, and 7967 peaks from 14 leaf samples were treated as valid peaks by MetaboAnalyst; 1587 peaks were treated as invalid because they showed a constant value across all samples or because they were detected only in one sample. To understand how these 14 samples from different localities responded to radiation, they were categorized into three groups depending on the ground radiation dose (high, low, and background levels) (Table 1). The PCA was performed using the 7967 peak area (intensity) data allocated to the 14 leaf samples (Figure 2a). PC1 and PC2 explained 24.0% and 16.9% of the variance, respectively. These percentages were not very high. The three dose-dependent groups were not well isolated from one another. Clustering by K-means (Figure 2b) and SOM (self-organizing map) (not shown) without predefining the dose-dependent groups showed that just a single group was statistically valid even when the number of groups was specified to be three. The single large group specified by K-means included nine Fukushima samples and a Miyagi sample and did not include two Fukushima (Iitate) samples (OC13 and OC14) and two Niigata samples (OC1 and OC2). This large cluster contained samples from all three dose-dependent groups.

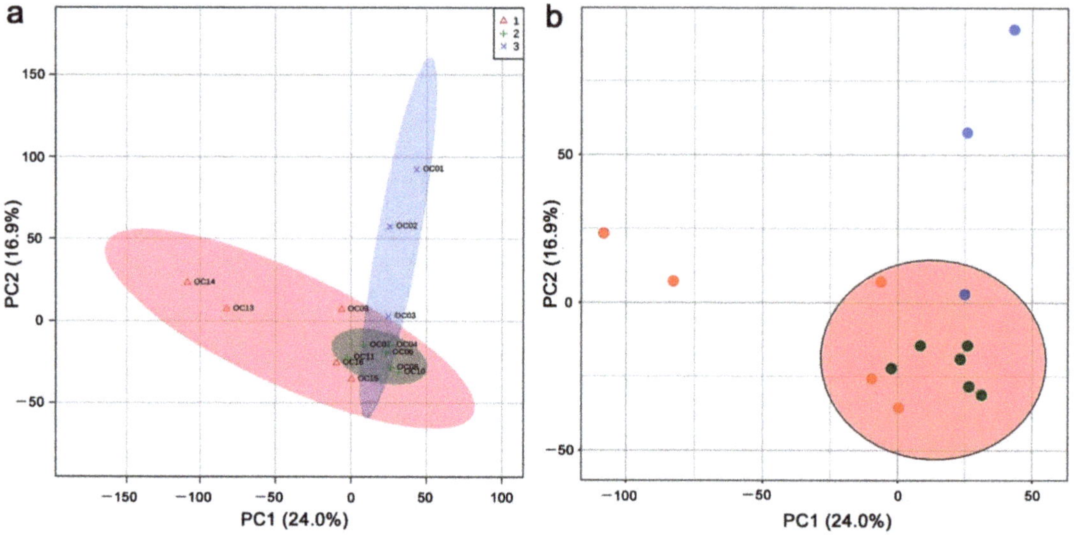

Figure 2. PCA using leaf samples from 14 localities in Fukushima, Miyagi, and Niigata prefectures: (**a**) Score plot, 95% confidence ranges are colored. Red, green, and blue sample dots and areas indicate high, low, and background dose-dependent groups, respectively; (**b**) K-means clustering analysis.

In the PCA plot, two spots in the negative area of PC1 (OC13 and OC14) were both from Iitate Village (Fukushima Prefecture, Japan), which is located at a relatively high altitude and, thus, geologically isolated from the rest of the Fukushima localities. Another two spots in the positive area of PC2 (OC1 and OC2) were both from Niigata Prefecture, which is located on the west side of Japan (Figure 1a,b). These four samples were likely genetically or environmentally different from the rest. These results suggest that the environmental radiation dose was not a primary factor influencing peak levels in LC–MS. In other words, in terms of overall peak dynamics, the plant may not respond strongly to environmental radiation.

A heatmap of all 7967 peaks also demonstrated that the three dose-dependent groups were not well justified (Figure 3a). An exception was the low-level group (shown in green bar at the top), which clustered together. However, the three groups were individually clustered when only the top 25 peaks were used to create the heatmap (Figure 3b). It appeared that a limited number of representative peaks responded to environmental radiation in a dose-dependent manner.

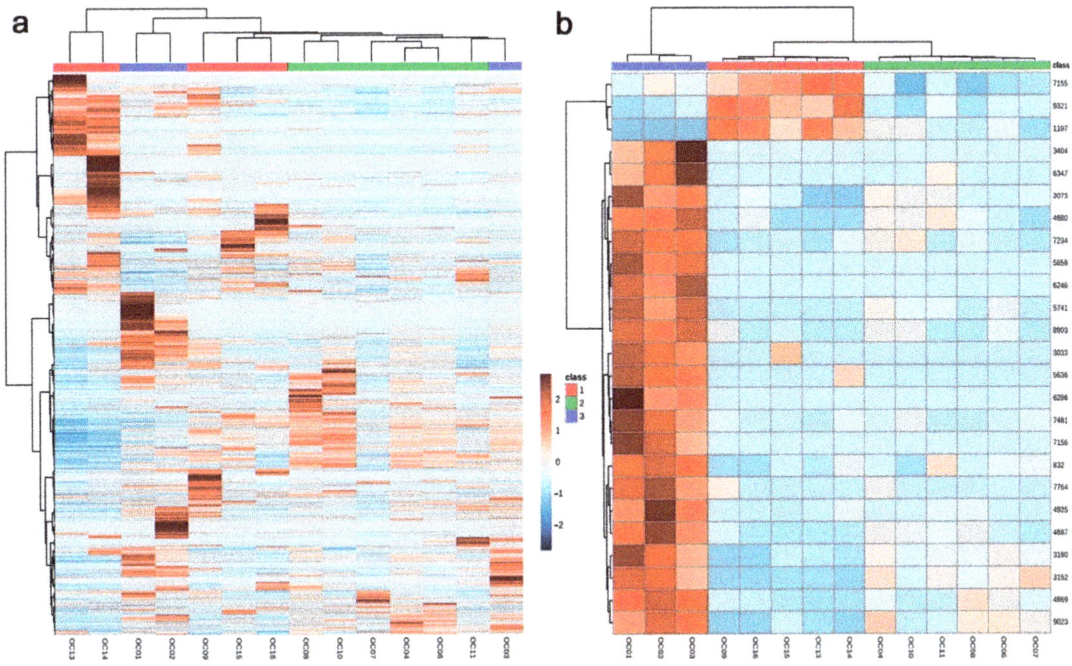

Figure 3. Heatmap analysis using samples from 14 localities in Fukushima, Miyagi, and Niigata prefectures. Red, green, and blue bars at the top of the heatmap indicate high, low, and background dose-dependent groups, respectively: (**a**) Heatmap using all peaks; (**b**) heatmap using the top 25 peaks.

3.2. Identification of Upregulated and Downregulated Peaks

Although no overall pattern justifying the three dose-dependent groups was observed in the PCA plot and heatmap using all peaks, there may have been some metabolites that were upregulated or downregulated in a dose-dependent manner. To examine this possibility, we performed one-way ANOVA with an adjusted p-value (FDR) cutoff at 0.05 (i.e., FDR < 0.05). We detected 93 significantly different peaks among the three groups; with FDR < 0.01, we detected 27 significantly different peaks and with FDR < 0.001, we detected two significantly different peaks (Supplementary Table S1 and Figure 4). After visual inspections of the peak values, the following numbers were obtained (Figure 5a): Among the 93 peaks, 15 peaks seemed to be dose-dependently upregulated; four peaks seemed to be upregulated (two peaks) or downregulated (two peaks) only at the low-level radiation, showing V-shaped or reversed V-shaped curves (i.e., irregular peaks); the rest (74 peaks) seemed to be downregulated.

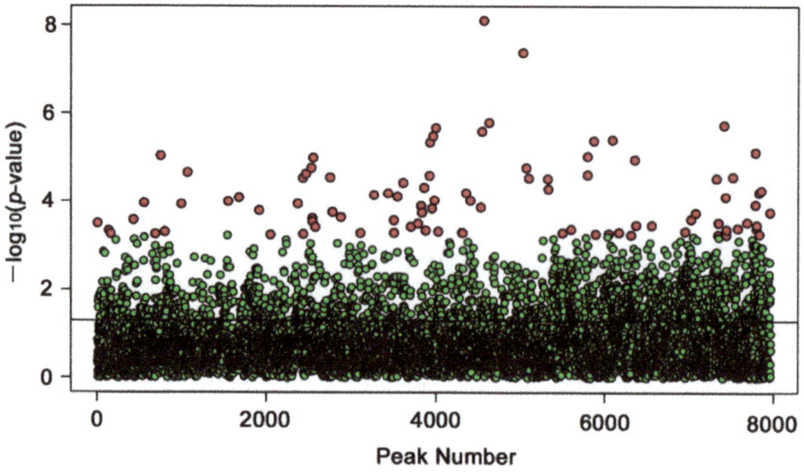

Figure 4. Plot of one-way ANOVA. A horizontal line at 1.301 on the y-axis indicates $p = 0.05$ (raw p-value). Red dots indicate peaks with FDR < 0.05. The peak number in the x-axis is adjusted for the valid number of peaks (i.e., 7967 peaks) by MetaboAnalyst.

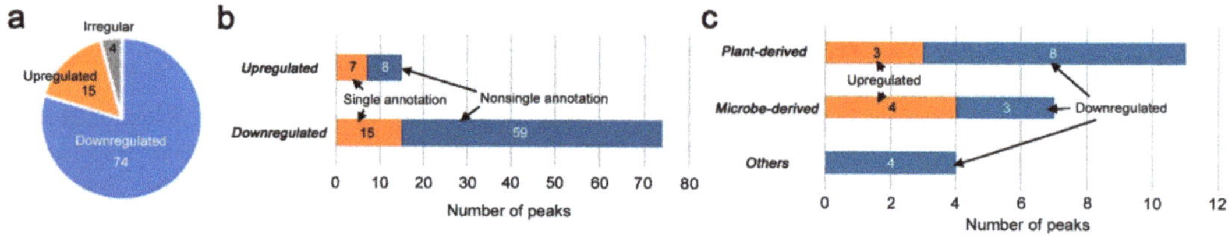

Figure 5. Number of upregulated and downregulated peaks of LC–MS with FDR < 0.05 (one-way ANOVA): (**a**) Pie chart; (**b**) number of upregulated and downregulated peaks with single or nonsingle annotations; (**c**) number of upregulated and downregulated peaks categorized into 3 groups, i.e., plant-derived compounds, microbe-derived compounds, and others.

Among the 15 upregulated peaks, seven peaks had single annotations (Figure 5b). Similarly, among the 74 downregulated peaks, 15 peaks had single annotations (Figure 5b). These peaks with single annotations were classified into three categories based on their origins: plant-derived compounds, microbe-derived compounds, and other compounds, including those of unknown or synthetic origin (Figure 5c). Plant-derived compounds were the most frequent, as expected, but unexpectedly, many microbe-derived compounds were found to be upregulated or downregulated.

3.3. Candidate Compounds for Upregulated and Downregulated Peaks

Upregulated peaks with single annotations (FDR < 0.05) were found, as shown in Table 2. Among them, only three upregulated peaks were considered to be plant derived. Candidate compounds for the upregulated peaks included kudinoside D, a triterpenoid saponin that has shown pharmacological activities in animal cells [52]; andrachcinidine, an alkaloid [53]; and pyridoxal phosphate, an active form of vitamin B6 that is involved in stress tolerance [54–63]. Other upregulated peaks were annotated as antibiotics, i.e., leptomycin B [64–66] and aldgamycin G [67], both from *Streptomyces*, a group of soil bacteria. Leptomycin B (No. 9321) was also found in the top 25 peaks for the heatmap (Figure 3b). Indeed, this peak showed the lowest FDR value (FDR = 0.0055) among the singularly annotated upregulated peaks. Additional compounds included carbamidocyclophane C, a cytotoxic compound derived from the cyanobacterium *Nostoc* sp. [68], and YM 47525, an antifungal compound of fungal origin [69].

Table 2. Summary of the upregulated peaks with single annotations.

No.	Formula	Exact Mass	Annotation (Compound Name)	Possible Function	Origin
9321	$C_{33}H_{48}O_6$	540.345	Antibiotic Cl 940, Antibiotic CL 1957A, Cl 940, Elactocin, Leptomycin B, Mantuamycin	Antibiotics	*Streptomyces* sp. (Microbe-derived)
9368	$C_{47}H_{72}O_{17}$	908.477	Kudinoside D	Triterpenoid saponin; Adipogenesis suppressor	*Ilex kudingucha* (Plant-derived)
8451	$C_{38}H_{56}O_8N_2Cl_2$	738.341	Carbamidocyclophane C, (+)-Carbamidocyclophane C	Cytotoxic compound	*Nostoc* sp. (Cyanobacteria) (Microbe-derived)
7968	$C_{37}H_{56}O_{15}$	740.362	Aldgamycin G	Antibiotics	*Streptomyces* sp. (Microbe-derived)
8935	$C_{13}H_{25}O_2N$	227.189	Andrachcinidine, (-)-Andrachcinidine	Alkaloid	*Andrachne aspera* (Plant-derived)
7563	$C_{33}H_{46}O_{11}$	618.304	YM 47525	Trichothecene; Fungicide	Fungus (Microbe-derived)
178	$C_8H_{10}O_6NP$	247.025	Pyridoxal phosphate	Activated vitamin B6	Plants and microbes (Plant-derived) (*1)

Note: This table lists candidate compounds in order of smaller FDR values. Raw *p*-values and FDR values for these peaks are listed in Supplementary Table S1. (*1) There is a possibility that No. 178 was derived from endophytic microbes because they also synthesize this compound (but see Section 4. Discussion).

Downregulated peaks with single annotations (FDR < 0.05) were found, as shown in Table 3. Among them, five peaks with the lowest FDR values (No. 4887, 7156, 6296, 3152, and 3073) were found in the top 25 peaks for the heatmap (Figure 3b). Candidate compounds for downregulated peaks included various types of chemicals. Plant-derived compounds were as follows: corchoionoside B (fatty acyl glucoside) [70], isoginkgetin-7-*O*-β-D-glucopyranoside (bioflavone glucoside) [71,72], sanjoinine A dialdehyde (cyclopeptide alkaloid) [73], zinolol (antioxidant) [74], acacetin-7-glucuronosyl-(1→2)-glucuronide (flavonoid) [75], tricalysioside N (*ent*-kaurane glucoside) [76], pregnadienolone-3-*O*-β-D-chacotrioside (saponin) [77], and silidianin (flavonolignan) [78]. Additionally, terreusinol, a metabolite of *Streptomyces* [79], and elloramycin E, an antibiotic from *Streptomyces* [80], were found. Trapoxin A (fungal cyclic peptide) [81,82] was also found. Other candidates included synthetic compounds [83–88]; thus, their annotations may not be accurate, although similar natural compounds may exist in the plant.

Table 3. Summary of the downregulated peaks with single annotations.

No.	Formula	Exact Mass	Annotation (Compound Name)	Possible Function	Origin
4887	$C_{19}H_{28}O_9$	400.173	Corchoionoside B	Fatty acyl glucoside; Membrane stabilizer	*Corchorus olitorius* (Plant-derived)
7156	$C_{38}H_{32}O_{15}$	728.174	Isoginkgetin-7-*O*-β-D-glucopyranoside	Bioflavone glucoside	*Ginkgo biloba* (Plant-derived)
6296	$C_{18}H_{22}O_5N_2$	346.153	Terreusinol, (+)-Terreusinol	Antibiotics	*Streptomyces* sp. (Microbe-derived)
3152	$C_{17}H_{16}ON_3Cl$	313.098	Amoxapine	GPCR (G-protein-coupled receptor) inhibitor	(Others, synthetic)
3073	$C_{32}H_{34}O_{15}$	658.190	Elloramycin E	Antibiotics	*Streptomyces* sp. (Microbe-derived)
8925	$C_{31}H_{42}O_6N_4$	566.310	Sanjoinine A dialdehyde	Alkaloid (Cyclopeptide)	*Zizyphus lotus* (Plant-derived) (*1)
609	$C_{14}H_{21}O_8N$	331.127	Zinolol	Antioxidant	*Anagallis onellin* (Plant-derived)

Table 3. *Cont.*

No.	Formula	Exact Mass	Annotation (Compound Name)	Possible Function	Origin
2963	$C_{28}H_{28}O_{17}$	636.133	Acacetin-7-glucuronosyl-(1→2)-glucuronide	Flavonoid	*Clerodendron trichotomum* (Plant-derived)
3171	$C_{12}H_{12}O_{3}N_{2}S$	264.057	Dapsone hydroxylamine	Dermatologically used drug	(Others, synthetic)
8804	$C_{34}H_{42}O_{6}N_{4}$	602.310	RF 1023A, Trapoxin A	Cyclic peptide; histone deacetylase inhibitor	*Helicoma ambiens* RF-1023 (Fungus) (Microbe-derived)
7781	$C_{28}H_{46}O_{11}$	558.304	Tricalysioside N, (-)-Tricalysioside N	*Ent*-kaurane glucoside	*Tricalysia dubia* (Plant-derived)
9091	$C_{39}H_{60}O_{15}$	768.393	Pregnadienolone-3-*O*-β-D-chacotrioside	Saponin	*Dioscorea panthaica* (Plant-derived)
4347	$C_{21}H_{26}ON_{3}SCl$	403.149	Perphenazine	Dopamine receptor D2 antagonist	(Others, synthetic)
3836	$C_{25}H_{24}O_{10}$	484.137	Silidianin	Flavonolignan	*Silybum marianum* (Plant-derived)
8800	$C_{22}H_{22}O_{2}N_{3}F$	379.170	Droperidol	Dopamine receptor antagonist	(Others, synthetic)

Note: This table lists candidate compounds in order of smaller FDR values. Raw *p*-values and FDR values for these peaks are listed in Supplementary Table S1. (*1) Sanjoinine A is a natural compound but sanjoinine A dialdehyde is a synthetically derived compound from sanjoinine A [73].

3.4. Correlation Analyses of Upregulated Peaks

We made scatter plots and performed a correlation analysis of the upregulated peaks with single annotations to examine dose dependence (Figure 6 and Table 4). In all seven cases, correlation coefficients were reasonably high ($r \geq 0.69$) with significantly small *p*-values. In three of seven cases, a logarithmic model fit better than a linear model, judging from correlation coefficients. Thus, these seven peaks may be upregulated in a dose-dependent manner in response to the ground radiation dose rate, confirming the ANOVA results. Correlation coefficients using the radioactivity concentration of ^{137}Cs showed lower values in all cases (Table 4).

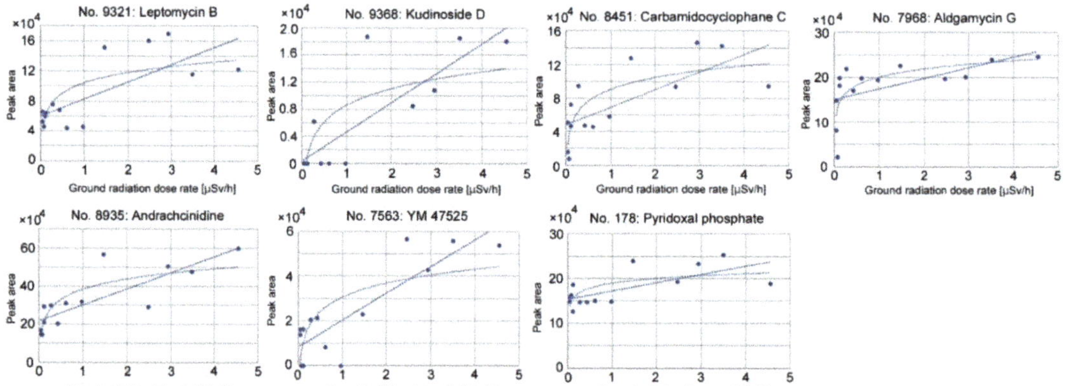

Figure 6. Scatter plots and linear and logarithmic fit curves of the upregulated peaks with single annotations against the ground radiation dose rate (μSv/h).

Table 4. Correlation coefficient *r* and its associated *p*-value of the upregulated peaks with single annotations.

No.	Brief Annotation	Ground Radiation Dose Rate [μSv/h]	Radioactivity Concentration of ^{137}Cs [Bq/kg]
9321	Leptomycin B	$r = 0.75, p = 0.0021$ ** (linear)	$r = 0.66, p = 0.0100$ * (linear)
9368	Kudinoside D	$r = 0.84, p = 0.0002$ *** (linear)	$r = 0.59, p = 0.027$ * (linear)
8451	Carbamidocyclophane C	$r = 0.72, p = 0.0037$ ** (linear) $r = 0.78, p = 0.0009$ *** (logarithmic)	$r = 0.51, p = 0.062$ (linear)
7968	Aldgamycin G	$r = 0.58, p = 0.030$ * (linear) $r = 0.72, p = 0.0037$ ** (logarithmic)	$r = 0.53, p = 0.051$ (linear)
8935	Andrachcinidine	$r = 0.83, p = 0.003$ *** (linear) $r = 0.84, p = 0.0001$ *** (logarithmic)	$r = 0.66, p = 0.0097$ ** (linear)
7563	YM 47525	$r = 0.87, p < 0.0001$ *** (linear)	$r = 0.72, p = 0.0035$ ** (linear)
178	Pyridoxal phosphate	$r = 0.69, p = 0.0088$ ** (linear)	$r = 0.42, p = 0.14$ (linear)

Note: When the coefficient was better in a logarithmic model in terms of *r* than in a linear model, both are shown. If not, only the coefficient of a linear model is shown. Asterisks indicate levels of statistical significance. *, $p < 0.05$; **, $p < 0.01$; ***, $p < 0.001$.

3.5. Correlation Analyses of Downregulated Peaks

We also made scatter plots and performed a correlation analysis of the downregulated peaks with single annotations (Figure 7 and Table 5). Overall, the absolute values of coefficients were reasonably high ($|r| \geq 0.59$), with reasonably small *p*-values. A logarithmic model fit better than a linear model, judging from correlation coefficients, in all cases examined except No. 8800, but in No. 7156 and No. 6296, peak levels were all or none. As seen in these two all-or-none cases, the downregulated response appeared to be very sensitive to the ground radiation dose rate, showing a steep logarithmic decline. It can be concluded that these 15 peaks were downregulated in a dose-dependent manner in response to the ground radiation dose rate, confirming the ANOVA results. Correlation coefficients using the radioactivity concentration of ^{137}Cs showed lower absolute values in all cases (Table 5).

Table 5. Correlation coefficient *r* and its associated *p*-value of the downregulated peaks with single annotations.

No.	Brief Annotation	Ground Radiation Dose Rate [μSv/h]	Radioactivity Concentration of ^{137}Cs [Bq/kg]
4887	Corchoionoside B	$r = -0.55, p = 0.041$ * (linear) $r = -0.76, p = 0.0017$ ** (logarithmic)	$r = -0.51, p = 0.065$ (linear)
7156	Isoginkgetin-7-*O*-β-D-glucopyranoside	$r = -0.41, p = 0.14$ (linear) $r = -0.66, p = 0.0102$ * (logarithmic)	$r = -0.45, p = 0.10$ (linear)
6296	Terreusinol	$r = -0.41, p = 0.15$ (linear) $r = -0.65, p = 0.012$ * (logarithmic)	$r = -0.45, p = 0.10$ (linear)
3152	Amoxapine	$r = -0.74, p = 0.0026$ ** (linear) $r = -0.92, p < 0.0001$ *** (logarithmic)	$r = -0.72, p = 0.0031$ ** (linear)
3073	Elloramycin E	$r = -0.62, p = 0.018$ * (linear) $r = -0.80, p = 0.0006$ *** (logarithmic)	$r = -0.51, p = 0.060$ (linear)
8925	Sanjoinine A dialdehyde	$r = -0.56, p = 0.039$ * (linear) $r = -0.76, p = 0.0015$ ** (logarithmic)	$r = -0.54, p = 0.046$ (linear)
609	Zinolol	$r = -0.34, p = 0.24$ (linear) $r = -0.59, p = 0.028$ * (logarithmic)	$r = -0.38, p = 0.18$ (linear)
2963	Acacetin-7-glucuronosyl-(1→2)-glucuronide	$r = -0.61, p = 0.020$ * (linear) $r = -0.79, p = 0.0009$ ***(logarithmic)	$r = -0.41, p = 0.14$ (linear)

Table 5. *Cont.*

No.	Brief Annotation	Ground Radiation Dose Rate [μSv/h]	Radioactivity Concentration of ^{137}Cs [Bq/kg]
3171	Dapsone hydroxylamine	$r = -0.39$, $p = 0.16$ (linear) $r = -0.60$, $p = 0.023$ * (logarithmic)	$r = -0.36$, $p = 0.21$ (linear)
8804	Trapoxin A	$r = -0.50$, $p = 0.066$ (linear) $r = -0.70$, $p = 0.0053$ ** (logarithmic)	$r = -0.54$, $p = 0.047$ * (linear)
7781	Tricalysioside N	$r = -0.65$, $p = 0.012$ * (linear) $r = -0.74$, $p = 0.0023$ ** (logarithmic)	$r = -0.59$, $p = 0.026$ * (linear)
9091	Pregnadienolone-3-*O*-β-D-chacotrioside	$r = -0.66$, $p = 0.0098$ ** (linear) $r = -0.84$, $p = 0.0102$ * (logarithmic)	$r = -0.68$, $p = 0.0079$ ** (linear)
4347	Perphenazine	$r = -0.52$, $p = 0.059$ (linear) $r = -0.66$, $p = 0.0002$ *** (logarithmic)	$r = -0.52$, $p = 0.056$ (linear)
3836	Silidianin	$r = -0.59$, $p = 0.027$ * (linear) $r = -0.72$, $p = 0.0034$ ** (logarithmic)	$r = -0.53$, $p = 0.052$ (linear)
8800	Droperidol	$r = -0.75$, $p = 0.0020$ ** (linear)	$r = -0.61$, $p = 0.019$ * (linear)

Note: When the coefficient was better in a logarithmic model in terms of r than in a linear model, both are shown. If not, only the coefficient of a linear model is shown. Asterisks indicate levels of statistical significance; *: $p < 0.05$, **: $p < 0.01$, ***: $p < 0.001$.

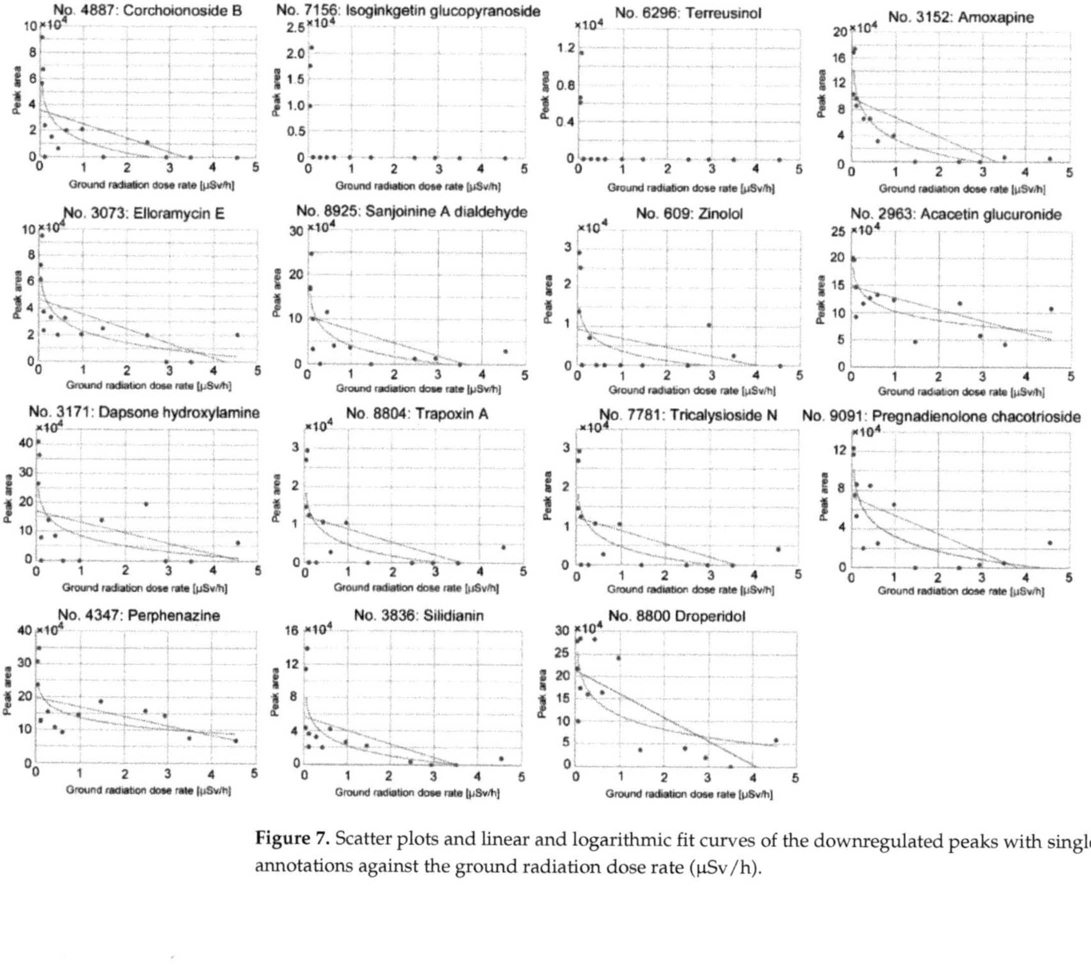

Figure 7. Scatter plots and linear and logarithmic fit curves of the downregulated peaks with single annotations against the ground radiation dose rate (μSv/h).

3.6. Peaks Upregulated or Downregulated at the "Low Level"

As mentioned before, among the peaks with FDR < 0.05 (ANOVA), there were four peaks that seem to be upregulated (No. 4745 and 7256) or downregulated (No. 8508 and 750) only at the "low level". Among them, No. 4745 lacked annotation due to unknown chemical formula, and No. 7256 was singularly annotated as $K_1R_1H_1$ (peptide), although its relevance to natural compounds in plants was unclear. Additionally, No. 8508 showed no database hit, and No. 750 was singularly annotated as DHAP(10:0), decanoyl dihydroxyacetone phosphate.

We made bar graphs and scatter plots for No. 7256 and No. 750 (Figure 8). In No. 7256 (Figure 8a), the low-level group was significantly larger than the background and high-level groups in terms of peak area. As expected, correlation coefficients were low, i.e., $r = -0.11$ ($p = 0.70$) for a linear model and $r = 0.24$ ($p = 0.41$) for a logarithmic model. In contrast, in No. 750 (Figure 8b), the low-level group was significantly smaller than the background and high-level groups in terms of peak area. Unexpectedly, correlation coefficients were not very low, i.e., $r = 0.54$ ($p = 0.048$) for a linear model and $r = 0.42$ ($p = 0.13$) for a logarithmic model.

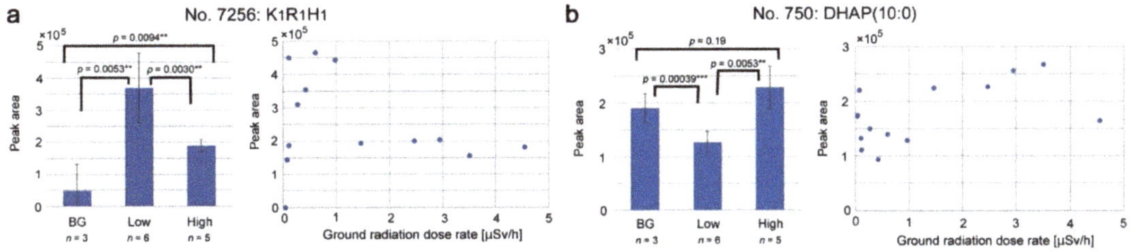

Figure 8. Bar graphs and scatter plots of singularly annotated peaks upregulated or downregulated at the "low level" of the ground radiation dose rate (μSv/h). Mean ± standard deviation values and results of *t*-test (raw *p*-values) are shown. Asterisks indicate levels of statistical significance; **, $p < 0.01$; ***, $p < 0.001$: (**a**) No. 7256, $K_1R_1H_1$; (**b**) No. 750, DHAP(10:0).

4. Discussion

In this study, we collected leaf samples of creeping wood sorrel, the host plant of the pale grass blue butterfly, from contaminated localities in Fukushima. These leaves had been chronically exposed to anthropogenic radiation in the field and were subjected to an LC–MS-based metabolomic analysis. Somewhat surprisingly, an overall dose-dependent trend for metabolomic changes in plants coping with radioactive environments was not observed in the PCA. One might think that this may be because the environmental pollution levels of the collection localities were not high enough for the plant to change its levels of many metabolites. However, this is not necessarily the case. In a previous study, the same plant species in Okinawa was subjected to acute external irradiation, and an overall irradiation response was clearly observed in GC–MS-based analyses despite low-level irradiation from the contaminated soil in Fukushima [29]. In contrast, such an overall response was less clear in an LC–MS-based analysis, suggesting that genetic background was a larger contributor to peak variations than external irradiation itself in a group of compounds amenable to LC–MS [29].

Notably, no identical compounds were annotated between the previous study [29] and the current study. It is also important to note that in the former study, the plant was exposed only externally, whereas in the latter study, the plant was exposed both externally and internally. Nevertheless, in both studies, compounds related to *Streptomyces* were found, i.e., three peaks in the present study and four peaks in the previous study [29]. In this sense, acute exposure under laboratory-based conditions (the previous study [29]) and chronic exposure under field conditions (the present study) may result in different outcomes in the plant but with some similarities. In the case of chronic exposure, the field plants may have already acclimated or adapted genetically to the radioactively contaminated environments by changing the levels of a relatively small number of key metabolites.

Although an overall dose-dependent response in plant samples from Fukushima was not observed, we identified some upregulated and downregulated peaks in response to ground radiation dose. In most upregulated and downregulated peaks, logarithmic fits were better than linear fits. Such a nonlinear response may be widely seen as a plant response to low-dose radiation. An all-or-none response was also observed in downregulated peaks.

There were three upregulated peaks annotated as plant-derived compounds: kudinoside D, andrachcinidine, and pyridoxal phosphate. Kudinoside D is a type of triterpenoid saponin and is known to have biological activities [52]. Importantly, when the ground radiation dose was close to zero, kudinoside D was rarely detected, showing nearly an all-or-none response. Thus, this compound may confer high stress tolerance against environmental radiation in plants. Andrachcinidine is an alkaloid [53]. These two compounds may be stress protectants for the plant and may also function to ward off herbivorous insects. They may cause abnormal and fatal phenotypes in pale grass blue butterfly larvae.

Interestingly, pyridoxal phosphate is an activated vitamin B6 known to function in response to salt stress and other types of stress in plants [54–61]. In addition to its function as a cofactor of stress protectant enzymes, vitamin B6 functions as an antioxidant [62,63]. Notably, this compound is also known to be upregulated in response to ultraviolet irradiation in plant acclimation [61]. The present study suggests that anthropogenic environmental irradiation in Fukushima may also cause upregulation of pyridoxal phosphate to cope with radiation stress in *O. corniculata*. Furthermore, based on the existing literature [54–58] and the present data, upon irradiation, sodium may be expelled from the plant more efficiently to induce salt tolerance due to the upregulation of pyridoxal phosphate. This speculation is consistent with the field-effect hypothesis that the sodium content in leaves of *O. corniculata* may decrease in response to radioactive pollutants, resulting in adverse effects on larvae of the pale grass blue butterfly due to sodium deficiency [28].

The above discussion can further be fortified by referring to KEGG (Kyoto Encyclopedia for Genes and Genomes) for metabolic reactions [89–91]. Among the upregulated metabolites, only pyridoxal phosphate was found in KEGG. Production of pyridoxal phosphate from pyridoxamine phosphate (R00277) or pyridoxine phosphate (R00278) also produces hydrogen peroxide. Its reverse reaction, thus, scavenges hydrogen peroxide when pyridoxal phosphate is provided from a different pathway, one of which is a reaction in pyridoxal and ATP (R00174). Interestingly, in other reactions, production of pyridoxal phosphate also produces D-alanine (R01147), D-glutamate (R01580), or L-glutamate (R07456). D-Alanine and pyridoxal phosphate are products from pyridoxamine phosphate and pyruvate, an important product of glycolysis, and D-glutamate and pyridoxal phosphate are together produced from pyridoxamine phosphate and 2-oxoglutarate, a key product in the TCA cycle [92], suggesting their involvement in a stress response associated with ATP production via glycolysis and the TCA cycle. L-Glutamate and pyridoxal phosphate are produced together by a reaction of D-glyceraldehyde-3-phosphate, D-ribulose-5-phosphate, and L-glutamine, suggesting their involvement in a stress response associated with photosynthesis. These amino acids, especially those of the D-configuration, may function as signaling molecules for a stress response [93–96].

In addition, the downregulated peaks contained various compounds, including plant-derived compounds (such as antioxidants, flavonoids, and saponins) and microbe-derived compounds (such as antibiotics). We do not know why some compounds were upregulated and functionally similar compounds were downregulated, but these compounds may be produced in different metabolic pathways and may respond to radiation stress independently.

We did not detect upregulation of antioxidants in this study other than pyridoxal phosphate, but we did detect downregulation of an antioxidant, zinolol. This is somewhat surprising because antioxidants function to nullify reactive oxygen species (ROS) that are generated by irradiation [97]. This result contrasts with a previous study of acute exposure, in which a few antioxidants were upregulated [29].

Nonetheless, there was a commonality between these studies, i.e., several peaks were annotated as compounds from soil microbes, especially antibiotics from *Streptomyces*, a group of soil bacteria. In a previous study, we thought that these microbe-derived compounds were contaminations of unrelated microbes from the soil [29]. However, leaves were washed well after collection, and no trace of contamination was seen visually. Even if a small amount of soil contamination occurred, its relative weight to the leaves of *O. corniculata* would be too small to contribute to the LC–MS results. Facing the fact that microbe-derived compounds were annotated frequently, we now think that these compounds were not from contamination but from endophytic microbes inside leaves. Indeed, many endophytic bacteria have widely been known in plants [98–101] and have been isolated from *O. corniculata* [102,103]. These microbes are probably of soil origin. These results suggest that *O. corniculata* may host various bacteria and fungi from the soil in its leaves and that compounds from these bacteria and fungi may contribute to plant functions when coping with radiation stress. Such cases of stress management appear to be common among plants [104,105]. To solidify this issue, PCR-based detection and isolated culture of endophytes from leaves may be necessary.

For the downregulated metabolites excluding synthetics, only "elloramycin" and "acacetin" were found in KEGG. The former is a bacterial metabolite, whereas the latter is a part of a plant metabolite, acacetin-7-glucuronosyl-(1→2)-glucuronide. Although elloramycin E was not found in KEGG, elloramycin A was found in the "biosynthesis of type II polyketide products" pathway (rn01057), and acacetin was found in the "flavone and flavonol biosynthesis" pathway (rn00944). These two metabolites seem to be unrelated at first glance. Interestingly, however, both reactions (R10959 and R03571) use *S*-adenosyl-L-methionine as a reactant and produce *S*-adenosyl-L-homocysteine. The present finding that elloramycin E and acacetin-7-glucuronosyl-(1→2)-glucuronide were downregulated together might indicate that the plant and its endophytes share *S*-adenosyl-L-methionine, which has important multiple roles in plant metabolism and signaling including ethylene biosynthesis and stress management [106–108]. This discussion supports possible functions of endophytic microbes in the *Oxalis* plant under radiation stress.

A possible function of these microbe-derived compounds may be to protect leaves from fungal infection. This may be relevant for the survival of *O. corniculata* because this plant is a small weed, and its leaves are very close to the ground. This means that leaves were placed under high humidity conditions, which may easily allow fungal infection to occur. Indeed, fungal infections on leaves have been observed in *O. corniculata* in our laboratory when humidity conditions were not well controlled. An interesting case was reported in which fungal damage to host plant leaves of a small weed, plantain, affected the relationship between the checkerspot butterfly and its host plant [109].

Additionally, we discovered $K_1R_1H_1$ and DHAP(10:0) as candidate compounds that responded most to the low-level radiation. The biological significance of the former is unknown, but the latter seems to be biologically significant. DHAP(10:0) is a derivative of dihydroxyacetone phosphate (DHAP), which is also called glycerone phosphate. DHAP is a metabolite in glycolysis and in the Calvin cycle. In the latter, DHAP is used to regenerate ribulose-1,5-bisphosphate, a key metabolite in the Calvin cycle. Importantly, DHAP is used for synthesis of vitamin B6 in plants but not in bacteria [110]. Thus, it is likely that the upregulation of pyridoxal phosphate detected in the present study occurred in plant cells. DHAP can be converted to glycerol-3-phosphate, which is known as a defense signaling molecule for systemic immunity in plants [111,112]. DHAP also produces methylglyoxal, a signaling molecule for abiotic stress in plants [113].

In the case of the pale grass blue butterfly in Fukushima, the high sensitivity of larvae to pollutants from the accident in the field is likely mediated by the physiological response of the host plant to the pollutants. The current study further suggested the involvement of endophytic soil microbes associated with the host plant. In the case of the monarch butterfly, larval sensitivity to neonicotinoid insecticides seems to be influenced by which host plant species larvae feed on [114]. These cases imply that the biological effects of any pollutants should be evaluated in the context of ecological interactions among plants, animals, and microbes.

5. Conclusions

In this study, we showed that the creeping wood sorrel likely responded to nuclear pollution in Fukushima by changing its levels of a limited number of key metabolites in a dose-dependent manner. The dose-dependent upregulated metabolites included not only plant-derived compounds (i.e., kudinoside D, andrachcinidine, and pyridoxal phosphate) but also microbe-derived compounds, some of which were antibiotics from *Streptomyces*. Pyridoxal phosphate is a stress-responding vitamer of vitamin B6 that may regulate leaf physiology, such as sodium contents. Other upregulated plant-derived compounds may function to ward off herbivorous animals, such as larvae of the pale grass blue butterfly. DHAP(10:0) is unique in that it was downregulated at the low-level radiation. DHAP(10:0) is a derivative of DHAP, which can produce vitamin B6 and stress signaling molecules. Microbe-derived compounds may also contribute to the stress response of the plant. Together, the contributions of these compounds (and their related microbes such as *Streptomyces*) to the radiation stress response should be investigated in the future and may demonstrate the importance of ecological field effects in understanding the biological impacts of the Fukushima nuclear accident. Other types of field effects [38,115–117] should also be investigated to understand the whole picture of the biological effects of the Fukushima nuclear accident.

Supplementary Materials: The following supporting information can be downloaded at: https://www.mdpi.com/article/10.3390/life12010115/s1, Supplementary File S1: LCMS_result field data KDRI, Supplementary File S2: LCMS peak data, Supplementary File S3 (PDF file) contains Supplementary Results and Discussion (including Supplementary Figure S1) and Supplementary Table S1. Note on Supplementary File S1: To open 2D Chromatogram in the "2DView" worksheet, select either "m/z(Detected)" or "Exact Mass" at the cell next to "Vertical axis", and select a sample at the cell next to "Group". To open MS/MS spectrogram in the "MS2" worksheet, double-click a peak number. The "PCA" and "t-Test" worksheets are not useful due to unrelated samples in the file.

Author Contributions: Conceptualization, J.M.O., W.T. and K.S.; methodology, W.T. and K.S.; software, W.T.; validation, J.M.O.; formal analysis and investigation, K.S., W.T. and J.M.O.; resources, K.S. and W.T.; data curation, K.S. and J.M.O.; writing—original draft preparation, J.M.O.; writing—review and editing, J.M.O., K.S. and W.T.; visualization, K.S. and J.M.O.; and supervision, project administration, and funding acquisition, J.M.O. All authors have read and agreed to the published version of the manuscript.

Funding: This research was funded by the Asahi Glass Foundation (Tokyo). The APC was funded by the Asahi Glass Foundation (Tokyo).

Institutional Review Board Statement: Not applicable.

Informed Consent Statement: Not applicable.

Data Availability Statement: The data presented in this study and the source data are available in this article and in the Supplementary Materials.

Acknowledgments: We thank K. Yoshida for a gamma monitor PM1710A, Y. Yona and Y. Iraha for their technical assistance using MetaboAnalyst, and other members of the BCPH Unit of Molecular Physiology for suggestions and technical help. We also thank the staff of the Kazusa DNA Research Institute for practical advice and technical assistance.

Conflicts of Interest: The authors declare no conflict of interest.

References

1. D'Mello, J.P.F. Preface. In *A Handbook of Environmental Toxicology: Human Disorders and Ecotoxicology*; D'Mello, J.P.F., Ed.; CAB International: Wallingford, UK, 2020; pp. xxv–xxxvi.
2. Arapis, G.D.; Karandinos, M.G. Migration of ^{137}Cs in the soil of sloping semi-natural ecosystems in Northern Greece. *J. Environ. Radioact.* **2004**, *77*, 133–142. [CrossRef]
3. Tahir, S.N.A.; Jamil, K.; Zaidi, J.H.; Arif, M.; Ahmed, N. Activity concentration of ^{137}Cs in soil samples from Punjab province (Pakistan) and estimation of gamma-ray dose rate for external exposure. *Radiat. Prot. Dosim.* **2006**, *118*, 345–351. [CrossRef] [PubMed]
4. Ambrosino, F.; Stellato, L.; Sabbarese, C. A case study on possible radiological contamination in the Lo Uttara landfill site (Caserta, Italy). *J. Phys. Conf. Ser.* **2020**, *1548*, 012001. [CrossRef]
5. Endo, S.; Kimura, S.; Takatsuji, T.; Nanasawa, K.; Imanaka, T.; Shizuma, K. Measurement of soil contamination by radionuclides due to the Fukushima Dai-ichi Nuclear Power Plant accident and associated estimated cumulative external dose estimation. *J. Environ. Radioact.* **2021**, *111*, 18–27. [CrossRef] [PubMed]
6. Hinton, T.G.; Alexakhin, R.; Balonov, M.; Gentner, N.; Hendry, J.; Prister, B.; Strand, P.; Woodhead, D. Radiation-induced effects on plants and animals: Findings of the United Nations Chernobyl Forum. *Health Phys.* **2007**, *93*, 427–440. [CrossRef] [PubMed]
7. Geras'kin, S.A.; Fesenko, S.V.; Alexakhin, R.M. Effects of non-human species irradiation after the Chernobyl NPP accident. *Environ. Int.* **2008**, *34*, 880–897. [CrossRef] [PubMed]
8. Beresford, N.A.; Scott, E.M.; Copplestone, D. Field effects studies in the Chernobyl Exclusion Zone: Lessons to be learnt. *J. Environ. Radioact.* **2020**, *211*, 105893. [CrossRef] [PubMed]
9. Møller, A.P.; Mousseau, T.A. Biological consequences of Chernobyl: 20 years on. *Trends Ecol. Evol.* **2006**, *21*, 200–207. [CrossRef] [PubMed]
10. Mousseau, T.A. The biology of Chernobyl. *Annu. Rev. Ecol. Evol. Syst.* **2021**, *52*, 87–109. [CrossRef]
11. Møller, A.P.; Hagiwara, A.; Matsui, S.; Kasahara, S.; Kawatsu, K.; Nishiumi, I.; Suzuki, H.; Mousseau, T.A. Abundance of birds in Fukushima as judges from Chernobyl. *Environ. Pollut.* **2012**, *164*, 36–39. [CrossRef] [PubMed]
12. Bonisoli-Alquati, A.; Koyama, K.; Tedeschi, D.J.; Kitamura, W.; Sukuzi, H.; Ostermiller, S.; Arai, E.; Møller, A.P.; Mousseau, T.A. Abundance and genetic damage of barn swallows from Fukushima. *Sci. Rep.* **2015**, *5*, 9432. [CrossRef]
13. Murase, K.; Murase, J.; Horie, R.; Endo, K. Effects of the Fukushima Daiichi nuclear accident on goshawk reproduction. *Sci. Rep.* **2015**, *5*, 9405. [CrossRef]
14. Hiyama, A.; Nohara, C.; Kinjo, S.; Taira, W.; Gima, S.; Tanahara, A.; Otaki, J.M. The biological impacts of the Fukushima nuclear accident on the pale grass blue butterfly. *Sci. Rep.* **2012**, *2*, 570. [CrossRef] [PubMed]
15. Hiyama, A.; Nohara, C.; Taira, W.; Kinjo, S.; Iwata, M.; Otaki, J.M. The Fukushima nuclear accident and the pale grass blue butterfly: Evaluating biological effects of long-term low-dose exposures. *BMC Evol. Biol.* **2013**, *13*, 168. [CrossRef]
16. Nohara, C.; Hiyama, A.; Taira, W.; Tanahara, A.; Otaki, J.M. The biological impacts of ingested radioactive materials on the pale grass blue butterfly. *Sci. Rep.* **2014**, *4*, 4946. [CrossRef] [PubMed]
17. Hiyama, A.; Taira, W.; Nohara, C.; Iwasaki, M.; Kinjo, S.; Iwata, M.; Otaki, J.M. Spatiotemporal abnormality dynamics of the pale grass blue butterfly: Three years of monitoring (2011–2013) after the Fukushima nuclear accident. *BMC Evol. Biol.* **2015**, *15*, 15. [CrossRef]
18. Akimoto, S. Morphological abnormalities in gall-forming aphids in a radiation-contaminated area near Fukushima Daiichi: Selective impact of fallout? *Ecol. Evol.* **2014**, *4*, 355–369. [CrossRef] [PubMed]
19. Akimoto, S.I.; Li, Y.; Imanaka, T.; Sato, H.; Ishida, K. Effects of radiation from contaminated soil and moss in Fukushima on embryogenesis and egg hatching of the aphid *Prociphilus oriens*. *J. Hered.* **2018**, *109*, 199–205. [CrossRef] [PubMed]
20. Hayama, S.; Tsuchiya, M.; Ochiai, K.; Nakiri, S.; Nakanishi, S.; Ishii, N.; Kato, T.; Tanaka, A.; Konno, F.; Kawamoto, Y.; et al. Small head size and delayed body weight growth in wild Japanese monkey fetuses after the Fukushima Daiichi nuclear disaster. *Sci. Rep.* **2017**, *7*, 3528. [CrossRef] [PubMed]
21. Ochiai, K.; Hayama, S.; Nakiri, S.; Nakanishi, S.; Ishii, N.; Uno, T.; Kato, T.; Konno, F.; Kawamoto, Y.; Tsuchida, S.; et al. Low blood cell counts in wild Japanese monkeys after the Fukushima Daiichi nuclear disaster. *Sci. Rep.* **2014**, *4*, 5793. [CrossRef] [PubMed]
22. Urushihara, Y.; Suzuki, T.; Shimizu, Y.; Ohtaki, M.; Kuwahara, Y.; Suzuki, M.; Uno, T.; Fujita, S.; Saito, A.; Yamashiro, H.; et al. Haematological analysis of Japanese macaques (*Macaca fuscata*) in the area affected by the Fukushima Daiichi Nuclear Power Plant accident. *Sci. Rep.* **2018**, *8*, 16748. [CrossRef]
23. Horiguchi, T.; Yoshii, H.; Mizuno, S.; Shiraishi, H. Decline in intertidal biota after the 2011 Great East Japan Earthquake and Tsunami and the Fukushima nuclear disaster: Field observations. *Sci. Rep.* **2016**, *6*, 20416. [CrossRef] [PubMed]
24. Hayashi, G.; Shibato, J.; Imanaka, T.; Cho, K.; Kubo, A.; Kikuchi, S.; Satoh, K.; Kimura, S.; Ozawa, S.; Fukutani, S.; et al. Unraveling low-level gamma radiation-responsive changes in expression of early and late genes in leaves of rice seedlings at Iitate Village, Fukushima. *J. Hered.* **2014**, *105*, 723–738. [CrossRef] [PubMed]
25. Watanabe, Y.; Ichikawa, S.; Kubota, M.; Hoshino, J.; Kubota, Y.; Maruyama, K.; Fuma, S.; Kawaguchi, I.; Yoschenko, V.I.; Yoshida, S. Morphological defects in native Japanese fir trees around the Fukushima Daiichi Nuclear Power Plant. *Sci. Rep.* **2015**, *5*, 13232. [CrossRef] [PubMed]

26. Yoschenko, V.; Nanba, K.; Yoshida, S.; Watanabe, Y.; Takase, T.; Sato, N.; Keitoku, K. Morphological abnormalities in Japanese red pine (*Pinus densiflora*) at the territories contaminated as a result of the accident at Fukushima Dai-ichi Nuclear Power Plant. *J. Environ. Radioact.* **2016**, *165*, 60–67. [CrossRef] [PubMed]
27. Rakwal, R.; Hayashi, G.; Shibato, J.; Deepak, S.A.; Gundimeda, S.; Simha, U.; Padmanaban, A.; Gupta, R.; Han, S.; Kim, S.T.; et al. Progress toward rice seed OMICS in low-level gamma radiation environment in Iitate Village, Fukushima. *J. Hered.* **2018**, *109*, 2089–2211. [CrossRef] [PubMed]
28. Sakauchi, K.; Taira, W.; Toki, M.; Tsuhako, M.; Umetsu, K.; Otaki, J.M. Nutrient imbalance of the host plant for larvae of the pale grass blue butterfly may mediate the field effect of low-dose radiation exposure in Fukushima: Dose-dependent changes in the sodium content. *Insects* **2021**, *12*, 149. [CrossRef] [PubMed]
29. Sakauchi, K.; Taira, W.; Otaki, J.M. Metabolomic response of the creeping wood sorrel *Oxalis corniculata* to low-dose radiation exposure from Fukushima's contaminated soil. *Life* **2021**, *11*, 990. [CrossRef] [PubMed]
30. Hancock, S.; Vo, N.T.K.; Omar-Nazir, L.; Batlle, J.V.I.; Otaki, J.M.; Hiyama, A.; Byun, S.H.; Seymour, C.B.; Mothersill, C. Transgenerational effects of historic radiation dose in pale grass blue butterflies around Fukushima following the Fukushima Dai-ichi Nuclear Power Plant meltdown accident. *Environ. Res.* **2019**, *168*, 230–240. [CrossRef] [PubMed]
31. Sakauchi, K.; Taira, W.; Hiyama, A.; Imanaka, T.; Otaki, J.M. The pale grass blue butterfly in ex-evacuation zones 5.5 years after the Fukushima nuclear accident: Contributions of initial high-dose exposure to transgenerational effects. *J. Asia-Pac. Entomol.* **2020**, *23*, 242–252. [CrossRef]
32. Nohara, C.; Taira, W.; Hiyama, A.; Tanahara, A.; Takatsuji, T.; Otaki, J.M. Ingestion of radioactively contaminated diets for two generations in the pale grass blue butterfly. *BMC Evol. Biol.* **2014**, *14*, 193. [CrossRef] [PubMed]
33. Taira, W.; Hiyama, A.; Nohara, C.; Sakauchi, K.; Otaki, J.M. Ingestional and transgenerational effects of the Fukushima nuclear accident on the pale grass blue butterfly. *J. Radiat. Res.* **2015**, *56*, i2–i18. [CrossRef]
34. Taira, W.; Nohara, C.; Hiyama, A.; Otaki, J.M. Fukushima's biological impacts: The case of the pale grass blue butterfly. *J. Hered.* **2014**, *105*, 710–722. [CrossRef] [PubMed]
35. Taira, W.; Toki, M.; Kakinohana, K.; Sakauchi, K.; Otaki, J.M. Developmental and hemocytological effects of ingesting Fukushima's radiocesium on the cabbage white butterfly *Pieris rapae*. *Sci. Rep.* **2019**, *9*, 2625. [CrossRef] [PubMed]
36. Nohara, C.; Hiyama, A.; Taira, W.; Otaki, J.M. Robustness and radiation resistance of the pale grass blue butterfly from radioactively contaminated areas: A possible case of adaptive evolution. *J. Hered.* **2018**, *109*, 188–198. [CrossRef] [PubMed]
37. Gurung, R.D.; Taira, W.; Sakauchi, K.; Iwata, M.; Hiyama, A.; Otaki, J.M. Tolerance of high oral doses of nonradioactive and radioactive caesium chloride in the pale grass blue butterfly *Zizeeria maha*. *Insects* **2019**, *10*, 290. [CrossRef]
38. Otaki, J.M. Understanding low-dose exposure and field effects to resolve the field-laboratory paradox: Multifaceted biological effects from the Fukushima nuclear accident. In *New Trends in Nuclear Science*; Awwad, N.S., AlFaify, S.A., Eds.; IntechOpen: London, UK, 2018; pp. 49–71. [CrossRef]
39. D'Mello, J.P.F. Phytotoxins. In *A Handbook of Environmental Toxicology: Human Disorders and Ecotoxicology*; D'Mello, J.P.F., Ed.; CAB International: Wallingford, UK, 2020; pp. 3–18.
40. Zagrobelny, M.; de Castro, É.C.P.; Møller, B.L.; Bak, S. Cyanogenesis in arthropods: From chemical warfare to nuptial gifts. *Insects* **2018**, *9*, 51. [CrossRef] [PubMed]
41. Brown, K.S., Jr.; Francini, R. Evolutionary strategies of chemical defense in aposematic butterflies: Cyanogenesis in Asteraceae-feeding American Acraeinae. *Chemoecology* **1990**, *1*, 52–56. [CrossRef]
42. Sakurai, N.; Ara, T.; Enomoto, M.; Motegi, T.; Morishita, Y.; Kurabayashi, A.; Iijima, Y.; Ogata, Y.; Nakajima, D.; Suzuki, H.; et al. Tools and databases of the KOMICS web portal for preprocessing, mining, and dissemination of metabolomics data. *BioMed Res. Int.* **2014**, *2014*, 194812. [CrossRef] [PubMed]
43. Sakurai, N.; Shibata, D. Tools and databases for an integrated metabolite annotation environment for liquid chromatography-mass spectrometry-based untargeted metabolomics. *Carotenoid Sci.* **2017**, *22*, 16–22.
44. Sakurai, N.; Narise, T.; Sim, J.-S.; Lee, C.-M.; Ikeda, C.; Akimoto, N.; Kanaya, S. UC2 search: Using unique connectivity of uncharged compounds for metabolite annotation by database searching in mass-spectrometry-based metabolomics. *Bioinformatics* **2018**, *34*, 698–700. [CrossRef]
45. Afendi, F.M.; Okada, T.; Yamazaki, M.; Hirai-Morita, A.; Nakamura, Y.; Nakamura, K.; Ikeda, S.; Takahashi, H.; Altaf-Ul-Amin, M.; Darusman, L.K.; et al. KNApSAcK family databases: Integrated metabolite-plant species databases for multifaceted plant research. *Plant Cell Physiol.* **2021**, *53*, e1. [CrossRef]
46. Wishart, D.S.; Tzur, D.; Knox, C.; Eisner, R.; Guo, A.C.; Young, N.; Cheng, D.; Jewell, K.; Arndt, D.; Sawhney, S. HMDB: The Human Metabolome Database. *Nucl. Acid Res.* **2007**, *35*, D521–D526. [CrossRef]
47. Wishard, D.S.; Feunang, Y.D.; Marcu, A.; Guo, A.C.; Liang, K.; Vázquez-Fresno, R.; Sajed, T.; Johnson, D.; Li, C.; Karu, N. HMDB 4.0: The human metabolome database for 2018. *Nucl. Acids Res.* **2018**, *46*, D608–D617. [CrossRef] [PubMed]
48. Sakurai, N.; Ara, T.; Kanaya, S.; Nakamura, Y.; Iijima, Y.; Enomoto, M.; Motegi, T.; Aoki, K.; Suzuki, H.; Shibata, D. An application of a relational database system for high-throughput prediction of elemental compositions from accurate mass values. *Bioinformatics* **2013**, *29*, 290–291. [CrossRef] [PubMed]
49. Xia, J.; Wishart, D.S. Web-based inference of biological patterns, functions and pathways from metabolomic data using MetaboAnalyst. *Nat. Protocol.* **2011**, *6*, 743–760. [CrossRef]

50. Xia, J.; Sinelnikov, I.; Han, B.; Wishart, D.S. MetaboAnalyst 3.0—making metabolomics more meaningful. *Nucl. Acids Res.* **2015**, *43*, W251–W257. [CrossRef]
51. Pang, Z.; Chong, J.; Zhou, G.; de Lima Morais, D.A.; Chang, L.; Barrette, M.; Gauthier, C.; Jacques, P.É.; Li, S.; Xia, J. MetaboAnalyst 5.0: Narrowing the gap between raw spectra and functional insights. *Nucl. Acids Res.* **2021**, *49*, gkab382. [CrossRef]
52. Che, Y.; Wang, Q.; Xiao, R.; Zhang, J.; Zhang, Y.; Gu, W.; Rao, G.; Wang, C.; Kuang, H. Kudinoside-D, a triterpenoid saponin derived from *Ilex kudingcha* suppresses adipogenesis through modulation of the AMPK pathway in 3T3-L1 adipocytes. *Fitoterapia* **2018**, *125*, 208–216. [CrossRef]
53. Mill, S.; Hootelé, C. Alkaloids of *Andrachne aspera*. *J. Nat. Prod.* **2000**, *63*, 762–764. [CrossRef] [PubMed]
54. Shi, H.; Xiong, L.; Stevenson, B.; Lu, T.; Zhu, J.-K. The Arabidopsis *salt overly sensitive 4* mutants uncover a critical role for vitamin B6 in plant salt tolerance. *Plant Cell* **2002**, *14*, 575–588. [CrossRef] [PubMed]
55. Titiz, O.; Tambasco-Studart, M.; Warzych, E.; Apel, K.; Amrheim, N.; Laloi, C.; Fitzpatrick, T.B. PDX_1 is essential for vitamin B6 biosynthesis, development and stress tolerance in Arabidopsis. *Plant J.* **2006**, *48*, 933–946. [CrossRef]
56. González, E.; Danehower, D.; Daub, M.E. Vitamer levels, stress response, enzyme activity, and gene regulation of Arabidopsis lines mutant in the pyridoxine/pyridoxamine 5′-phosphate oxidase (PDX3) and pyridoxal kinase (SOS4) genes involved in the vitamin B6 salvage pathway. *Plant Physiol.* **2007**, *145*, 985–996. [CrossRef]
57. Mahajan, S.; Pandey, G.K.; Tuteja, N. Calcium- and salt-stress signaling in plants: Shedding light on SOS pathway. *Arch. Biochem. Biophys.* **2008**, *471*, 146–158. [CrossRef]
58. Hussain, S.; Huang, J.; Zhu, C.; Zhu, L.; Cao, X.; Hussain, S.; Ashraf, M.; Khaskheli, M.A.; Kong, Y.; Jin, Q.; et al. Phyridoxal 5′-phosphate enhances the growth and morpho-physiological characteristics of rice cultivars by mitigating the ethylene accumulation under salinity stress. *Plant Physiol. Biochem.* **2020**, *154*, 782–795. [CrossRef]
59. You, J.; Hu, H.; Xiong, L. An ornithine δ-aminotransferase gene OsOAT confers drought and oxidative stress tolerance in rice. *Plant Sci.* **2021**, *197*, 59–69. [CrossRef]
60. Singh, R.P.; Shelke, G.M.; Kumar, A.; Jha, P.N. Biochemistry and genetics of ACC deaminase: A weapon to "stress ethylene" produced in plants. *Front. Microbiol.* **2015**, *6*, 937. [CrossRef] [PubMed]
61. Czégény, G.; Kőrösi, L.; Strid, Å.; Hideg, É. Multiple roles for vitamin B_6 in plant acclimation to UV-B. *Sci. Rep.* **2019**, *9*, 1259. [CrossRef]
62. Havaux, M.; Ksas, B.; Szewczyk, A.; Rumeau, D.; Franck, F.; Caffarri, S.; Triantaphylidès, C. Vitamin B6 deficient plants display increased sensitivity to high light and photo-oxidative stress. *BMC Plant Biol.* **2009**, *9*, 130. [CrossRef] [PubMed]
63. Vanderschuren, H.; Boycheva, S.; Li, K.-T.; Szydlowski, N.; Gruissem, W.; Fitzpatrick, T.B. Strategies for vitamin B6 biofortification of plants: A dual role as a micronutrient and a stress protectant. *Front. Plant Sci.* **2013**, *4*, 143. [CrossRef] [PubMed]
64. Komiyama, K.; Okada, K.; Tomisaka, S.; Umezawa, I.; Hamamoto, T.; Beppu, T. Antitumor activity of leptomycin B. *J. Antibiotics* **1985**, *38*, 427–429. [CrossRef] [PubMed]
65. Tunac, J.B.; Graham, B.D.; Dobson, W.E.; Lenzini, M.D. Novel antitumor antibiotics, CI-940 (PD 114,720) and PD 114,721. Taxonomy, fermentation and biological activity. *J. Antibiotics* **1985**, *38*, 460–465. [CrossRef]
66. Hamamoto, T.; Uozumi, T.; Beppu, T. Leptomycins A and B, new antifungal antibiotics. III. Mode of action of leptomycin B on *Schizosaccharomyces pombe*. *J. Antibiotics* **1985**, *38*, 1573–1580. [CrossRef] [PubMed]
67. Mizobuchi, S.; Mochizuki, J.; Soga, H.; Tanba, H.; Inoue, H. Aldgamycin G, a new macrolide antibiotic. *J. Antibiotics* **1986**, *39*, 1776–1778. [CrossRef] [PubMed]
68. Bui, H.T.N.; Jansen, R.; Pham, H.T.L.; Mundt, S. Carbamidocyclophanes A-E, chlorinated paracyclophanes with cytotoxic and antibiotic activity from the Vietnamese cyanobacterium *Nostoc* sp. *J. Nat. Prod.* **2007**, *70*, 499–503. [CrossRef]
69. Sugawara, T.; Tanaka, A.; Nagai, K.; Suzuki, K.; Okada, G. New member of the trichothecene family. *J. Antibiotics* **1997**, *50*, 778–780. [CrossRef]
70. Yoshikawa, M.; Shimada, H.; Saka, M.; Yoshizumi, S.; Yamahara, J.; Matsuda, H. Medicinal foodstuffs. V. Moroheiya. (1): Absolute stereostructures of corchoionosides A, B, and C, histamine release inhibitors from the leaves of Vietnamese *Corchorus olitorius* L. (Tiliaceae). *Chem. Pharm. Bull.* **1997**, *45*, 464–469. [CrossRef]
71. Hyun, S.K.; Kang, S.S.; Son, K.H.; Chung, H.Y.; Choi, J.S. Biflavone glucosides from *Ginkgo biloba* yellow leaves. *Chem. Pharm. Bull.* **2005**, *53*, 1200–1201. [CrossRef]
72. Zhao, Y.; Wu, Y.; Wang, M. Bioactive substances of plant origin. In *Handbook of Food Chemistry*; Cheung, P.C.K., Mehta, B.M., Eds.; Springer: Berlin/Heidelberg, Germany, 2015; pp. 967–1008. [CrossRef]
73. Tan, N.-H.; Zhou, J. Plant cyclopeptides. *Chem. Rev.* **2006**, *106*, 840–895. [CrossRef]
74. Ammar, S.; Mahjoub, M.A.; Charfi, N.; Skandarani, I.; Chekir-Ghedira, L.; Mighri, Z. Mutagenic, antimutagenic and antioxidant activities of a new class of β-glucoside hydroxyhydroquinone from *Anagallis monelli* growing in Tunisia. *Chem. Pharm. Bull.* **2007**, *55*, 385–388. [CrossRef]
75. Kim, M.; Jang, Y. Phytochemical analysis of *Clerodendron trichotomum* by UHPLC-ESI-MS. *Planta Med.* **2013**, *79*, PL20. [CrossRef]
76. He, D.-H.; Matsunami, K.; Otsuka, H.; Shinzato, T.; Aramoto, M.; Bando, M.; Takeda, Y. Tricalysiosides H-O: *Ent*-kaurane glucosides from the leaves of *Tricalysia dubia*. *Phytochemisty* **2005**, *66*, 2857–2864. [CrossRef] [PubMed]
77. Dong, M.; Feng, X.Z.; Wu, L.J.; Wang, B.X.; Ikejima, T. Two new steroidal saponins from the rhizomes of *Dioscorea panthaica* and their cytotoxic activity. *Planta Med.* **2001**, *67*, 853–857. [CrossRef] [PubMed]

78. Bosisio, E.; Benelli, C.; Pirola, O. Effect of the flavonolignans of *Silybum marianum* L. on lipid peroxidation in rat liver microsomes and freshly isolated hepatocytes. *Pharmacol. Res.* **1992**, *25*, 147–165. [CrossRef]
79. Li, X.; Lee, S.M.; Choi, H.D.; Kang, J.S.; Son, B.W. Microbial transformation of terreusione, an ultraviolet-A (UV-A) protecting dipyrroloquinone, by *Streptomyces* sp. *Chem. Pharm. Bull.* **2003**, *51*, 1458–1459. [CrossRef]
80. Fiedler, H.-P.; Rohr, J.; Zeeck, A. Minor congeners of the elloramycin producer *Streptomyces olivaceus*. *J. Antibiot.* **1986**, *39*, 856–859. [CrossRef]
81. Itazaki, H.; Nagashima, K.; Sugita, K.; Yoshida, H.; Kawamura, Y.; Yasuda, Y.; Matsumoto, K.; Ishii, K.; Uotani, N.; Nakai, H.; et al. Isolation and structural elucidation of new cyclotetrapeptides, trapoxins A and B, having detransformation activities as antitumor agents. *J. Antibiotics* **1990**, *43*, 1524–1532. [CrossRef]
82. Wang, X.; Lin, M.; Xu, D.; Lai, D.; Zhou, L. Structural diversity and biological activities of fungal cyclic peptides, excluding cyclodipeptides. *Molecules* **2017**, *22*, 2069. [CrossRef]
83. Lydiard, R.B.; Gelenberg, A.J. Amoxapine—An antidepressant with some neuroleptic properties?: A review of its chemistry, animal pharmacology and toxicology, human pharmacology, and clinical efficacy. *Pharmacotherapy* **1981**, *1*, 163–178. [CrossRef]
84. Wozel, G.; Blasum, C. Dapsone in dermatology and beyond. *Arch. Dermatol. Res.* **2014**, *306*, 103–124. [CrossRef]
85. Smith, H.S.; Cox, L.R.; Smith, B.R. Dopamine receptor antagonists. *Ann. Palliat. Med.* **2012**, *1*, 137–142. [CrossRef]
86. Yelnosky, J.; Katz, R.; Dietrich, E.V. A study of some of the pharmacologic actions of droperidol. *Toxicol. Appl. Pharmacol.* **1964**, *6*, 37–47. [CrossRef]
87. Freye, E.; Kuschinsky, E. Effects of fentanyl and droperidol on the dopamine metabolism of the rat striatum. *Pharmacology* **1976**, *14*, 1–7. [CrossRef]
88. Bradshaw, H.; Pleuvry, B.J.; Sharma, H.L. Effect of droperidol on dopamine-induced increase in effective retinal plasma flow in dogs. *British J. Anaesthesia* **1980**, *52*, 879–883. [CrossRef] [PubMed]
89. Kanehisa, M.; Goto, S. KEGG: Kyoto Encyclopedia of Genes and Genomes. *Nucl. Acids Res.* **2000**, *28*, 27–30. [CrossRef] [PubMed]
90. Kanehisa, M. KEGG bioinformatics resource for plant genomics and metabolomics. *Methods Mol. Biol.* **2016**, *1374*, 55–70. [CrossRef]
91. Kanehisa, M.; Furumichi, M.; Sato, Y.; Ishiguro-Watanabe, M.; Tanabe, M. KEGG: Integrating viruses and cellular organisms. *Nucl. Acids Res.* **2021**, *49*, D545–D551. [CrossRef] [PubMed]
92. Araújo, W.L.; Martins, A.O.; Femie, A.R.; Tohge, T. 2-Oxoglutarate: Linking TCA cycle function with amino acid, glucosinolate, flavonoid, alkaloid, and gibberellin. *Front. Plant Sci.* **2014**, *5*, 552. [CrossRef] [PubMed]
93. Chen, Y.; Zhang, X.; Sun, T.; Tian, Q.; Zhang, W.-H. Glutamate receptor homolog3.4 is involved in regulation of seed germination under salt stress in Arabidopsis. *Plant Cell Physiol.* **2018**, *59*, 978–988. [CrossRef] [PubMed]
94. Li, Z.-G.; Ye, X.-Y.; Qiu, X.-M. Glutamate signaling enhances the heat tolerance of maize seedling by plant glutamate receptor-like channels-mediated calcium signaling. *Protoplasma* **2019**, *256*, 1165–1169. [CrossRef]
95. Yamasaki, H.; Ogura, M.P.; Kingjoe, K.A.; Cohen, M.F. D-Cysteine-induced rapid root abscission in the water fern *Azolla pinnata*: Implications for the linkage between D-amino acid and reactive sulfur species (RSS) in plant environmental responses. *Antioxidants* **2019**, *8*, 411. [CrossRef]
96. Qiu, X.-M.; Sun, Y.-Y.; Ye, X.-Y.; Li, Z.G. Signaling role of glutamate in plants. *Front. Plant Sci.* **2020**, *10*, 1743. [CrossRef] [PubMed]
97. Sakihama, Y.; Yamasaki, H. Phytochemical antioxidants: Past, present and future. In *Antioxidants—Benefits, Sources, Mechanisms of Action*; Waisundara, V., Ed.; IntechOpen: London, UK, 2021. [CrossRef]
98. Qin, S.; Xing, K.; Jiang, J.-H.; Xu, L.-H.; Li, W.-J. Biodiversity, bioactive natural products and biotechnological potential of plant-associated endophytic actinobacteria. *Appl. Microbiol. Biotechnol.* **2011**, *89*, 457–473. [CrossRef]
99. Shimizu, M. Endophytic actinomycetes: Biocontrol agents and growth promoters. In *Bacteria in Agrobiology: Plant Growth Responses*; Maheshwari, D., Ed.; Springer: Berlin/Heidelberg, Germany, 2011; pp. 201–220. [CrossRef]
100. Golinska, P.; Wypij, M.; Agarkar, G.; Rathod, D.; Dahm, H.; Rai, M. Endophytic actinobacteria of medicinal plants: Diversity and bioactivity. *Antonie Leeuwenhoek* **2015**, *108*, 267–289. [CrossRef] [PubMed]
101. Grover, M.; Bodhankar, S.; Maheswari, M.; Srinivasarao, C. Actinomycetes as mitigators of climate change and abiotic stress. In *Plant Growth Promoting Actinobacteria*; Subramaniam, G., Arumugam, S., Rajendran, V., Eds.; Springer: Singapore, 2016; pp. 203–212. [CrossRef]
102. Peng, A.; Liu, J.; Gao, Y.; Chen, Z. Distribution of endophytic bacteria in *Alopecurus aequalis* Sobol and *Oxalis corniculata* L. from soils contaminated by polycyclic aromatic hydrocarbons. *PLoS ONE* **2013**, *8*, e83054. [CrossRef]
103. Mufti, R.; Amna Rafique, M.; Haq, F.; Munis, M.F.H.; Masood, S.; Mumtaz, A.S.; Chaudhary, H.J. Genetic diversity and metal resistance assessment of endophytes isolated from *Oxalis corniculata*. *Soil Environ.* **2015**, *34*, 89–99.
104. Kuldau, G.; Bacon, C. Clavicipitaceous endophytes: Their ability to enhance resistance of grasses to multiple stresses. *Biol. Control* **2008**, *46*, 57–71. [CrossRef]
105. Rodriguez, R.J.; Henson, J.; Volkenburgh, E.V.; Hoy, M.; Wright, L.; Beckwith, F.; Kim, Y.-O.; Redman, R.S. Stress tolerance in plants via habitat-adapted symbiosis. *ISME J.* **2008**, *2*, 404–416. [CrossRef]
106. Roje, S. *S*-Adenosyl-L-methionine: Beyond the universal methyl group donor. *Phytochemistry* **2006**, *67*, 1686–1698. [CrossRef] [PubMed]
107. Anjum, N.A.; Gill, R.; Kaushik, M.; Hasanuzzaman, M.; Pereira, E.; Ahmad, I.; Tuteja, N.; Gill, S.S. ATP-sulfurylase, sulfur-compounds, and plant stress tolerance. *Front. Plant Sci.* **2015**, *6*, 210. [CrossRef]

108. Pattyn, J.; Vaughan-Hirsch, J.; Van de Poel, B. The regulation of ethylene biosynthesis: A complex multilevel control circuitry. *New Phytol.* **2021**, *229*, 770–782. [CrossRef]
109. Severns, P.M.; Guzman-Martinez, M. Plant pathogen invasion modifies the eco-evolutionary host plant interactions of an endangered checkerspot butterfly. *Insects* **2021**, *12*, 246. [CrossRef] [PubMed]
110. Tambasco-Studart, M.; Titiz, O.; Raschle, T.; Forster, G.; Amrhein, N.; Fitzpatrick, T.B. Vitamin B6 biosynthesis in higher plants. *Proc. Natl. Acad. Sci. USA* **2005**, *102*, 13687–13692. [CrossRef] [PubMed]
111. Venugopal, S.C.; Chanda, B.; Vaillancourt, L.; Kachroo, A.; Kachroo, P. The common metabolite glycerol-3-phosphate is a novel regulator of plant defense signaling. *Plant Signal. Behav.* **2009**, *4*, 746–749. [CrossRef] [PubMed]
112. Mandal, M.K.; Chanda, B.; Xia, Y.; Yu, K.; Sekine, K.; Gao, Q.; Selote, D.; Kachroo, A.; Kachroo, P. Glycerol-3-phosphate and systemic immunity. *Plant Signal. Behav.* **2011**, *6*, 1871–1874. [CrossRef] [PubMed]
113. Hoque, T.; Hossain, M.A.; Mostofa, M.G.; Burritt, D.J.; Fujita, M.; Tran, L.-S.P. Methylglyoxal: An emerging signaling molecule in plant abiotic stress responses and tolerance. *Front. Plant Sci.* **2016**, *7*, 1341. [CrossRef]
114. Prouty, C.; Barriga, P.; Davis, A.K.; Krischik, V.; Altizer, S. Host plant species mediates impact of neonicotinoid exposure to Monarch butterflies. *Insects* **2021**, *12*, 999. [CrossRef] [PubMed]
115. Otaki, J.M. Fukushima's lessons from the blue butterfly: A risk assessment of the human living environment in the post-Fukushima era. *Integr. Environ. Assess. Manag.* **2016**, *12*, 667–672. [CrossRef]
116. Otaki, J.M.; Taira, W. Current status of the blue butterfly in Fukushima research. *J. Hered.* **2018**, *109*, 178–187. [CrossRef]
117. Otaki, J.M. Fukushima Nuclear Accident: Potential health effects inferred from butterfly and human cases. In *A Handbook of Environmental Toxicology: Human Disorders and Ecotoxicology*; D'Mello, J.P.F., Ed.; CAB International: Wallingford, UK, 2020; pp. 497–514.

life

MDPI

Article

Radiocaesium Contamination of Mushrooms at High- and Low-Level Chernobyl Exposure Sites and Its Consequences for Public Health

Ondřej Harkut, Petr Alexa * and Radim Uhlář

Department of Physics, Faculty of Electrical Engineering and Computer Science, VŠB-Technical University of Ostrava, 17. listopadu 2172/15, 708 00 Ostrava, Czech Republic; ondrej.harkut.st@vsb.cz (O.H.); radim.uhlar@vsb.cz (R.U.)

* Correspondence: petr.alexa@vsb.cz

Abstract: We compare the specific activities of ^{137}Cs and ^{40}K in stipes and caps of three different common mushroom species (*Xerocomus badius*, *Russula ochroleuca* and *Armillariella mellea*) measured at the Czech Chernobyl hot spot in the Opava area (Silesia) and at a low-exposed site at the Beskydy mountains in 2011. The highest values of ^{137}Cs were found in caps of *Xerocomus badius* and *Russula ochroleuca* in the Opava area (11.8 and 8.77 kBq/kg, respectively). The source of ^{137}Cs was verified by the measurement of the ^{134}Cs/^{137}Cs ratio. Based on our results, we estimate an effective dose per year due to radiocaesium intake in the two investigated areas for *Xerocomus badius*, one of the most popular edible mushrooms in the Czech Republic. In 2011, the effective dose reached the maximum value of 0.102 mSv in the Opava area and 0.004 mSv at the low-exposed site at the Beskydy mountains. Therefore, it does not represent a significant risk for public health.

Keywords: radiocaesium; mushrooms; Chernobyl accident

Citation: Harkut, O.; Alexa, P.; Uhlář, R. Radiocaesium Contamination of Mushrooms at High- and Low-Level Chernobyl Exposure Sites and Its Consequences for Public Health. *Life* **2021**, *11*, 1370. https://doi.org/10.3390/life11121370

Academic Editors: Fabrizio Ambrosino and Supitcha Chanyotha

Received: 6 November 2021
Accepted: 4 December 2021
Published: 9 December 2021

1. Introduction

Wild fungi and their fruiting bodies—so-called mushrooms—tend to accumulate radiocaesium that represents a problematic environmental issue particularly due to a relatively long half-life, emission of gamma radiation and high risk of incorporation into living organisms [1–4]. This effect has already been examined closely over decades after pollution events [2,5]. Due to the impact of nuclear weapon tests, the Chernobyl accident in 1986 [4,6,7] and the Fukushima accident in 2011 [8], soils around the globe are contaminated by radiocaesium [9].

The soil-to-fungi transfer causes the accumulation of larger amounts of ^{137}Cs in wild mushrooms depending on type of soil and its surface activity of radiocaesium [10–14]. Wild edible mushrooms in Czech forest ecosystems are commonly picked and eaten by dwellers, which represents a risk of receiving additional effective doses by ingesting higher levels of ^{137}Cs than recommended by IAEA [15].

The IAEA recommendation suggests the generic action level for ^{137}Cs of 1 kBq/kg. If the specific activity exceeds the level, an action of some sort should be taken. Simultaneously, the IAEA recommendation states that classes of food that are consumed in small quantities, e.g., less than 10 kg per person per year, which represent a very small fraction of the total diet and would make very small additions to individual exposures, may have action levels ten-times higher than those for major foodstuffs.

Measurements of ^{137}Cs specific activity in different parts of mushrooms (caps and stipes—in some cases gills and pores) have been carried out at particular areas in the Slovak Republic [16], in Poland [1,17,18], in Austria [19] and in southern Germany [20]. The studies [21,22] dealt with this issue in the Czech Republic. All these countries have been affected by a radioactive cloud from Chernobyl. Mountain areas are susceptible to rainfalls that are able to release particulates from radioactive clouds into a forest environment [23].

The measured specific activities greatly depend on the amount of precipitations that were absorbed by soil. This effect created high- and low-level Chernobyl exposure sites across the country.

Mushrooms are characterized by a high ability to accumulate radiocaesium and work well as bioindicators of radioactivity in nature [24]. The reason lies within their structure, which consists of gentle fibres. The genetic constitution of mushrooms differs from green plants that absorb caesium less efficiently than its nutrient element, potassium. The so-called Cs/K discriminator factor (DF) at mushrooms indicates the transportation efficiency of these elements within the mushroom structure, e.g., from stipe to cap [16,17,25–30].

The aim of our paper is to compare specific activities of ^{137}Cs in mushrooms from two areas in the eastern part of the Czech Republic with different total precipitation amounts from the radioactive Chernobyl cloud that passed the areas on 30 April/1 May 1986. In the Opava region (Silesia) the total precipitation amount exceeded 15 mm, while in the Ostravice river valley in the Beskydy mountains, it was lower than 0.5 mm [6]. This resulted in a different initial surface activity in both areas.

The Chernobyl hot spot in the Opava region has not yet been examined in terms of the content of ^{137}Cs in mushrooms in spite of the fact that the fallout from the radioactive cloud from the Chernobyl accident was one of the largest in the Czech Republic. Activity levels of ^{137}Cs reached up to 52 kBq/m^2 in soil samples [6]. We also tested a possibility to determine both ^{137}Cs and ^{40}K activity in caps and stipes for small samples (masses around 1 g and less) using a low-background well HPGe spectrometer.

Different parts of the fruitbody (caps and stipes) of the collected specimens (*Xerocomus badius, Russula ochroleuca* and *Armillariella mellea*) were analysed. The species *Xerocomus badius* was chosen as a commonly used reference edible mushroom for its high ability to accumulate radiocaesium [1], and a potential radiation risk due to high consumption of this species was determined.

2. Materials and Methods

In October and November 2011, fruiting bodies of three commonly used reference edible mushrooms (*Xerocomus badius, Russula ochroleuca* and *Armillariella mellea*) were collected from a square area of approximately 2.6 km^2 in the Opava area (GPS coordinates of the centre of the area: 49°52′26.432″ N, 18°0′30.972″ E) and from a similar square area in the Ostravice river valley in the Beskydy mountains (GPS coordinates of the centre of the area: 49°30′0.825″ N, 18°26′48.283″ E).

The Chernobyl hot spot area in the Opava region where the collecting of mushrooms took place is located on a geological bedrock consisting of paleozoic predominantly sedimentary rocks (shale, greywacke, quartzite and limestone) whereas the geological bedrock in the Ostravice river valley consists of mezozoic sedimentary rocks (sandstone and shale). In the Opava region, sandy-loam brown soils prevail, while, in the Ostravice river valley, acid loam brown soils dominate.

In Figure 1 the precipitation in mm at the area of the former Czechoslovakia is depicted in the time span of 24 h on 30 April/1 May 1986, shortly after the Chernobyl accident. During that time, the radioactive cloud from the Chernobyl accident crossed the former Czechoslovakian border at the Opava region. It is clear that most of the precipitation fell on this area. In Figure 2 surface activities of ^{137}Cs in the soil measured after the Chernobyl accident are presented.

As a consequence of high precipitation, the initial values of surface activities in the Opava region measured on 17 June 1986 [31], exceeded 10 kBq/m^2 and spanned the interval from 23 kBq/m^2 to 52 kBq/m^2; whereas, in the Ostravice river valley, only 0.59 kBq/m^2 were obtained, i.e., at least a 40-times lower value. It is interesting to point out that the Ostravice river valley (Staré Hamry) belongs to the network of localities where mushrooms are regularly checked for their ^{137}Cs content by the National Radiation Protection Institute of the Czech Republic, while the hot spot area at the Opava region is not checked [32].

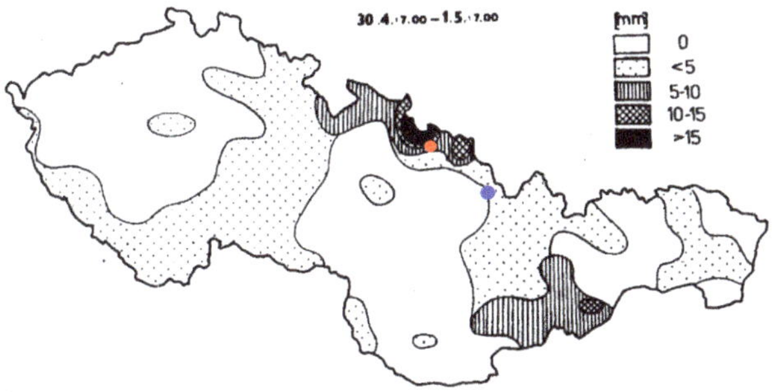

Figure 1. Precipitation in mm that fell on the area of the former Czechoslovakia in the time span from 30.4.1986 7AM CET to 1.5.1986 7AM CET shortly after the Chernobyl accident [6]. The red dot represents the hot spot in the Opava region, whereas the blue dot represents the second investigated area in the Ostravice river valley.

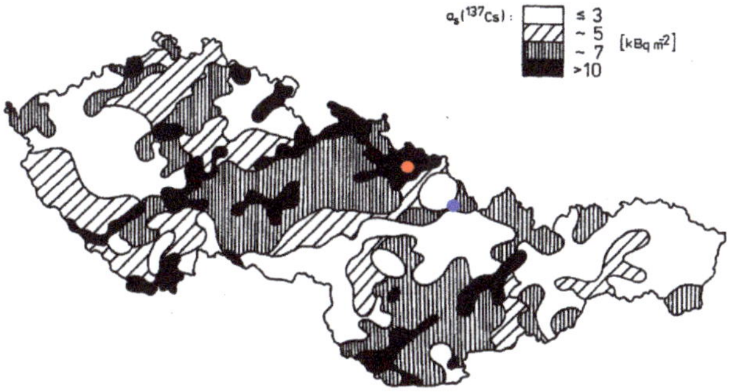

Figure 2. Distribution of the ^{137}Cs surface activity on the area of the former Czechoslovakia after the Chernobyl accident in 1986 [6]. The red dot represents the hot spot in the Opava region whereas the blue dot represents the second investigated area in the Ostravice river valley.

The collected mushrooms were cleaned, divided into caps and stipes and then sliced and dried for 4 days in air. After 4 days, they were dried in a laboratory dryer for 22 h at 105 °C. The individual parts of the fruiting bodies were chopped in a blender and filled into 3 mL plastic vials that fit the well of a 30% relative efficiency low-background well-type HPGe spectrometer (GWD-3023, Baltic Scientific Instruments, Riga, Latvia).

The well detector dimensions were 16 mm in diameter and 50 mm in depth. The ultra-low background cryostat was made from ultra pure Al (5N5 AlSi 1%), OFE-OK electrolytic copper and its uranium and thorium content is less than 1 ppb. The detector was placed in a 10 cm lead shielding with an 8 mm radiopure copper liner. Activity of natural occurring radionuclides in the 2 cm inner chamber of the lead shielding was less than 5 Bq/kg.

The detector operates in a shallow underground laboratory at VŠB-Technical University of Ostrava, Czech Republic, at about 4 meters below the ground level. The resulting gamma background represents 0.0023 and 0.0029 cps in the regions of interest of 661.66 keV ^{137}Cs and 1460.82 keV ^{40}K gamma peaks, respectively.

As heights of the mushroom samples in the 3 mL vials differ and span the interval from 8 mm to 16 mm, the efficiency calibration for the ^{137}Cs 661.66 keV gamma line was performed for four different heights of a standard ^{137}Cs solution provided by Eurostandard, Czech Republic (4.3, 8.5, 12.7 and 17.0 mm). The efficiency curve for the GWD-3023 spectrometer as a function of the sample height was approximated by a quadratic function fitting the measured values obtained for the standard solutions thus enabled to determine the efficiency for an arbitrary sample height (see Figure 3). The resulting relative standard uncertainty of the efficiency introduced by the fitting procedure is less than 0.07%.

A similar procedure was applied to the efficiency calibration for the ^{40}K 1460.82 keV gamma peak. Here, three samples of a powder 99.5% pure KCl provided by Penta, Czech Republic, of different heights were prepared (8.8, 15.3 and 20.6 mm), and the efficiency curve was approximated by a linear function (see Figure 3). The resulting relative standard uncertainty of the efficiency introduced by the fitting procedure is less than 1.2%.

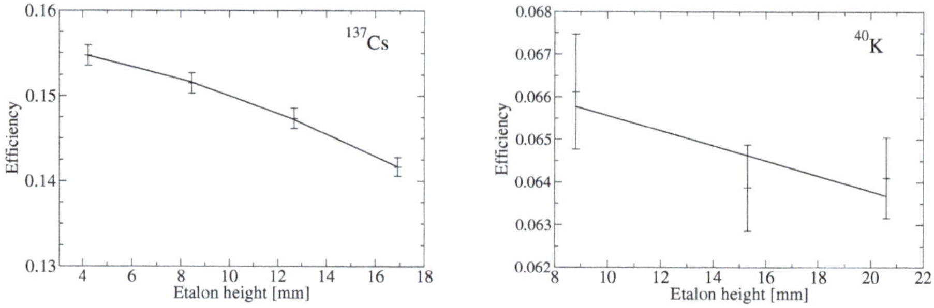

Figure 3. Efficiency calibration curves for different sample heights in the 3 mL vials for ^{137}Cs (**left**) and for ^{40}K (**right**) for the low-background well-type HPGe spectrometer GWD-3023.

In addition to ^{137}Cs, there exists another radiocaesium isotope in nature, ^{134}Cs. The activity of ^{134}Cs can be calculated from the 604.72 keV peak. The specific activity ratio of ^{134}Cs and ^{137}Cs, a_{134}/a_{137}, can help to track the source of radiocaesium. Taking into account different half-lives of ^{137}Cs and ^{134}Cs, $T_{137} = 30.08(9)$ year [33] and $T_{134} = 2.0652(4)$ year [34], respectively, we can calculate the initial ratio of the specific activities of ^{134}Cs and ^{137}Cs, a_{134_0}/a_{137_0}, for April 1986 (the Chernobyl accident) under the assumption that all radiocaesium has the Chernobyl origin:

$$a_{134_0}/a_{137_0} = a_{134}/a_{137} \times \exp\left[\ln 2 \times t(1/T_{134} - 1/T_{137})\right], \qquad (1)$$

where t is the time between the initial deposition and measurement. If the assumption is correct a_{134_0}/a_{137_0} should coincide (within the error bars) with the reported Chernobyl experimental values 0.5–0.6 [35,36] and also with the value $(a_{134_0}/a_{137_0})_{\mathrm{exp}} = 0.515(15)$ calculated from the ratios of the surface activities of ^{134}Cs and ^{137}Cs in the Opava region measured on 17 June 1986 [31].

If the ratio obtained from Equation (1) is higher than the reported Chernobyl initial experimental value, this indicates an additional post-Chernobyl radiocaesium source; if it is lower, a pre-Chernobyl radiocaesium source plays a non-negligible role. This is the case of the second investigated area at the Ostravice river valley where the initial ratio of the surface activities of ^{134}Cs and ^{137}Cs measured on 17 June 1986 equals 0.22 [31]. Therefore, to analyse suspected additional non-Chernobyl sources of radiocaesium, it is useful to define a radiocaesium enhancement factor F_{enh}:

$$F_{\mathrm{enh}} = \frac{a_{134_0}/a_{137_0}}{(a_{134_0}/a_{137_0})_{\mathrm{exp}}} \qquad (2)$$

To determine the ratio a_{134}/a_{137}, a large amount of material is necessary in order to detect ^{134}Cs after more than 20 years after the Chernobyl accident. The samples of *Xerocomus badius* containing both caps and stipes from the Chernobyl hot spot in the Opava area collected in October and November 2011 and in October 2012 underwent the same procedure as the small samples and finally were placed into a Marinelli beaker (volume 450 mL) and measured on the top of a 30% relative efficiency coaxial HPGe spectrometer (GC-3018, Canberra).

The detector was shielded by a massive shielding (100 mm Pb + 1 mm Cd + 1 mm Cu). The efficiency curve for the GC-3018 HPGe spectrometer in the Marinelli geometry was obtained from the MBSS2 standard containing isotopes ^{241}Am, ^{109}Cd, ^{57}Co, ^{139}Ce, ^{203}Hg, ^{113}Sn, ^{85}Sr, ^{137}Cs, ^{88}Y and ^{60}Co provided by Eurostandard, Czech Republic (see Figure 4).

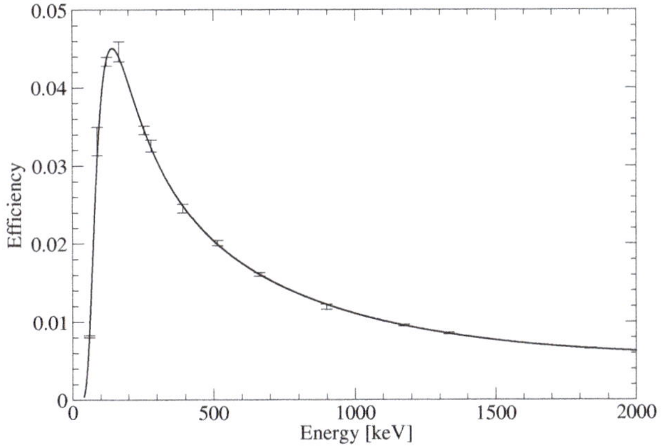

Figure 4. Efficiency calibration curve for the 30% relative efficiency coaxial HPGe spectrometer GC-3018 in the Marinelli geometry.

The effect of selfabsorption was estimated for the GC-3018 HPGe spectrometer and the 3 mL vials and was found to represent less than 4% for ^{137}Cs and less than 2% for ^{40}K. Spectra of the samples were collected with and without calibration point sources provided by Eurostandard, Czech Republic, which were placed separately above each sample.

A ^{137}Cs point source was used to determine the selfabsorption for the 661.66 keV gamma line and the gamma line of 1408.01 keV from a ^{152}Eu point source was used to estimate the effect for the 1460.82 keV ^{40}K gamma line. The selfabsorption effect for the well-type HPGe spectrometer decreases rapidly due to geometry of the well.

To estimate a committed effective dose E caused by the consumption of mushrooms containing a higher amount of ^{137}Cs, the following formula can be applied [37]:

$$E = m \times a_{137f} \times h_{137} \, , \tag{3}$$

where m is the annual intake of fresh mushrooms (kg per person), a_{137f} the ^{137}Cs specific activity of fresh mushrooms (Bq/kg), and h_{137} stands for the conversion factor for ingestion intake of ^{137}Cs (1.3×10^{-8} Sv/Bq) [38].

3. Results and Discussion

Spectra of the samples in the 3 mL vials were measured in October 2016 using the low-background GWD-3023 HPGe spectrometer. Measurement times spanned the interval from 3.5 to 87 h. The obtained specific activities of ^{137}Cs and ^{40}K, recalculated for 1 November 2011 (middle of the collection period), are presented in Table 1. It is clearly seen that the specific activities of ^{137}Cs are higher in the Opava region, while the specific activities of ^{40}K are almost the same in both investigated areas.

A slightly higher level of both ^{137}Cs and ^{40}K is observed in the caps with an exception of *Armillariella mellea* in the Ostravice river valley for ^{137}Cs. The highest values of the specific activity in Table 1 are close to the mean values for fruiting bodies of fungi in the Opole Anomaly collected in 2019 [39]. The Opole Anomaly is well known for extreme levels of ^{137}Cs in Poland (surface activity exceeded 50 kBq/m^2 in 1986) [40]. The Opole Anomaly is quite close to the Opava region investigated in this study.

Table 2 compares the ratios of the specific activities in caps and stipes in the two investigated localities. One can see that both *Russula ochroleuca* and *Xerocomus badius* are highly sensitive to the ^{137}Cs soil content whereas caps of *Armillariella mellea* are about seven to eight times less sensitive, and its stipes are even 10- to 17-times less sensitive. Similar results were reported, e.g., in [16].

Table 1. Specific activities of ^{137}Cs and ^{40}K, a_{137} and a_{40}, for caps and stipes of the investigated dried mushrooms at the two localities and their ratio. Ratios of specific activities a_{137} and a_{40} in caps and stipes, R_{137} and R_{40}, are also displayed. Calculated standard uncertainties are shown in parentheses.

Locality	Species	Part	a_{137}(Bq/kg)	a_{40}(Bq/kg)	a_{137}/a_{40}	R_{137}	R_{40}
Opava area	*Russula ochroleuca*	cap	8772(89)	1123(59)	7.81(42)	2.199(43)	1.15(11)
		stipe	3990(67)	975(76)	4.09(33)		
	Xerocomus badius	cap	11,810(160)	1250(160)	9.4(12)	1.132(22)	1.23(18)
		stipe	10,430(150)	1017(75)	10.26(77)		
	Armillariella mellea	cap	217.0(62)	1717(82)	0.1264(70)	2.129(98)	1.115(70)
		stipe	101.9(37)	1541(63)	0.0662(36)		
Ostravice area	*Russula ochroleuca*	cap	406.8(69)	1078(47)	0.377(18)	1.626(54)	1.119(95)
		stipe	250.2(7.2)	963(70)	0.260(21)		
	Xerocomus badius	cap	428.7(91)	1005(63)	0.427(28)	1.075(35)	1.27(14)
		stipe	398.8(97)	789(67)	0.505(45)		
	Armillariella mellea	cap	62.8(18)	1687(44)	0.0372(14)	0.962(40)	1.299(53)
		stipe	65.2(20)	1299(40)	0.0502(21)		

Table 2. Ratios of the specific activities a_{137} and a_{40} for caps and stipes in the investigated areas. Standard uncertainties are shown in parentheses.

Part	Species	$a_{137_{Opava}}/a_{137_{Ostravice}}$	$a_{40_{Opava}}/a_{40_{Ostravice}}$
Cap	*Russula ochroleuca*	21.56(43)	1.042(70)
	Xerocomus badius	27.54(69)	1.25 (18)
	Armillariella mellea	3.46(14)	1.018(56)
Stipe	*Russula ochroleuca*	15.95(53)	1.01(11)
	Xerocomus badius	26.15(74)	1.29(15)
	Armillariella mellea	1.563(73)	1.186(61)

To estimate the strength of the linear relationship between the ratios of the specific activities of ^{137}Cs and ^{40}K (in Table 1 in the column a_{137}/a_{40}) for stipes and caps in both areas, a Pearson correlation test was applied. The Pearson correlation coefficient was equal to 0.93 and the p-value was less than 0.01 ($p = 0.0074$) indicating a strong linear relationship between the ratios of the specific activities in agreement with [1], which supports the hypothesis that the transport of ^{137}Cs from stipe to cap depends directly on ^{40}K concentration for all three investigated species.

A mixed sample of the total dry weight of 29.591(15) g containing both caps and stipes of *Xerocomus badius* collected in October and November 2011 from the Chernobyl hot spot (Opava region) was measured in the Marinelli geometry in April 2012. The measurement time comprised 654,037 s. We found that the ^{137}Cs specific activity $a_{137} = 9400(200)$ Bq/kg is compatible with our results obtained from the measurement of the small samples of *Xerocomus badius* caps and stipes in the Opava region. The obtained ^{134}Cs specific activity a_{134} equalled 2.27(57) Bq/kg.

In October 2012, we collected caps of *Xerocomus badius* from the same place in the Opava region that underwent the same procedure as the previous sample and measured them for a longer time of 1,800,000 s. The initial ratios of the specific activities of ^{134}Cs and ^{137}Cs calculated from Equation (1) for both samples for April 1986 are summarized in Table 3 and compared to the initial reported Chernobyl experimental values in the Opava region.

The slightly higher values of the radiocaesium enhancement factor F_{enh} may indicate an additional contribution from the Fukushima accident in March 2011, but the final conclusion cannot be drawn because F_{enh} does not exceed 1 by more than 2σ.

Table 3. Initial ratios of a_{134_0}/a_{137_0} calculated from Equation (1) for two samples of *Xerocomus badius* from the Opava region and the radiocaesium enhancement factors F_{enh} calculated from Equation (2) for the experimental initial ratio in the Opava region, $(a_{134_0}/a_{137_0})_{exp} = 0.515(15)$. Standard uncertainties are shown in parentheses.

Sample Collection	a_{134_0}/a_{137_0}	F_{enh}
October–November 2011	0.75(23)	1.46(45)
October 2012	0.74(17)	1.44(34)

The highest value of the ^{137}Cs specific activity was observed for the species of *Xerocomus badius* (see Table 1) collected in the Opava region. Supposing the moisture content of mushrooms to be at 90% [17], the specific activity of the whole fresh mushroom is, in this case, at the value of 1119 Bq/kg, which already exceeds the limit in foodstuff recommended by IAEA (1000 Bq/kg fresh weight) [15].

The share of the ^{137}Cs in the annual committed effective dose has been significantly increasing since the Chernobyl accident [41], and thus it is important to focus on mushroom consumers with special dietary habits.

The mean consumption of mushrooms calculated for the period 1986–2014 was 1.7 kg per year for the general population in the Czech Republic [41]. The annual consumption of wild mushrooms by dwellers has been estimated by Šišák [42] to be 7 kg per person. Based on Equation (3) the annual committed effective dose E for *Xerocomus badius* for dwellers equalled 0.102 mSv in the Opava region in 2011.

In the second examined location (the Ostravice river valley), the specific activity of ^{137}Cs for the whole fresh mushroom $a_{137} = 41.5$ Bq/kg resulted in the annual committed effective dose of $E = 0.004$ mSv for dwellers. Therefore, the radiation risk in the Opava region is about 26 times higher. If we take into account a 50% decrease of the ^{137}Cs activity due to cooking reported in [41], the annual committed effective dose becomes even lower.

4. Conclusions

The highest levels of ^{137}Cs were found in caps of the species *Xerocomus badius* and *Russula ochroleuca* in the Opava region (11.8 kBq/kg and 8.77 kBq/kg, respectively). *Armillariella mellea* shows very low accumulation of radiocaesium in both locations. Furthermore, the dominant Chernobyl origin of radiocaesium at the hot spot in the Opava region was confirmed by means of the ^{134}Cs/^{137}Cs activity ratio. The linear relationship between the ratios of specific activities of ^{137}Cs and ^{40}K for stipes and caps was validated as well.

The potential risk from the consumption of *Xerocomus badius* in the Opava region is about 26-times higher than in the Ostravice river valley and represented the annual committed effective dose of 0.102 mSv at maximum in 2011. We also showed that the low-background well HPGe detector GWD-3023 equipped with ultra-low background shielding can be efficiently used for routine investigation of the ^{137}Cs content in small mushroom samples with a dry weight of less than 1 g and a volume lower than 2–3 mL, which fit in the detector well.

Author Contributions: Conceptualization, P.A. and O.H.; methodology, O.H. and R.U.; validation, O.H., P.A. and R.U.; formal analysis, O.H. and R.U.; investigation, O.H.; resources, O.H.; data curation, O.H.; writing—original draft preparation, O.H.; writing—review and editing, O.H., P.A. and R.U.; visualization, O.H.; supervision, P.A.; project administration, P.A.; funding acquisition, P.A. All authors have read and agreed to the published version of the manuscript.

Funding: This research was funded by the Ministry of Education, Youth and Sports of the Czech Republic, project number SP2021/64.

Institutional Review Board Statement: Not applicable.

Informed Consent Statement: Not applicable.

Data Availability Statement: https://homel.vsb.cz/~ale02/mushrooms_life2021.xls, accessed on 4 December 2021.

Acknowledgments: The authors thank Petra Količová for mushroom collecting and identifying and Petr Jandačka for initiating the study.

Conflicts of Interest: The authors declare no conflict of interest.

Abbreviations

The following abbreviations are used in this manuscript:

IAEA	International Atomic Energy Agency
GPS	Global Positioning System
CET	Central European Time
OFE-OK	Oxygen Free Copper of Very High Purity
HPGe	High Purity Germanium Spectrometer

References

1. Bazala, M.A.; Bystrejewska-Piotrowska, G.; Čipáková, A. Bioaccumulation of ^{137}Cs in wild mushrooms collected in Poland and Slovakia. *Nukleonika* **2005**, *50* (Suppl. S1), 15–18.
2. Falandysz, J.; Saniewski, M.; Fernandes, A.R.; Meloni, D.; Cocchi, L.; Struminska-Parulska, D.; Zalewska, T. Radiocaesium in Tricholoma spp. from the Northern Hemisphere in 1971–2016. *Sci. Total Environ.* **2022**, *802*. 149829. [CrossRef]
3. Guillen, J.; Baeza, A. Radioactivity in mushrooms: A health hazard? *Food Chem.* **2014**, *154*, 14–25. [CrossRef] [PubMed]
4. Zarubina, N.E. The content of accidental radionuclides in mushrooms from the 30-km zone of Chernobyl Nuclear Power Station. *Mikol. I Fitopatol.* **2004**, *38*, 36–40.
5. Beňová, K.; Dvořák, P.; Tomko, M.; Falis, M. Artificial environmental radionuclides in Europe and methods of lowering their foodstuff contamination—A review. *Acta Vet. Brno* **2016**, *85*, 105–112. [CrossRef]
6. Bučina, I.; Dvořák Z.; Malátová, I.; Vrbová, H.; Drábová, D. Radionuclides from the Chernobyl accident in soil over the Czechoslovak Territory, their origin, deposition and distribution. In Proceedings of the XV Regional Congress of IRPA, The Radioecology of Natural and Artificial Radionuclides, Visby, Sweden, 10–14 September 1989; pp. 170–175.
7. Savino, F.; Pugliese, M.; Quarto, M.; Adamo, P.; Loffredo, F.; De Cicco, F.; Roca, V. Thirty years after Chernobyl: Long-term determination of ^{137}Cs effective half-life in the lichen Stereocaulon vesuvianum. *J. Environ. Radioact.* **2017**, *172*, 201–206. [CrossRef] [PubMed]
8. Hashimoto, S.; Imamura, N.; Kawanishi, A.; Komatsu, M.; Obashi, S.; Nishina, K.; Kaneko, S.; Shaw, G.; Thiry, Y. A dataset of ^{137}Cs activity concentration and inventory in forests contaminated by the Fukushima accident. *Sci. Data* **2020**, *7*, 431. [CrossRef]
9. Ambrosino, F.; Stellato, L.; Sabbarese, C. A case study on possible radiological contamination in the Lo Uttaro landfill site (Caserta, Italy). *J. Phys. Conf. Ser.* **2020**, *1548*, 012001. [CrossRef]
10. Červínková, A.; Poschl, M.; Pospíšilová, L. Radiocaesium transfer from forest soils to wild edible fruits and radiation dose assessment through their ingestions in Czech Republic. *J. For. Res.* **2017**, *22*, 91–96. [CrossRef]
11. Ivanic, M.; Fiket, Z.; Medunic, G.; Turk, M.F.; Marovic, G.; Sencar, J.; Kniewald, G. Multi-element composition of soil, mosses and mushrooms and assessment of natural and artificial radioactivity of a pristine temperate rainforest system (Slavonia, Croatia). *Chemosphere* **2019**, *215*, 668–677. [CrossRef] [PubMed]
12. Daillant, O.; Boilley, D.; Josset, M.; Hettwig, B.; Fischer, H.W. Evolution of radiocaesium contamination in mushrooms and influence of treatment after collection. *J. Radioanal. Nucl. Chem.* **2013**, *297*, 437–441. [CrossRef]
13. Calmon, P.; Thiry, Y.; Zibold, G.; Rantavaara, A.; Fesenko, S. Transfer parameter values in temperate forest ecosystems: A review. *J. Environ. Radioact.* **2009**, *100*, 757–766. [CrossRef]
14. Čipáková, A. Migration of radiocaesium in individual parts of the environment. *Nukleonika* **2005**, *50*, 19–23.
15. IAEA. Intervention criteria in a nuclear or radiation emergency. In *IAEA Safety Series No. 109*; International Atomic Energy Agency: Viena, Austria, 1994.
16. Čipáková, A. ^{137}Cs content in mushrooms from localities in eastern Slovakia. *Nukleonika* **2004**, *49*, 25–29.
17. Falandysz, J.; Zalewska, T.; Saniewski, M.; Fernandes, A.R. An evaluation of the occurrence and trends in ^{137}Cs and ^{40}K radioactivity in King Bolete Boletus edulis mushrooms in Poland during 1995–2019. *Environ. Sci. Pollut. Res.* **2021**, *28*, 405–415. [CrossRef]
18. Zalewska, T.; Cocchi, L.; Falandysz, J. Radiocaesium in Cortinarius spp. mushrooms in the regions of the Reggio Emilia in Italy and Pomerania in Poland. *Environ. Sci. Pollut. Res.* **2016**, *23*, 169–174. [CrossRef] [PubMed]
19. Heinrich, G. Distribution of Radiocesium in the different parts of mushrooms. *J. Environ. Radioact.* **1993**, *18*, 229–245. [CrossRef]

20. Kocadag, M.; Exler, V.; Christopher, B.S.; Baumgartner, A.; Stietka, M.; Landstetter, C.; Korner, M.; Maringer, F.J. Environmental radioactivity study of Austrian and Bavarian forest ecosystems: Long-term behaviour of contamination of soil, vegetation and wild boar and its radioecological coherences. *Appl. Radiat. Isot.* **2017**, *126*, 106–111. [CrossRef] [PubMed]
21. Kalač, P. A review of edible mushrooms radioactivity. *Food Chem.* **2001**, *75*, 29–35. [CrossRef]
22. Horyna, J.; Řanda, Z. Uptake of radiocesium and alkali metals by mushrooms. *J. Radioanal. Nucl. Chem.* **1988**, *127*, 107–120. [CrossRef]
23. Kita, K.; Igarashi, Y.; Kinase, T.; Hayashi, N.; Ishizuka, M.; Adachi, K.; Koitabashi, M.; Sekiyama, T.T.; Onda, Y. Rain-induced bioecological resuspension of radiocaesium in a polluted forest in Japan. *Sci. Rep.* **2020**, *10*, 15330. [CrossRef]
24. Marovic, G.; Franic, Z.; Sencar, J.; Bituh, T.; Vugrinec, O. Mosses and Some Mushroom Species as Bioindicators of Radiocaesium Contamination and Risk Assessment. *Coll. Antropol.* **2008**, *32*, 109–114. [PubMed]
25. Baeza, A.; Guillen, F.J.; Salas, A.; Manjon, J.L. Distribution of radionuclides in different parts of a mushroom: Influence of the degree of maturity. *Sci. Total Environ.* **2006**, *359*, 255–266. [CrossRef]
26. Falandysz, J.; Zhang, J.; Saniewski, M. ^{137}Cs, ^{40}K and K in raw and stir-fried mushrooms from the Boletaceae family from the Midu region in Yunnan, Southwest China. *Environ. Sci. Pollut. Res.* **2020**, *27*, 509–517. [CrossRef] [PubMed]
27. Falandysz, J.; Zalewska, T.; Fernandes, A.R. ^{137}Cs and ^{40}K in Cortinarius caperatus mushrooms (1996–2016) in Poland—Bioconcentration and estimated intake: ^{137}Cs in Cortinarius spp. from the Northern Hemisphere from 1974 to 2016. *Environ. Pollut.* **2019**, *255*, 113208. [CrossRef] [PubMed]
28. Falandysz, J.; Saniewski, M.; Zhang, J.; Zalewska, T.; Liu, H.G.; Kluza, K. Artificial ^{137}Cs and natural ^{40}K in mushrooms from the subalpine region of the Minya Konka summit and Yunnan Province in China. *Environ. Sci. Pollut. Res.* **2018**, *25*, 615–627. [CrossRef]
29. Kalač, P. Chemical composition and nutritional value of European species of wild growing mushrooms: A review. *Food Chem.* **2009**, *113*, 9–16. [CrossRef]
30. Baeza, A.; Guillén, J.; Hernández, S.; Salas, A.; Bernedo, M.; Manjón, J.L.; Moreno, G. Influence of the nutritional mechanism of fungi (mycorrhize/saprophyte) on the uptake of radionuclides by mycelium. *Radiochim. Acta* **2005**, *93*, 233–238. [CrossRef]
31. Surface Activity of ^{137}Cs, ^{134}Cs and ^{103}Ru Obtained for Soil Samples during the Nationwide Survey on 17 June 1986. Available online: https://www.suro.cz/cz/publikace/cernobyl/plosna-aktivita-radionuklidu-zjistena-ve-vzorcich-odebranych-pud/pruzkum_pud_1986.pdf/at_download/file (accessed on 26 October 2021). (In Czech)
32. Monitoring of Radiation Situation, State Office for Nuclear Safety. Available online: https://www.sujb.cz/aplikace/monras/?lng=cs_CZ (accessed on 26 October 2021). (In Czech)
33. Brown, E.; Tuli, J.K. Nuclear Data Sheets for A = 137. *Nucl. Data Sheets* **2007**, *108*, 2173–2318. [CrossRef]
34. Sonzogni, A.A. Nuclear Data Sheets for A = 134. *Nucl. Data Sheets* **2004**, *103*, 1–182. [CrossRef]
35. Kawada, Y.; Yamada, T. Radioactivity ratios of ^{134}Cs/^{137}Cs released by the nuclear accidents. *Isot. News* **2012**, *697*, 16–20.
36. Taylor, H.W.; Svoboda, J.; Henry, G.H.R.; Wein, R.W. Post-Chernobyl ^{134}Cs and ^{137}Cs levels at some localities in Northern Canada. *Arctic* **1988**, *41*, 293–296. [CrossRef]
37. ICRP. Age-dependent Doses to the Members of the Public from Intake of Radionuclides—Part 5 Compilation of Ingestion and Inhalation Coefficients (ICRP Publication 72). *Ann. ICRP* **1996**, *26*, 1–91.
38. ICRP. Compendium of Dose Coefficients based on ICRP Publication 60 (ICRP Publication 119). *Ann. ICRP (Suppl.)* **2012** *41*, 40.
39. Ronda, O.; Grządka, E.; Ostolska, I.; Orzeł, J.; Cieślik, B.M. Accumulation of radioisotopes and heavy metals in selected species of mushrooms. *Food Chem.* **2022**, *367*, 130670. [CrossRef] [PubMed]
40. Oloś, G.; Dolhańczuk-Śródka, A. Levels of ^{137}Cs in game and soil in Opole Anomaly, Poland in 2012–2020. *Ecotoxicol. Environ. Saf.* **2021**, *223*, 112577. [CrossRef]
41. Škrkal, J.; Fojtík, P.; Malátová, Bartusková, M. Ingestion intakes of ^{137}Cs by the Czech population: Comparison of different approaches. *J. Environ. Radioact.* **2017**, *171*, 110–116. [CrossRef] [PubMed]
42. Šišák, L. The importance of forests as a source of mushrooms and berries in the Czech Republic. *Mykol. Sborník* **1996**, *73*, 98–101. (In Czech)

life

MDPI

Article

A Potential Serum Biomarker for Screening Lung Cancer Risk in High Level Environmental Radon Areas: A Pilot Study

Narongchai Autsavapromporn [1,*], Pitchayaponne Klunklin [1], Imjai Chitapanarux [1], Churdsak Jaikang [2], Busyamas Chewaskulyong [3], Patumrat Sripan [4], Masahiro Hosoda [5] and Shinji Tokonami [6]

1 Division of Radiation Oncology, Department of Radiology, Faculty of Medicine, Chiang Mai University, Chiang Mai 50200, Thailand; pklunklin@gmail.com (P.K.); imjai.chitapanarux@cmu.ac.th (I.C.)
2 Toxicology Section, Department of Forensic Medicine, Faculty of Medicine, Chiang Mai University, Chiang Mai 50200, Thailand; churdsak.j@cmu.ac.th
3 Division of Oncology, Department of Internal Medicine, Faculty of Medicine, Chiang Mai University, Chiang Mai 50200, Thailand; busyamas.chewask@cmu.ac.th
4 Research Institute for Health Science, Chiang Mai University, Chiang Mai 50200, Thailand; patumrat.sripan@cmu.ac.th
5 Graduate School of Health Science, Hirosaki University, Hirosaki, Aomori 036-8564, Japan; m_hosada@hirosaki-u.ac.jp
6 Institute of Radiation Emergency Medicine, Hirosaki University, Hirosaki, Aomori 036-8564, Japan; tokonami@hirosaki-u.ac.jp
* Correspondence: narongchai.a@cmu.ac.th

Citation: Autsavapromporn, N.; Klunklin, P.; Chitapanarux, I.; Jaikang, C.; Chewaskulyong, B.; Sripan, P.; Hosoda, M.; Tokonami, S. A Potential Serum Biomarker for Screening Lung Cancer Risk in High Level Environmental Radon Areas: A Pilot Study. *Life* **2021**, *11*, 1273. https://doi.org/10.3390/life11111273

Academic Editors: Fabrizio Ambrosino and Supitcha Chanyotha

Received: 24 October 2021
Accepted: 19 November 2021
Published: 21 November 2021

Abstract: Radon is a major cause of lung cancer (LC) deaths among non-smokers worldwide. However, no serum biomarker for screening of LC risk in high residential radon (HRR) areas is available. Therefore, the aim of this study was to determine diagnostic values of serum carcinoembryonic antigen (CEA), cytokeratin 19 fragment (Cyfra21-1), human epididymis protein 4 (HE4), interleukin 8 (IL-8), migration inhibitory factor (MIF), tumor nuclear factor-alpha (TNF-α) and vascular endothelial growth factors (VEGF) occurring in high radon areas. Seventy-five LC non-smoker patients and seventy-five healthy controls (HC) were enrolled in this study. Among the HC groups, twenty-five HC were low residential radon (LRR) and fifty HC were HRR. Significantly higher ($p < 0.0004$) serum levels of CEA, Cyfra21-1, IL-8 and VEGF were found in the LC compared with the LRR and HRR groups. More importantly, significantly higher levels ($p < 0.009$) of serum CEA, Cyfra21-1 and IL-8 were observed in HRR compared with the LRR group. Likewise, a ROC curve demonstrated that serum CEA and Cyfra21-1 could better distinguish LC risk from HRR groups than IL-8. These results indicated that serum CEA and Cyfra21-1 were significantly increased in the HRR group and may be considered as potential biomarkers for individuals at high-risk to develop LC.

Keywords: radon; serum biomarker; lung cancer; CEA; Cyfra21-1

1. Introduction

Lung cancers (LC) are the most aggressive malignant solid tumor causes of cancer-related deaths for both men and women worldwide. Approximately 15–20% of LCs are small cell lung cancers (SCLCs) and other 80–85% of LC are non-small cell lung cancer (NSCLCs). NSCLC can be subdivided into three histological subtypes, namely squamous cell carcinoma, adenocarcinoma and large-cell carcinoma. The treatment of LC includes surgery, chemotherapy and radiation therapy [1,2]. LC progresses quietly and the majority of LC patients are typically diagnosed at an advanced or late stage, with only 15% of LC patients begin diagnosed at an early stage [3]. The median survival of LC patients after treatment is only about 1 year (or less) and the 5-year survival rate is approximately 20% [1]. Over 70% of LC patients are diagnosed in advanced stages because there remains no practical way to identify high-risk individuals. Thus, detection of LC at an early stage could help to improve the survival, prediction of prognosis and treatment outcome of LC patients.

In Chiang Mai province, in upper northern Thailand, LC is the second most common cancer in both men and women, according to the World Health Organization (WHO) Report in 2020 [4]. The main factors identified as responsible for increased LC incidence rate were demographic characteristics, tobacco smoke, secondhand smoke, environmental exposure and indoor radon exposure [5–7]. It is considered that 3–20% of all LC deaths worldwide are attributable to indoor radon [8]. Radon (^{222}Rn) is a radioactive noble gas from the decay product of uranium-238 (^{238}U). It has a half-life of 3.82 days and possesses the capacity of damaging respiratory epithelium cells through the emission of an alpha particle (high linear energy transfer radiation). Radon is present in rock, soil, groundwater, natural gas, and building materials found in dwellings [8,9]. Residential radon exposure depends not only on factors related to housing, but also on the geological structures, the ventilation of radon in air and environmental conditions [10]. According to the WHO, exposure to high levels of radon for a long period of time is the second most common risk factor for LC after tobacco smoke and the major risk factor for LC in non-smokers [8,9]. In addition, the latency of LC is between 5 and 25 years for indoor radon exposure [11]. Thus, long-term exposure to radon and its decay products within dwellings could play an important role in LC risk during a lifetime of exposure in both non-smokers and smokers.

In our previous study, the concentration of indoor radon in Chiang Mai province (57 Bq/m^3) was considerably higher than the worldwide average value of 39 Bq/m^3. Within the district of San Pa Tong, the indoor radon activity concentration reached 219 Bq/m^3, exceeding the WHO reference level of 100 Bq/m^3 [8]. The annual effective dose was found to be 5.5 mSv, a value higher than the global average of 1 mSv [12]. Therefore, the identification of a useful biomarker for screening the early-stage LC in high residential radon exposure is particularly important for improving LC prognosis and treatment outcomes in Chiang Mai province.

To date, serum biomarkers represent the non-invasive blood test for the screening of LC. Several serum tumor markers for LC have been studied extensively, such as carcinoembryonic antigen (CEA), cytokeratin 19 fragment (Cyfra21-1), human epididymis protein 4 (HE4), interleukin 8 (IL-8), migration inhibitory factor (MIF), tumor nuclear factor-alpha (TNF-α) and vascular endothelial growth factor (VEGF) [2,13–18]. However, there is currently no serum biomarker specifically for the detection of LC risk in environmentally high radon areas. Therefore, it is crucial to explore potential serum biomarkers that can detect the diagnosis of LC induced by high radon exposure. In this study, we investigated the serum levels of CEA, Cyfra21-1, HE4, IL-8, MIF, TNF-α and VEGF in LC patients and residential radon exposure, and we evaluated the diagnostic ability of those serum for LC risk in high radon areas.

2. Materials and Method

2.1. Study Area

Thailand is a country located in the middle of mainland south-east Asia (Figure 1a). It has a total area of 198,120 square miles with a population of 68 million people [19]. It is bounded to the north by Myanmar and Laos, to the west with the Andaman Sea and Myanmar, to the east by Cambodia and Laos, and to the south by the Gulf of Thailand and Malaysia. Thailand has 77 provinces that are further divided into six geographical regions —Northern, Northeast, Central, Eastern, Western and Southern Thailand—based on natural features: Thailand has a tropical climate, characterized by monsoons [20]. Chiang Mai is the largest city in the upper northern region of Thailand. It is located on the Ping River and surrounded by the mountain ranges of the Thai highlands whose geological and geochemical characteristics increase the levels of natural background radiation from sources such as radon. The city is subdivided into 25 districts. The Hang Dong, Muang, Saraphi and San Pha Tong districts of Chiang Mai were selected as the study area based on the higher mortality rate of lung cancer in upper northern Thailand than in other areas [5,6]. Based on our previous study, the radon levels in the study area are divided into three groups (Figure 1b): "low" (<44 Bq/m^3), "moderate" (44–70 Bq/m^3) and "high" (>70 Bq/m^3) based on indoor radon concentration in the dwellings [12].

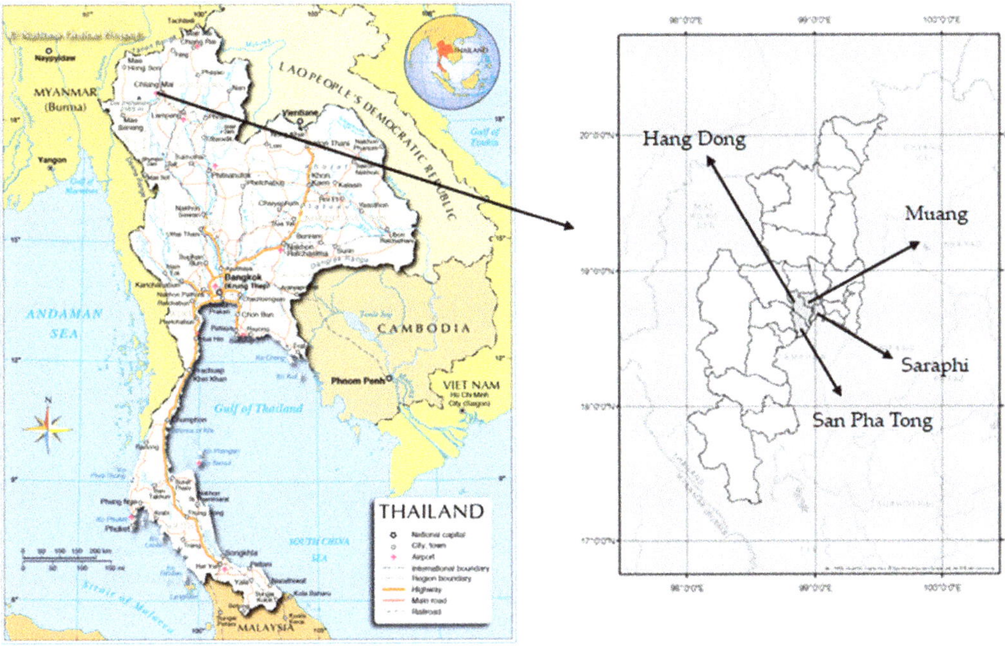

(a) **(b)**

Figure 1. Geological map of Thailand. (**a**) Map of Thailand; (**b**) The study area location in Chiang Mai. Geological map of Thailand obtained from the Nations Online Project (Available online: https://www.nationsonline.org/oneworld/map/thailand_map.htm (accessed on 15 September 2021).

2.2. Study Design

The transitional study was conducted on selected individuals in the following Chiang Mai districts: Hang Dong, Muang, Saraphi and San Pha Tong (Figure 1b). A total of 150 non-smokers was examined including 75 LC patients (38 males and 37 females), aged from 38 to 87 years, with the median age of 60.3 ± 10.8 years, and 75 healthy controls (HC, 38 males and 37 females), aged from 37 to 86 years, with the median age of 59.6 ± 8.3 years (Table 1). The recruitment period of LC patients took place at Maharaj Nakon Chiang Mai University Hospital and Saraphi Hospital, in Chiang Mai between 2016 and 2020 All LC patients were diagnosed as NSCLC and non-smokers or former smokers (never smoked or stopped smoking for more than 15 years). Then, we randomly selected HC groups that comprised 25 low residential radon or LRR areas (15 males and 10 females) and 50 high residential radon or HRR areas (23 males and 27 females), who had lived in during the past 10 years (or more) in the measured dwellings. All HC groups were individuals without a past history of cancer, minor surgery and non-smokers (never smoked or less than 100 cigarettes smoked in his or her lifetime). Participants were interviewed by trained interviewers using a questionnaire that collected information on possible confounding factors (such as smoking status, lifestyle, environmental tobacco smoke, occupational/environmental/medical exposure to radiation and alcohol consumption).

Table 1. Characteristics of lung cancer (LC) patients and healthy controls (HC).

Characteristics.	LC (n = 75)	HC		
		LRR (n = 25)	HRR (n = 50)	Total (n = 75)
Age in years, mean (SD)	60.3 (10.8)	61.2 (7.1)	58.8 (8.8)	59.6.(8.3)
Gender				
Male	38	15	23	38
Female	37	10	27	37

2.3. Sample Collection

A 10 mL of blood samples were collected from both LC and HC groups in a serum-separating sterile tubes. The samples were then centrifuged at 3000× g for 10 min at 4 °C and stored at −80 °C for further analysis.

2.4. Biochemical Analyses

The serum levels of Cyfra21-1, CEA, HE4, IL-8, MIF, TNF-α and VEGF were performed using a Milliplex map kit assay (Millipore, Billerica, MA, USA) according to the manufacturer's instructions [21]. All samples were analyzed in duplicate with the xPONENT software (Luminex) and expressed in picograms (pg) per milliliter (mL). The intra-assay and inter-assay variabilities were ≤5%.

2.5. Statistical Analysis

The statistical analyses were performed with the software Sigma Plot 10 (Systat Software Inc, San Jose, CA USA). The values of serum (CEA, Cyfra21-1, HE4, IL-8, MIF, TNF-α and VEGF) were summarized as mean ± SD. The significance between the two groups were evaluated by Mann-Whitney U test. To determine the diagnostic value of these analyses, the receiver operating characteristic (ROC) curve was plotted and relevant results including the area under the curve (AUC) combined with sensitivity and specificity were estimated. A p values < 0.05 were considered as statistically significant.

3. Results

3.1. Characteristics of LC and HC Groups

Overall, 75 LC patients and 75 from the HC groups were enrolled in this study. Among the HC groups, 25 individuals (33.3%) were from LRR areas and 50 (66.7%) were from HRR areas. All subjects were non-smokers and all LC patients were diagnosed as NSCLC. There were no statistically significant differences between the two groups in the age (p = 0.65) or gender. The median age of LC groups at diagnosis was 60.3 ± 10.8 years. In the HC groups, the median age was 59.6 ± 8.3 years. Thirty-eight (50.7%) were males and thirty-seven (49.3%) were females in both LC and HC groups. The detailed information is shown in Table 1.

3.2. Levels of Serum Analytes in LC and HC Groups

The serum levels of CEA, Cyfra21-1, HE4, IL-8, MIF, TNF-α and VEGF in LC and HC groups were presented in Figure 2. The levels of serum CEA, Cyfra21-1, IL-8 and VEGF were significantly significant differences (p < 0.0001) between LC and HC groups (Figure 2a,b,d,g). However, no significant differences (p > 0.05) were observed in serum HE4, MIF and TNF-α levels between LC and HC groups (Figure 2c,e,f). These results illustrate that serum CEA, Cyfra21-1, IL-8 and VEGF are potential biomarkers for detection of LC risk in HC groups as well as residential radon exposure.

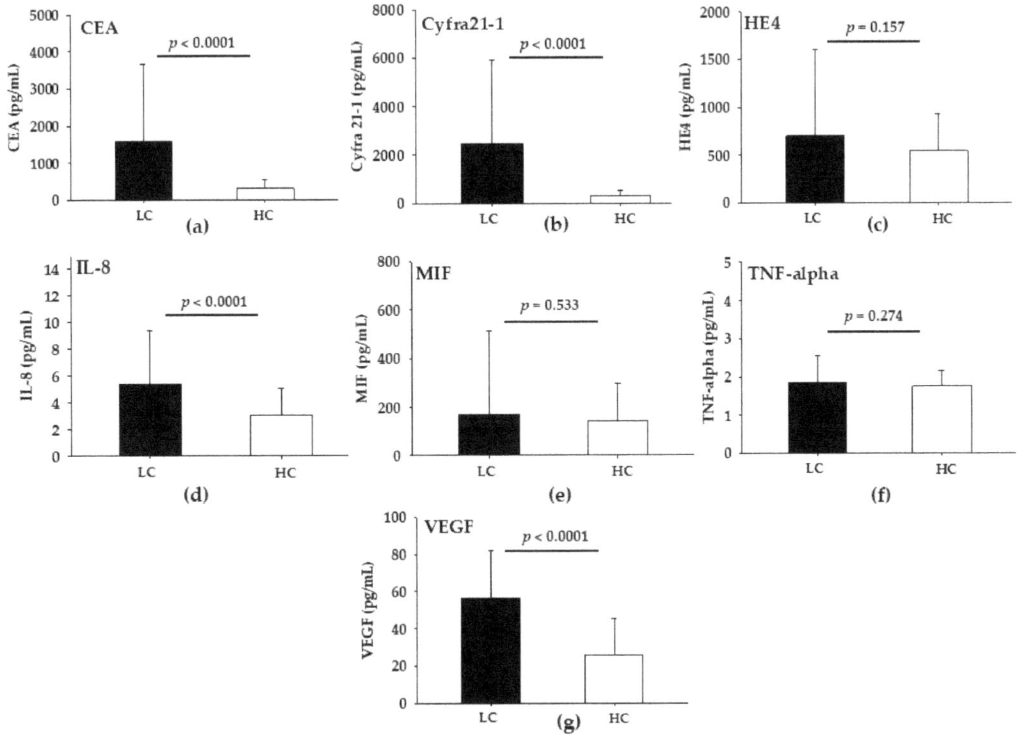

Figure 2. Levels of serum in lung cancer (LC) patients and healthy controls (HC). (**a**) CEA; (**b**) Cyfra21-1; (**c**) HE4; (**d**) IL-8; (**e**) MIF; (**f**) TNF-α; (**g**) VEGF.

3.3. Levels of Serum Analytes in LC, LRR and HRR Groups

To further verify the potential serum biomarker for screening LC risk in high radon areas, the HC groups were divided into LRR and HRR groups according to the radon concentration in their dwellings. As shown in Figure 3, significantly higher ($p < 0.05$) serum levels of CEA, Cyfra21-1, IL-8 and VEGF were observed for the LC group in a comparison between LRR and HRR groups. However, there were no statistically significant differences ($p > 0.05$) in serum HE4, MIF and TNF-α. Furthermore, the levels of serum CEA, Cyfra21-1 and IL-8 were significantly higher ($p < 0.05$) in HRR than LRR groups, but there were no statistically significant differences ($p > 0.05$) between LRR and HRR groups for serum HE4, MIF, TNF-α and VEGF. These results indicated that serum CEA, Cyfra21-1 and IL-8 possess potential ability to distinguish high risk of LC from HC groups.

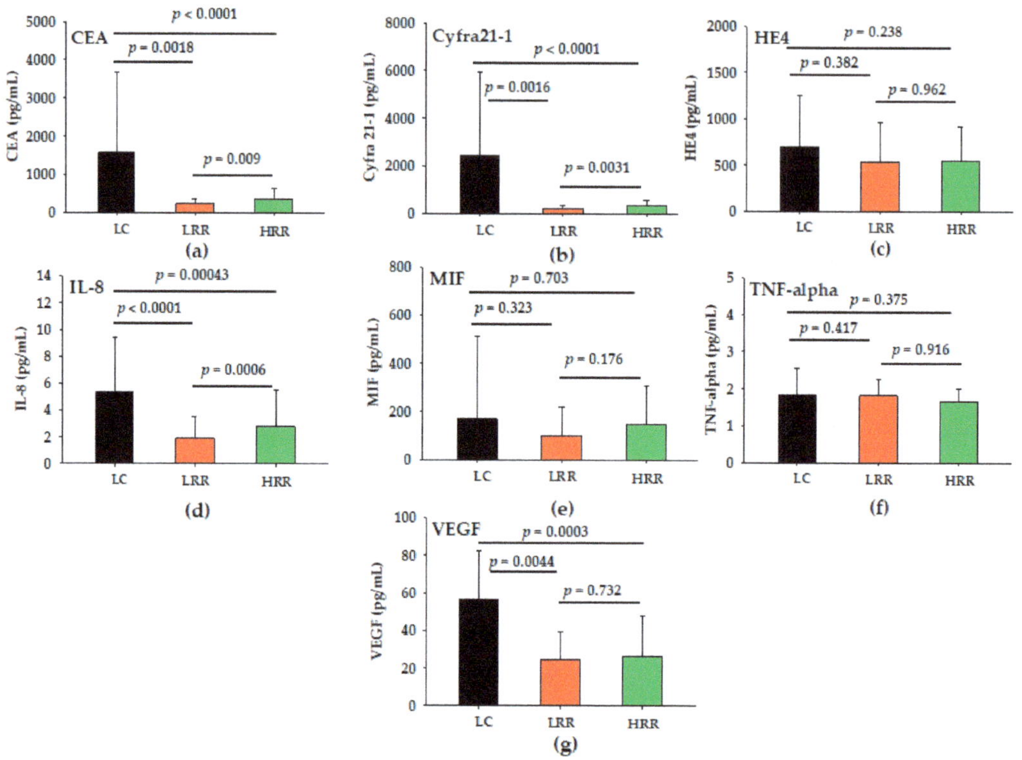

Figure 3. Levels of serum in lung cancer (LC) patients, low residential radon (LRR) and high residential radon (HRR). (**a**) CEA; (**b**) Cyfra21-1; (**c**) HE4; (**d**) IL-8; (**e**) MIF; (**f**) TNF-α; (**g**) VEGF.

3.4. Diagnostic Ability of Serum Biomarker for LC Risk in High Level Environmental Radon Areas

After having confirmed that serum CEA, Cyfra21-1 and IL-8 could be better biomarkers to distinguish between LRR and HRR groups, the predictive power as a screening tool to distinguish LC risk from HRR groups was then evaluated. For this purpose, the ROC curves were calculated the diagnostic efficacy of serum CEA, Cyfra21-1 and IL-8 as potential biomarkers of LC risk in high level environmental radon areas. The area under the ROC (AUC-ROC) curve, sensitivity, specificity and all cut-off values of serum were determined using ROC analysis and summarized in Table 2. The AUC-ROC curve for discriminating LC from HRR groups were 0.782, 0.797 and 0.606 for serum CEA, Cyfra21-1 and IL-8, respectively, relative to the HRR groups (Figure 4). The comparison of ROC demonstrated that serum CEA and Cyfra21-1 performed better in identifying LC risk in HRR groups compared with IL-8. Then, we evaluated the sensitivity and specificity of serum CEA, Cyfra21-1 and IL-8 levels in LC patients compared to HRR groups. The sensitivity of serum CEA, Cyfra21-1 and IL-8 were 57.3%, 58.6% and 48% and the specificity were 98%, 94% and 76%. The cut off values of serum CEA Cyfra21-1 and IL-8 were 890.4 pg/mL, 682.5 pg/mL and 5 pg/mL (Table 2). Based on this result, it seems that serum CEA and Cyfra21-1 were better diagnostic markers for early detection of LC risk in high radon areas.

Table 2. The diagnostic sensitivity and specificity of serum CEA, Cyrfra21-1 and IL-8 in LC patients compared to HRR groups.

Biomarker	Sensitivity (%)	Specificity (%)	AUC
CEA	57.3	98	0.7821
Cyfra21-1	58.6	94	0.7968
IL-8	48	76	0.6063

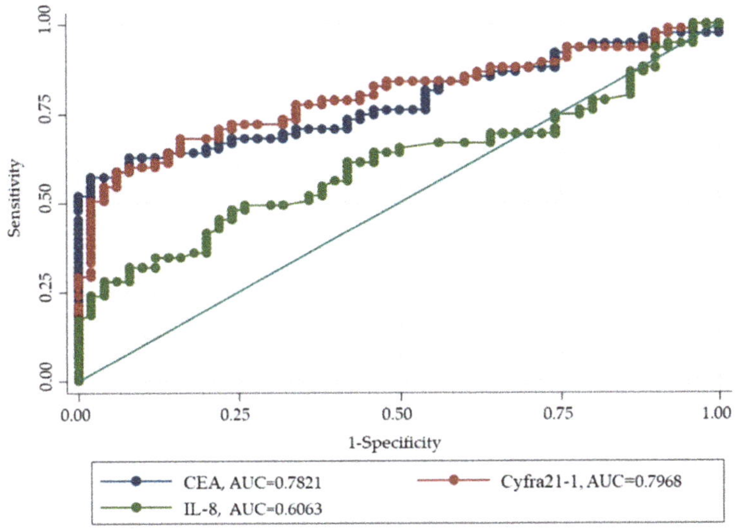

Figure 4. ROC curves for the diagnosis of LC risk in LC patients compared to HRR groups.

4. Discussion

According to the global cancer statistical analysis, LC is one of the main health problems worldwide, showing the highest rates of incidence and death and being the most common cancer among the population in Chiang Mai (Thailand) [1,2,4]. Radon is the seconding leading cause of LC after tobacco smoking and the major risk to non-smokers [5–9,11]. In a previous study we demonstrated that the values of indoor radon concentration in Chiang Mai were considerably higher than the corresponding global average values (39 Bq/m^3), ranging between 35 to 219 Bq/m^3, with an average value of 57 Bq/m^3 [12]. It has been considered that the risk of LC development is increased by 16% per 100 Bq/m^3 [8,11]. Since the high risk of developing LC is due to long-term toxic effects of radon and its decay products in HRR group, early diagnosis is vital for the prevention, diagnosis and treatment of LC. Our previous study showed that short telomere length and high level of expression of PARP1, WT1, TRERF1 and NOLOC4 serve as biomarkers to screen populations with high risks of LC in high radon exposure areas [12,22]. However, neither method is practical for early screening of LC risk for a large-scale general population. Therefore, finding serum biomarkers as a noninvasive diagnostic method and more rapid technique would improve the diagnosis and treatment of LC for a larger population. There appear to have been no previous studies of serum biomarkers that can be used as LC biomarkers in areas subject to high radon levels.

We evaluated the serum CEA, Cyfra21-1, HE4, IL-8, MIF, TNF-α and VEGF in all non-smoking LC patients with NSCLC and HC groups. In addition, HC groups were divided into LRR and HRR groups according to the radon activity concentration recorded in their dwellings. The results show that serum levels of CEA, Cyfra21-1, IL-8 and VEGF in LC patients were significantly higher ($p < 0.0001$) than in HC groups. This finding is in agreement with previous studies [13,15,16,23–26]. Further markers such as HE4, MIF and TNF-α are likely not useful as serum biomarkers for detection of LC within this study. The reason for this fact may be due to the different clinical stage, histologic type and smoking status [14,16–18,24]. Furthermore, the results showed that serum CEA, Cyfra21-1, IL-8 and VEGF in LC were higher ($p < 0.05$) than in the LRR and HRR groups. Interestingly, significantly higher levels ($p < 0.05$) of serum CEA, Cyfra21-1 and IL-8 were observed in HRR compared with LRR groups. This indicates that a high level of serum CEA, Cyfra21-1 and IL-8 in HRR groups may be a better biomarker of LC risk for differentiating between LRR and HRR groups.

In this study, we also evaluated the diagnostic criteria for predicting LC risk in LC patients compared to HRR groups based on sensitivity, specificity and ROC for serum CEA, Cyfra21-1 and IL-8. We found that the respective sensitivity and specificity were as follows: 57.3% and 98% for CEA; 58.6% and 94% for Cyfra21-1 and 48% and 76% for IL-8. It appears that serum CEA and Cyfra21-1 levels are more accurate, sensitive and specific than that of IL-8. These results further indicated that serum CEA and Cyfra21-1 had a relatively high ability to distinguish LC risk in HRR groups. In addition, the AUC value of serum CEA and Cyfra21-1 were 0.7821 and 0.7968, respectively, and further confirm the ability of these serum to have diagnostic value for LC risk in HRR groups. Based on the findings reported here, this study is the first to establish that serum CEA and Cyfra21-1 were able to select high-risk individuals with LC in high level radon areas, thus having the potential biomarkers to aid in the early screening and diagnosis of those at high-risk of LC. However, these serum markers are relatively limited due to their inadequate sensitivity (~57.3–58.6%). Thus, combined detection of serum CEA, Cyfra21-1 and other markers may improve the early diagnostic sensitivity and decreased specificity, which can lead to faster detection of high-risk groups. These will be the purpose of our future study to provide and improve the evidence for this study.

Nevertheless, a few limitations should be considered when interpreting of research results of this study. Firstly, only gender, age, histologic type and smoking status were included in this study, while other factors such as stage of cancer, alcohol consumption, genetic factors, lung disease, estrogens and occupational/environmental/medical exposure to radiation were not further studied. Secondly, since the sample size was limited, our findings may not be generalizable to other populations. Thirdly, due to the limited number of non-smoking LC patients in the study area, we were not able to divide the group into LC-LRR and LC-HRR groups. However, the results of previous studies have shown that the telomere length, protein expression [12,22] were different in LC patients compared to LRR and HRR groups and similarly our current study also found difference in serum biomarkers among those groups. Finally, this is a preliminary observational study to determine serum CEA and Cyfra21-1 as biomarkers for the diagnosis of LC risk in HRR groups; more longitudinal studies are needed to evaluate and validate the prognostic values in HRR groups with LC and to confirm these findings.

5. Conclusions

In summary, the results of the current study show that serum levels of CEA, Cyfra21-1, IL-8 and VEGF were significantly higher in LC patients than residential radon exposure (LRR and HRR groups). Among those biomarkers, serum CEA and Cyfra21-1 performed better in identifying LC risk in HRR groups with satisfactory specificity and sensitivity according to the AUC-ROC. These may be considered as potential serum biomarkers for indicating individuals at high-risk to develop LC. However, further studies in a larger population sample using multiple serum markers are necessary to confirm our current data before serum CEA and Cyfra21-1 can be used clinically as a tumor biomarker for the risk of high radon exposure-induced LC.

Author Contributions: Conceptualization, N.A.; Formal analysis, N.A., P.K., I.C., C.J., B.C., P.S., M.H. and S.T.; Investigation, N.A.; Writing—original draft preparation, N.A.; Writing—review and editing, P.S. and N.A.; Visualization, N.A.; Project administration, N.A.; Funding acquisition, N.A. and S.T. All authors have read and agreed to the published version of the manuscript.

Funding: This research has been funded by the National Research Council of Thailand, and Thailand Science Research and Innovation (RSA6280045); Faculty of Medicine, Chiang Mai University (MT01/2562), the International Atomic Energy Agency (Research contract number 21062) and in part by Grant for Environmental Research Projects from The Sumitomo Foundation.

Institutional Review Board Statement: The study was approved by the Research Ethics Committee in the Faculty of Medicine, Chiang Mai University, Thailand (Research ID: 2559-04011, 2559-04252 and 2562-06213).

Informed Consent Statement: Informed consent was obtained from all the participants in this study before blood sample collection.

Data Availability Statement: All data are available in this article.

Acknowledgments: The authors are grateful to Supamat Amphol, Chanis Pronnumpa, Kantaphat Krandrod, Thamaborn Ploykrathok, Tengku Ahbrizal Farizal Tengku Ahmad, Takahito Suzuki, Hiromi Kudo, Kochakorn Phantawong and Satit Inta for all their support. We would like to thank the Maharaj Nakon Chiang Mai University Hospital and Saraphi Hospital staffs for their assistance in the handling of blood samples. Finally, we would like to thank all participants in this study.

Conflicts of Interest: The authors have no conflict of interest to declare.

References

1. Sung, H.; Ferlay, J.; Siegel, R.L.; Laversanne, M.; Soerjomataram, I.; Jemal, A.; Bray, F. Global cancer statistic 2020: GLOBOCAN estimates of incidence and mortality worldwide for 36 cancers in 185 countries. *CA Cancer J. Clin.* **2021**, *71*, 209–249. [CrossRef]
2. Zhao, H.; Shi, X.; Liu, J.; Chen, Z.; Wang, G. Serum Cyfra21-1 as a biomarker in patients with nonsmall cell lung cancer. *J. Can. Res. Ther.* **2014**, *10*, 215–217.
3. Mao, Y.; Yang, D.; He, J.; Krasna, M.J. Epidemiology of lung cancer. *Surg. Oncol. Clin. N. Am.* **2016**, *25*, 439–445. [CrossRef]
4. Thailand-Global Cancer Observatory. Available online: https://gco.iarc.fr/today/data/factsheets/populations/764-thailand-fact-sheets.pdf (accessed on 20 September 2021).
5. Wiwatanadate, P. Lung cancer related to environmental and occupational hazards and epidemiology in Chiang Mai, Thailand. *Gene Environ.* **2011**, *33*, 120–127. [CrossRef]
6. Rankantha, A.; Chitapanarux, I.; Pongnikorn, D.; Prasitwattanaseree, S.; Bunyatisai, W.; Sripan, P.; Traisathit, P. Risk patterns of lung cancer mortality in northern Thailand. *BMC Public Health.* **2018**, *18*, 1138. [CrossRef]
7. Reungwetwattana, T.; Oranratnachai, S.; Puataweepong, P.; Tangsujarivijit, V.; Cherntanomwong, P. Lung cancer in Thailand. *J. Thorac. Oncol.* **2020**, *15*, 1714–1721. [CrossRef] [PubMed]
8. World Health Organization. *Handbook on Indoor Radon*; World Health Organization: Geneva, Switzerland, 2009.
9. United Nations. Scientific Committee on the Effect of Atomic Radiation. In *Source and Effects of Ionizing Radiation: United Nations Scientific Committee on the Effect of Atomic Radiation: UNSCEAR 2008 Report to the Genera; Assembly with Scientific Annexes*; United Nations: New York, NY, USA, 2010.
10. Sabbarese, C.; Ambrosino, F.; D'Onofrio, A. Development of radon transport model in different types of dwellings to assess indoor activity concentration. *J. Environ. Radioact.* **2021**, *227*, 106501. [CrossRef] [PubMed]
11. Al-Zoughool, M.; Krewski, D. Health effects of radon: A review of the literature. *Int. J. Radiat. Biol.* **2009**, *85*, 57–69. [CrossRef] [PubMed]
12. Autsavapromporn, N.; Klunklin, P.; Threeatana, C.; Tuntiwechapikul, W.; Hosada, M.; Tokonami, S. Short telomere length as a biomarker risk of lung cancer development induced by high radon levels: A pilot study. *Int. J. Environ. Res. Public Health* **2018**, *15*, 2152. [CrossRef]
13. Dong, Y.; Zheng, X.; Yang, Z.; Sun, M.; Zhang, G.; An, X.; Pan, L.; Zhang, S. Serum carcinoembryonic antigen, neuron-specific enolase as biomarkers for diagnosis of nonsmall cell lung cancer. *J. Can. Res. Ther.* **2016**, *12*, C34–C36.
14. Zeng, Q.; Liu, M.; Zhou, N.; Liu, L.; Song, X. Serum human epididymis protein4 (HE4) may be a better tumor marker in early lung cancer. *Clin. Chim. Acta* **2016**, *455*, 102–106. [CrossRef]
15. Tas, F.; Duranyildiz, D.; Oguz, H.; Camlica, H.; Yasasever, V.; Topuz, E. Serum vascular endothelial growth factor (VEGF) and interleukin-8 (IL-8) levels is small cell lung cancer. *Cancer Investig.* **2006**, *24*, 492–496. [CrossRef]
16. Balla, M.M.; Desai, S.; Purwar, P.; Kumar, A.; Bhandarkar, P.; Shejul, Y.K.; Pramesh, C.S.; Laskar, S.; Pandey, B.N. Differential diagnosis of lung cancer, its metastasis and chronic obstructive pulmonary disease based on serum VEGF, Il-8 and MMP-9. *Sci. Rep.* **2016**, *6*, 36065. [CrossRef] [PubMed]

17. Gámez-Pozo, A.; Sánchez-Navarro, I.; Calvo, E.; Teresa Agulló-Ortuño, M.T.; López-Vacas, R.; Díaz, E.; Camafeita, E.; Nistal, M.; Madero, R.; Espinosa, E.; et al. PTRF/Cavin-1 and MIF proteins are identified as non-small cell lung cancer biomarkers by label-free proteomics. *PLoS ONE* **2012**, *7*, e33752. [CrossRef]
18. Chen, Z.; Xu, Z.; Sun, S.; Yu, Y.; Lv, D.; Cao, C.; Deng, Z. TGF-β1, IL-6 and TNF-α in bronchoalveolar lavage fluid: Useful markers for lung cancer? *Sci. Rep.* **2014**, *4*, 5595. [CrossRef]
19. National Statistical Office Thailand. Population and Housing Census 2019. Available online: http://web.nso.go.th/index.htm (accessed on 20 September 2021).
20. Department of Mineral Resources. Geology of Thailand. Available online: http://www.dmr.go.th/main.php?filename=Mineral_re2015_EN (accessed on 15 September 2021).
21. Oyanagi, J.; Koh, Y.; Sato, K.; Mori, K.; Teraoka, S.; Akamatsu, H.; Kanai, K.; Hayata, A.; Tokudome, N.; Akamatsu, K.; et al. Predictive value of serum protein levels in patients with advanced non-small cell lung cancer treated with nivolumab. *Lung Cancer* **2019**, *132*, 107–113. [CrossRef] [PubMed]
22. Autsavapromporn, N.; Dukaew, N.; Wongnoppavicvich, A.; Chewaskulyong, B.; Roytrakul, S.; Klunklin, P.; Phantawong, K.; Chitapanarux, I.; Sripan, P.; Kritsananuwat, R.; et al. Identification of novel biomarkers for lung cancer risk in high levels of radon by proteomics: A pilot study. *Radiat. Prot. Dosimetry* **2019**, *184*, 496–499. [CrossRef] [PubMed]
23. Hanagiri, T.; Sugaya, M.; Takenaka, M.; Oka, S.; Baba, T.; Shigematsu, Y.; Nagata, Y.; Shimokawa, H.; Uramoto, H.; Takenoyama, M.; et al. Preoperative Cyfra21-1 and CEA as prognostic factors in patients with srage I non-small cell lung cancer. *Lung Cancer* **2011**, *74*, 112–117. [CrossRef] [PubMed]
24. Dalaveris, E.; Kerenidi, T.; Katsabeki-katsafli, A.; Kiropoulos, T.; Tanou, K.; Gourgoulianis, K.I.; Kostikas, K. VEGF, TNF-α and 8-isoprostane levels in exhaled breath condensate and serum of patients with lung cancer. *Lung Cancer* **2009**, *64*, 219–225. [CrossRef] [PubMed]
25. Cai, D.; Xu, Y.; Ding, R.; Qiu, K.; Zhang, R.; Wang, H.; Huang, L.; Xie, X.; Yan, H.; Deng, Y.; et al. Extensive serum biomarker analysis in patients with non-small-cell lung carcinoma. *Lung Cancer* **2020**, *126*, 154868. [CrossRef]
26. Balla, M.M.S.; Patwardhan, S.; Melwani, P.K.; Purwar, P.; Kumar, A.; Pramesh, C.S.; Laskar, S.; Pandey, B.N. Prognosis of metastasis based on age and serum analytes after follow-up of non-metastatic lung cancer patients. *Transl. Oncol.* **2021**, *14*, 100933. [CrossRef] [PubMed]

life

MDPI

Article

Metabolomic Response of the Creeping Wood Sorrel *Oxalis corniculata* to Low-Dose Radiation Exposure from Fukushima's Contaminated Soil

Ko Sakauchi [1], Wataru Taira [1,2] and Joji M. Otaki [1,*]

1 The BCPH Unit of Molecular Physiology, Department of Chemistry, Biology and Marine Science, Faculty of Science, University of the Ryukyus, Okinawa 903-0213, Japan; yamatoshijimi@sm1044.skr.u-ryukyu.ac.jp (K.S.); wataira@lab.u-ryukyu.ac.jp (W.T.)

2 Center for Research Advancement and Collaboration, University of the Ryukyus, Okinawa 903-0213, Japan

* Correspondence: otaki@sci.u-ryukyu.ac.jp; Tel.: +81-98-895-8557

Abstract: The biological consequences of the Fukushima nuclear accident have been intensively studied using the pale grass blue butterfly *Zizeeria maha* and its host plant, the creeping wood sorrel *Oxalis corniculata*. Here, we performed metabolomic analyses of *Oxalis* leaves from Okinawa to examine the plant metabolites that were upregulated or downregulated in response to low-dose radiation exposure from Fukushima's contaminated soil. The cumulative dose of radiation to the plants was 5.7 mGy (34 μGy/h for 7 days). The GC-MS analysis revealed a systematic tendency of downregulation among the metabolites, some of which were annotated as caproic acid, nonanoic acid, azelaic acid, and oleic acid. Others were annotated as fructose, glucose, and citric acid, involved in the carbohydrate metabolic pathways. Notably, the peak annotated as lauric acid was upregulated. In contrast, the LC-MS analysis detected many upregulated metabolites, some of which were annotated as either antioxidants or stress-related chemicals involved in defense pathways. Among them, only three metabolite peaks had a single annotation, one of which was alfuzosin, an antagonist of the α_1-adrenergic receptor. We conclude that this *Oxalis* plant responded metabolically to low-dose radiation exposure from Fukushima's contaminated soil, which may mediate the ecological "field effects" of the developmental deterioration of butterflies in Fukushima.

Keywords: metabolome; GC-MS; LC-MS; Fukushima nuclear accident; low-dose radiation exposure; plant physiology; *Oxalis corniculata*; radioactive pollution; field effect; alfuzosin

Citation: Sakauchi, K.; Taira, W.; Otaki, J.M. Metabolomic Response of the Creeping Wood Sorrel *Oxalis corniculata* to Low-Dose Radiation Exposure from Fukushima's Contaminated Soil. *Life* **2021**, *11*, 990. https://doi.org/10.3390/life11090990

Academic Editors: Fabrizio Ambrosino and Supitcha Chanyotha

Received: 17 August 2021
Accepted: 17 September 2021
Published: 20 September 2021

1. Introduction

Radioactive pollution caused by anthropogenic radionuclides has been widespread worldwide since the middle of the twentieth century. At present, anthropogenic ^{137}Cs can be detected globally [1–4]. One of the most severe pollution events was the Fukushima nuclear accident in 2011, which was the second-largest nuclear accident after the Chernobyl nuclear accident in 1986. The biological impacts of the Fukushima nuclear accident have been studied in contaminated fields, which have focused on many organisms, including birds such as the barn swallow and goshawk [5–7], Japanese monkeys [8–10], intertidal invertebrates [11], gall-forming aphids [12,13], and plants [14–18]. However, to the best of our knowledge, the most intensively studied species in both field and laboratory experiments is the pale grass blue butterfly *Zizeeria maha* [19–38]. This small butterfly is popular in Japan (except for in Hokkaido) and has been established as an excellent field indicator species for environmental assessments and evolutionary studies [39–41] and as an excellent model organism in the laboratory for developmental and physiological studies [42–45]. Collectively, these studies have concluded that the butterflies in Fukushima have been affected both genetically by high-level initial exposure and physiologically by low-level chronic exposure through "field effects", even though this butterfly is resistant to low-level radiation exposure from ingested ^{137}Cs under experimental conditions; that is, this butterfly is dosimetrically resistant in the laboratory but vulnerable in the field, possibly due to ecological interactions.

What kind of ecological interactions in the field does this butterfly species have? The pale grass blue butterfly likely has relatively simple ecological interactions, because its life history is simple. Its larvae monophagously eat the leaves of the creeping wood sorrel *Oxalis corniculata* [20,21,42]. Adult butterflies fly around this plant and do not travel over long distances from their original location unless they are blown by a strong wind [41]. We therefore hypothesized that this host plant may change its biochemical contents after irradiation stress, even at a low level of exposure, which could then affect the butterfly larvae [27,31,32,37]. This field effect hypothesis is reasonable, considering that some plants have shown morphological and gene expression changes in response to radiation exposure due to the Fukushima nuclear accident [14–18]. Along this line, we discovered that the sodium content in the leaves of the host plant was inversely correlated with both the ^{137}Cs radioactivity concentration in the leaves and the ground radiation dose [38]. Since the sodium deficiency in herbivore animals generally results in serious pathological consequences, this could be a possible mechanism of the ecological field effect on these butterflies. However, there may be multiple plant-mediated pathways that affect the larval physiology.

In this study, we performed metabolomic analyses of irradiated and nonirradiated *Oxalis* plants using both GC-MS (gas chromatography-mass spectrometry) and LC-MS (liquid chromatography-mass spectrometry) analyses to identify candidate metabolites with changed levels in the leaves upon irradiation. GC-MS can analyze gaseous or volatile compounds with relatively small molecular weights and relatively high heat resistance, including carbohydrates, amino acids, organic acids, and fatty acids, whereas LC–MS can analyze a wide range of compounds that can dissolve in solvents, including aromatic glycosides, terpenoid derivatives, and amino acid derivatives. Many plant primary metabolites belong to the former group of compounds, whereas many plant secondary metabolites belong to the latter. In this study, targeted and nontargeted GC-MS analyses were treated separately, because they detected two different sets of metabolite peaks.

Radiation metabolomics has often been employed to diagnostically or therapeutically search for relevant biochemicals after exposure to ionizing radiation [46–50]. Metabolomics was used to investigate metabolite changes in rice seeds [15,18] and calf blood plasma [51] after the Fukushima nuclear accident. To demonstrate that the creeping wood sorrel responds to the low dose of radiation that is relevant in the field, we used contaminated soil collected from Fukushima as the radiation source. For this irradiation experiment, we used whole plants collected in the field from Okinawa, the least contaminated prefecture in Japan. In this way, we reproduced in our laboratory (in Okinawa) the possible acute radioactive environment of this plant in Fukushima immediately after the nuclear accident, although the present system focused only on external exposure.

2. Materials and Methods

2.1. Plant and Culture Soil

Whole plants of the creeping wood sorrel *O. corniculata*, including the roots, were obtained from 3 localities on Okinawa-jima Island, Okinawa Prefecture, Japan: Nishihara Town (Uehara-Takadai Park), Yomitan Village (near Zakimi Castle Ruin), and Yaese Town (Kochinda Undo Park) (Figure 1). This plant has color variants, but a typical green variant was used (Figure 2a). To generate a pair of genetically identical samples, each individual plant was separated into two batches with the same identification number (1: Yaese, 2: Nishihara, 3: Yomitan). That is, a pair of two batches were a clone. The clones were potted individually in cylindrical pots (100 mm in diameter × 150 mm in height) using a package of commercially available culture soil for flowers and vegetables, Hanasaki Monogatari (Akimoto Tensanbutsu, Iga, Mie, Japan). This soil was analyzed for its radioactivity concentration in a way similar to that used for the contaminated soil from Fukushima (see below). We detected 2.14 Bq/kg ^{137}Cs (n = 1; measured on 8 August, 2018), but we considered this amount to be negligible in comparison to the contaminated soil from Fukushima (see below). The potted plants were grown outside under natural conditions for approximately 2 weeks until many leaves were produced. The plants were watered every day before and during irradiation treatment.

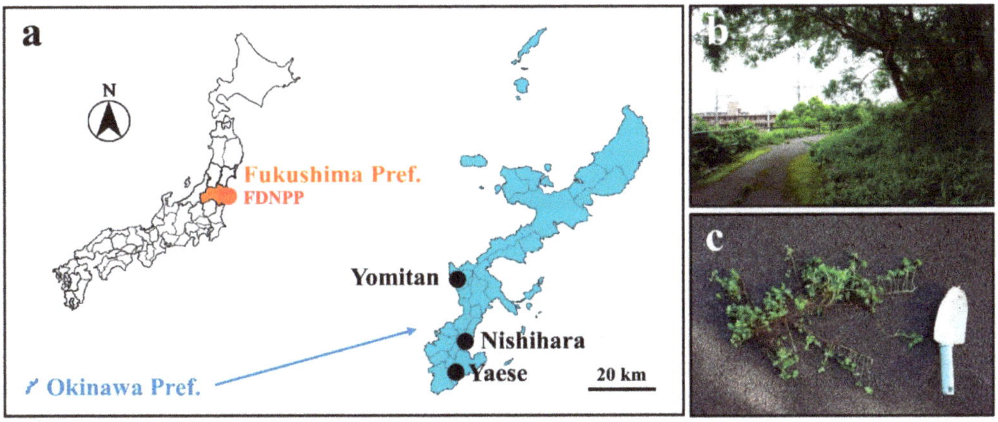

Figure 1. Sampling localities. (**a**) Maps of Japan and Okinawa. Shown are Fukushima Prefecture in orange (left); Fukushima Dai-ichi Nuclear Power Plant (FDNPP) in the red circle (left); Okinawa Prefecture in blue (left and right); and the 3 sampling localities (Yomitan Village, Nishihara Town, and Yaese Town) in Okinawa in black circles (right). (**b**) A landscape of the sampling site in Nishihara Town. (**c**) The creeping wood sorrel *O. corniculata* with its creeping stem just obtained at a sampling site in Yaese Town.

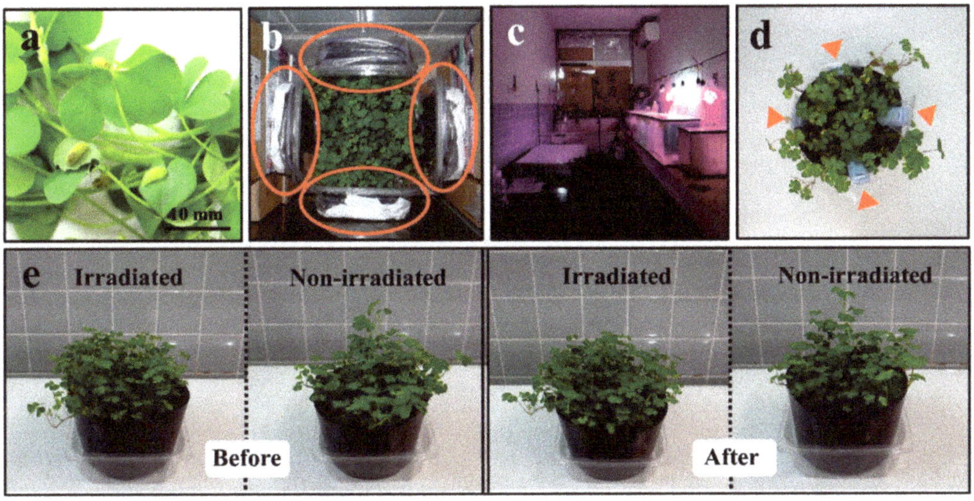

Figure 2. Irradiation treatment. (**a**) Typical green variant of *Oxalis* leaves used in this experiment. Larvae of the pale grass blue butterflies are also pictured (but the larvae were not irradiated in this experiment). (**b**) Top-down view of the irradiation setup. A plant pot was placed at the center, surrounded by 4 packages of contaminated soil as the radiation sources (circled). (**c**) Overview of the irradiation room. The irradiation system was set up under horticulture lights on the right and surrounded by aquarium tanks filled with water. (**d**) Four dosemeters in a pot (arrowheads). (**e**) Irradiated and nonirradiated *Oxalis* plants before and after treatment.

2.2. Irradiation Treatment

Plants of the experimental group (irradiated; IR) were treated with external radiation for 7 days (168.0 h) at room temperature (27 °C ± 1 °C) (Figure 2b), and plants of the control group (nonirradiated control; NC) were placed under the same conditions but without irradiation. To do so, plants were kept under 18 L:6 D-long day conditions using Derlights horticulture LED light bulbs (40-W equivalence) (Figure 2c). The plants in the irradiation group were surrounded by 4 transparent plastic square containers (135 mm × 135 mm × 57 mm) containing contaminated soil, which were then surrounded by a concrete wall, two walls of several lead blocks (a single block: 100 mm × 200 mm × 50 mm), and aquarium tanks filled with water (Figure 2b). The water tanks and lead blocks were further aligned to minimize the radiation exposure of the researchers (Figure 2c). The plants in the nonirradiated control group were surrounded similarly but without the soil containers. The irradiation periods for the Nishihara, Yomitan, and Yaese plant samples were 25 September–2 October, 8–15 October, and 18–25 November 2018, respectively.

Contaminated soil as the plant radiation source described above was collected in Minamisoma City, Fukushima Prefecture on 28 November 2014 after an evacuation order for that area was partially lifted. Surface soil at a depth of 0–50 mm was collected. The soil was packed in plastic bags, which were then contained in transparent plastic square containers, as described before. For evaluation of the cumulative absorbed dose under these conditions, 4 dosemeters were put in a pot (see below) (Figure 2d).

We confirmed that the plants looked equally healthy before and after treatment; no signs of leaf necrosis, chlorosis, or other abnormalities were found by visual inspection (Figure 2e). After the 7-day exposure period, the plant leaves (5–10 g per plant sample) were handpicked with disposable gloves. Relatively young leaves were preferably collected, and relatively old ones were excluded, because young leaves are likely preferred by butterfly larvae and because secondary metabolites may be more abundant in young leaves. The leaf samples were then washed with Evian bottled natural mineral water (Evian les Bains, France). After that, the water was completely drained from the leaves. The leaves were then pat-dried and quickly frozen in liquid nitrogen. The frozen samples were packed with dry ice and sent to the Kazusa DNA Research Institute, Kisarazu, Chiba, Japan (www.kazusa.or.jp, accessed on 15 September 2021) for GC-MS and LC-MS analyses (see below).

2.3. Cumulative Dose and Dose Rate

To measure the cumulative dose of radiation to the plants, we used a wide-range glass badge dosemeter for the environment (type code: ES) (60 mm × 28 mm × 16 mm) from Chiyoda Technol (Tokyo, Japan). This dosemeter is the latest-generation dosemeter, which takes advantage of radiophotoluminescence (RPL). According to the manufacturer's specifications, this dosemeter can detect X-rays (10 keV–10 MeV), γ-rays (10 keV–10 MeV), and β-rays (130 keV–3 MeV) and can measure 0.1 mSv–10 Sv of radiation. In the present study, there was no X-ray source, and the β-rays were probably shielded by the plastic containers but only to some extent. Thus, we considered that the dosemeters mainly detected γ-rays, as well as β-rays. We put each dosemeter in a plastic bag for waterproofing, and four of them were placed in a pot for 7 days (168.0 h) (Figure 2d). Two new pots were prepared from a plant collected from Nanjo City (Fusozaki Park), Okinawa: one for irradiation and the other for the nonirradiated control, as explained above. After 7 days (168.0 h) of exposure (16–23 October 2018) under the experimental conditions above, the dosemeters were sent back to Chiyoda Technol, where image development and quantification were performed. The outputs were given as dose equivalents at a depth of 70 μm from the human body surface in sieverts (Sv). We assumed that the process of dose absorption on the surface is similar between humans and *Oxalis* plants and that Sv values in humans and Gy values in plants may be interchangeable.

2.4. Radioactivity Concentration of the Contaminated Soil from Fukushima

The contaminated soil from Fukushima was placed in a cylindrical columnar plastic vial (15 mm in diameter and 50 mm in height) to make an 8-mm sample height, and its radioactivity concentration was measured using a Canberra GCW-4023 germanium semiconductor radiation detector (Meriden, CT, USA) until an error rate of less than 2% was reached. The counting efficiency values for ^{40}K and ^{137}Cs were 7.92% and 16.62%, respectively. The branching ratios for ^{40}K and ^{137}Cs used here were 10.67% and 85.21%, respectively. The half-lives of ^{40}K and ^{137}Cs used here were 1.251×10^9 years and 30.17 years, respectively. Three soil samples were measured, and their outputs were averaged. The measurements were performed from 9 May to 1 July, 2016. All calculations of the radioactivity concentrations were set at the first exposure date, 25 September, 2018, for simplicity.

2.5. GC-MS Sample Preparation

The leaf samples were first mixed with methanol to make a 75–80% methanol solution, which was homogenized using zirconia beads. After centrifugation at 15,000 rpm for 5 min, the supernatants were collected and subjected to a MonoSpin C18 column (GL Sciences, Tokyo, Japan). The column was pretreated with 100% methanol and centrifuged at 5000× *g* for 2 min, and 70% methanol was further added. The column was again centrifuged at 5000× *g* for 2 min. The sample was then applied to this pretreated column, which was then centrifuged at 3000× *g* for 2 min. The eluted samples were collected and subsequently treated with nitrogen gas and methoxamine hydrochloride with pyridine for methoxime derivatization of the compounds and then with MSTFA (*N*-methyl-*N*-(trimethylsilyl)trifluoroacetamide) for trimethylsilyl (TMS) derivatization.

2.6. GC-MS Analysis

The samples were analyzed using a SHIMADZU gas chromatograph quadrupole mass spectrometer GCMS-QP2010 Ultra (Kyoto, Japan) with a SHIMADZU autosampler AOC-5000 Plus and an Agilent Technologies DB-5 column (30 m in length, 0.250 mm in internal diameter, and 1.00 μm in membrane thickness) (Santa Clara, CA, USA). The electron ionization (EI) method was used to ionize the samples. The following setting values were employed: chamber temperature, 280 °C; oven temperature, 100 °C for 4 min at a rate of 4 °C/min, followed by holding at 320 °C for 8 min; connection temperature, 280 °C; ionization source temperature, 200 °C; flow rate, 390 mm/s (1.1 mL/min); scanning speed, 2000 u/s; mass detection range, 45–600 (*m/z*); and sample injection volume, 0.5 μL. An autotuning function was used for machine tuning and validation, in which a standard calibration sample of PFTBA (perfluorotributylamine) was used for tuning the resolution, sensitivity, mass calibration, and vacuum. For time adjustment, mixed alkanes C_7–C_{33} were analyzed. An internal standard was not used.

2.7. GC-MS Peak Detection, Alignment, and Annotation

The targeted and nontargeted (untargeted) methods were performed independently. For the targeted methods, SHIMADZU analysis software GCMSsolution and its associated GC-MS Metabolite Database ver. 2 were employed for compound annotation and comparisons between samples. Peaks were detected in reference to those in the database based on specific fragment ions and retention times and were annotated based on the following three points between the peaks detected in the sample and those in the compound library: similarity in the mass spectral patterns, similarity in the intensity ratios of the specific mass fragments, and similarity in the retention times. When two or more peaks were annotated as the same chemical compound, the area values and retention time values of those peaks were extracted and aligned.

For the nontargeted methods (analysis of all peaks), SpectraWorks MS spectrum data mining software AnalyzerPro (Runcorn, Cheshire, UK) and SHIMADZU GC-MS peak alignment software FragmentAlign (www.kazusa.or.jp/komics/software/FragmentAlign, accessed on 15 September 2021) [52] were employed. Peaks were detected automatically by AnalyzerPro based on the peak area, peak width, signal-to-noise ratio, and specific thresholds. Whether the ions originated from the same chemical compound was examined by a deconvolution treatment based on the peak retention time and peak shape. When it was concluded that the ions had an identical origin, they were bundled together. Alignments were performed based on the scores (using Pearson product-moment correlation and cosine similarity) and retention times. Chemical compound annotations were determined based on the similarity between the mean mass spectrum patterns of the samples and those of the chemical compound library. The results of both the targeted and nontargeted methods were compiled in a Microsoft Excel file, "GC-MS_Result Kazusa DNA Institute" (Supplementary Table S1).

2.8. LC-MS Sample Preparation

The leaf samples were put in methanol (final concentration of 75%), homogenized with zirconia beads, and centrifuged at 15,000 rpm for 10 min. The supernatant was collected. Additionally, a MonoSpin C18 column (GL Sciences) was treated with 100% methanol and centrifuged at 5000× *g* for 2 min. The column was again treated with 75% methanol and centrifuged at 5000× *g* for 2 min. The prepared sample was then applied to the column, which was centrifuged at 5000× *g* for 2 min. The eluted solution was collected and filtered through a 0.2-μm filter. Each sample was analyzed 3 times for reproducibility, and the results were averaged before the statistical analyses.

2.9. LC-MS Analysis

The samples were analyzed using a SHIMADZU Nexera X2 high-performance liquid chromatography (HPLC) instrument connected to a Thermo Fisher Scientific Q Exactive Plus high-resolution mass analyzer (Waltham, MA, USA). An InertSustain AQ-C18 column (2.1 mm × 150 mm, 3-μm particle size) (GL Sciences) was used as the reversed-phase HPLC column. A Nexera X2 system was run under the following conditions: column temperature, 40 °C; mobile phase A, 0.1% formic acid in water; mobile phase B, acetonitrile; flow of mobile phase, 0.2 mL/min; and sample injection volume, 2 μL. The Q Exactive Plus instrument was run under the following conditions using electrospray ionization (ESI): measurement time, 3–30 min; measurement range, 80–1200 (*m/z*); full-scan resolution, 70,000; and MS/MS scan resolution, 17,500. For MS/MS precursor selection, a Data-Dependent Scan (Top 10) was performed, in which the top 10 precursor ions detected by the full scan were subjected to MS/MS analysis. The dynamic exclusion was 20 s, in which the precursor ions that were previously measured were excluded from the MS/MS analysis to measure as many precursor ions as possible.

An LCMS QC Reference Material (Waters, Milford, MA, USA) containing 9 known standard chemical compounds with known concentrations was analyzed before and after the sample analysis to confirm that the retention times and detection sensitivity were set within acceptable ranges. The maximum ion intensity from each sample was confirmed to be sufficient for the peak signal detection (10^8–10^9 cps). The LCMS QC Reference Material contained the following compounds: acetaminophen ($C_8H_9NO_2$) at 152.0712 (10 μg/mL), caffeine ($C_8H_{10}N_4O_2$) at 195.0882 (1.5 μg/mL), sulphaguanidine ($C_7H_{10}N_4O_2S$) at 215.0603 (5 μg/mL), sulfadimethoxine ($C_{12}H_{14}N_4O_4S$) at 311.0814 (1 μg/mL), Val-Tyr-Val ($C_{19}H_{29}N_3O_5$) at 380.2185 (2.5 μg/mL), verapamil ($C_{27}H_{38}N_2O_4$) at 455.2910 (0.2 μg/mL), terfenadine ($C_{32}H_{41}NO_2$) at 472.3216 (0.2 μg/mL), Leu-enkephalin ($C_{28}H_{37}N_5O_7$) at 556.2771 (2.5 μg/mL), and reserpine ($C_{33}H_{40}N_2O_9$) at 609.2812 (0.6 μg/mL).

2.10. LC-MS Peak Detection, Alignment, and Annotation

The LC-MS data obtained above were converted to mzXML format using ProteoWizard (http://proteowizard.sourceforge.net, accessed on 15 September 2021). Peak detection, determination of the ionizing states, and peak alignments were performed automatically using the data analysis software PowerGetBatch (http://www.kazusa.or.jp/komics/software/PowerGetBatch/ja, accessed on 15 September 2021) developed by the

Kazusa DNA Research Institute [53]. Each detected peak was associated with a retention time, a peak area (intensity), an exact mass, and a MS/MS spectrogram. At this point, a collection of detected peaks included not only those from compounds within the samples but, also, those from noise and false positives. This is partly because a single compound is detected as multiple peaks, depending on the ionization state after ESI. Thus, the ionization states (adducts) were determined based on differences in the exact masses of the peaks detected with similar retention times. Each pair of peaks that was consistent with the known adduct pairs was identified based on the assumption that the basic cation adduct was $[M+H]^+$ and the basic anion adduct was $[M-H]^-$. When pairing was not possible, either $[M+H]^+$ or $[M-H]^-$ was assigned, because these ions were produced most frequently. All the detected peaks were aligned using all samples based on their *m/z* and retention time values, resulting in a peak intensity matrix. The noise peaks and false-positive peaks were then eliminated based on the negative controls. The detected peaks were considered valid when they were reproducibly detected in each of three trials of an identical sample.

Exact mass values of the nonionized compounds calculated from the adducts were searched against the following chemical mass databases developed by the Kazusa DNA Research Institute: (1) UC2 (http://webs2.kazusa.or.jp/mfsearcher/uc2/, accessed on 15 September 2021) [54], which contains metabolites recorded in the following two databases: KNApSAcK (http://knapsackfamily.com/KNApSAcK_Family/, accessed on 15 September 2021) [55] and Human Metabolome Database (https://hmdb.ca, accessed on 15 September 2021), (2) a theoretical chemical composition formula database (containing chemically reasonable compounds and peptides less than molecular weight 1000), and (3) an in-house database. The database searches were performed using the search program MFSearcher [56] developed by the Kazusa DNA Research Institute. When the alignments had a compound that matched its own mass (±1 ppm) from these databases, a chemical composition formula was assigned. When a matching compound was not found within ±5 ppm of its own mass, its mass was adjusted by adding ±1 ppm, and a new search was performed.

Valid alignments (peaks) were further examined against the reference material standards that were analyzed previously under the same conditions at the Kazusa DNA Research Institute. The reference standards included plant metabolite data that were provided by the collaborative study between the Kazusa DNA Research Institute and Tokiwa Phytochemical (Sakura, Chiba, Japan). The mean *m/z* values and MS/MS spectrograms of a set of alignments were compared with those of reference material standards. Within a mean retention time ±1.5 min, the cosine similarity of the MS/MS spectrograms must be >0.9, and the *m/z* error must be within ±5 ppm to be considered likely identical. The LC-MS/MS results were compiled in the Microsoft Excel file "LC-MS_Result Kazusa DNA Research Institute" (Supplementary Table S2).

2.11. Statistical Analysis of the Peak Area Data

The output peak data of GC-MS were compiled in Microsoft Excel files (Supplementary Table S3 and S4). The output peak data of LC-MS were also treated similarly (Supplementary Table S5). The peak area (intensity) data were subjected to statistical analyses using MetaboAnalyst 5.0 [57–59]. In the process of uploading the data into MetaboAnalyst, the peak data with a constant or single value across the samples were deleted automatically. The data were subjected to autoscaling, a mean-centered normalization process divided by the standard deviation of each sample. However, box plots and graphs related to the *t*-test and fold change analysis were made using the original peak area values without normalization to understand the original levels, although the original levels did not necessarily reflect the levels in the leaf samples in these analyses. During upload, data filtering was not performed. Sample normalization to adjust for differences among the samples was not performed.

We performed Student's *t*-tests (unpaired and bi-sided) and used $p < 0.05$ (without adjustment) as the criterion to consider statistical significance, and the peaks that met this criterion were examined independently. The statistical contribution of pairs was ignored, because MetaboAnalyst cannot perform such calculations when the number of peaks exceeds 1000, and because we confirmed that the p-values did not change much with or without pairing. We also performed a fold change analysis and referred to fold change (FC) values, and peaks with $FC > 2.0$ for upregulation and $FC < -2.0$ for downregulation were considered important. When calculating the FC values, MetaboAnalyst uses the equations $FC = a/c$ when $a > c$ and $FC = -c/a$ when $a < c$, where a is an experimental group and c is a control group. However, we also manually calculated a/c when $a < c$ to examine the candidate peaks independently. In this case, $FC < 0.5$ was considered important. A principal component analysis (PCA) and heat map analysis were performed to obtain possible relationships among the samples. In the latter, Euclidean distance and the Ward linkage method were employed for clustering.

2.12. Comparison of the LC-MS/MS Spectrograms for Alfuzosin

To clarify the identity of peak No. 4746, we referred to public LC-MS/MS spectrogram records of alfuzosin in the Human Metabolome Database (HMDB) ver. 4.0 (https://hmdb.ca) (accessed on 21 July 2021) [60,61]. Alfuzosin (HMDB0014490) had 10 experimental $[M+H]^+$ spectrograms available as of July 2021, which were referred to in this study.

3. Results

3.1. Cumulative Dose, Dose Rate, and Radioactivity Concentration

The cumulative absorbed doses after 7 days (168.0 h) from the four dosemeters were 7.1, 4.6, 5.5, and 5.5 mGy, resulting in a value of 5.7 ± 1.0 mGy (mean $\pm$ standard deviation). That is, the dose was on the order of milligrays in this system. The dose rates from four dosemeters were calculated to be 42, 27, 33, and 33 μGy/h, resulting in a value of 34 ± 6 μGy/h. In contrast, all four dosemeters in the nonirradiated control showed no detection.

The contaminated soil contained 363.54 ± 11.25 Bq/kg ^{40}K and 1.529 ± 0.013 MBq/kg ^{137}Cs. Since these are the major radionuclides in the soil, and because ^{137}Cs outnumbered ^{40}K, it is likely that the plant mostly received γ-rays and β-rays from anthropogenic ^{137}Cs from the Fukushima Dai-ichi Nuclear Power Plant.

3.2. GC-MS: Targeted Method

In the targeted GC-MS method, 428 compounds were targeted; the shortest retention time was 7.444 min (boric acid), and the largest retention time was 62.738 min (cholesterol). Using six plant samples (three irradiated and three nonirradiated), 61 peaks were detected in total. To understand the relationships among the six plant samples, we first performed a principal component analysis (PCA) using the output peak area (intensity) data (Figure 3a). In the score plot, the nonirradiated samples were scattered, but the irradiated samples were relatively clustered together, suggesting that these plants might have responded to irradiation in a stereotypical manner despite genetic differences. However, PC1 and PC2 explained only 33.0% and 28.1% of the variation, respectively. In the loading plot, the 61 peaks were mostly scattered (Figure 3b).

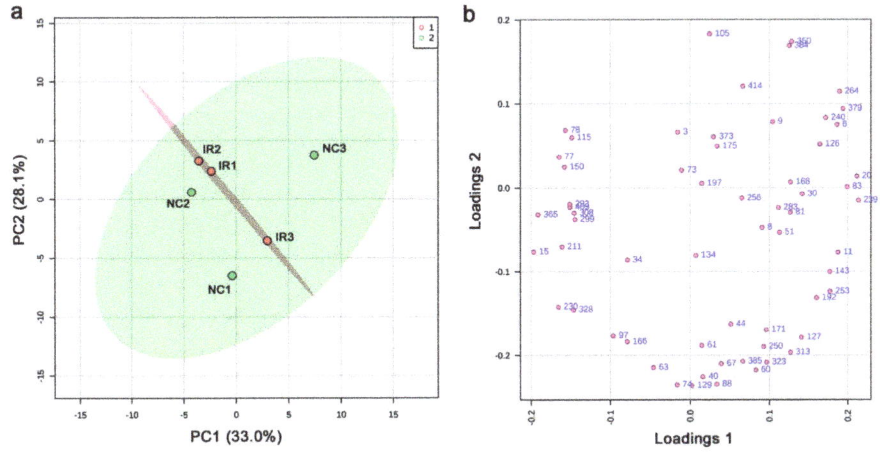

Figure 3. PCA results from the targeted GC-MS analysis. (**a**) Score plot; 95% confidence ranges are colored. IR, irradiated; NC, nonirradiated. (**b**) Loading plot; 61 peaks are located in a 2D plane.

A heat map using all 61 peaks revealed some overall trends. All three nonirradiated samples (NC1, NC2, and NC3) had high levels of peaks at certain locations on the map, whereas two irradiated samples (IR1 and IR2) had only relatively low levels of peaks (Figure 4a). All three nonirradiated samples had different patterns, and their corresponding irradiated samples showed different patterns from those of the nonirradiated samples. Using the top 25 peaks, the heat map clearly indicated that the irradiated and nonirradiated groups were differently clustered (Figure 4b). The three irradiated samples were similar to one another, and the three nonirradiated samples were less similar to one another. Overall, the nonirradiated samples were more diverse and robust in metabolite levels than the irradiated samples, at least as determined by the targeted GC-MS method. These results suggest that the plant responded to irradiation at the metabolome level despite the genetic differences, probably by slowing down some metabolic pathways, although the response patterns were not very consistent among the three irradiated plant samples.

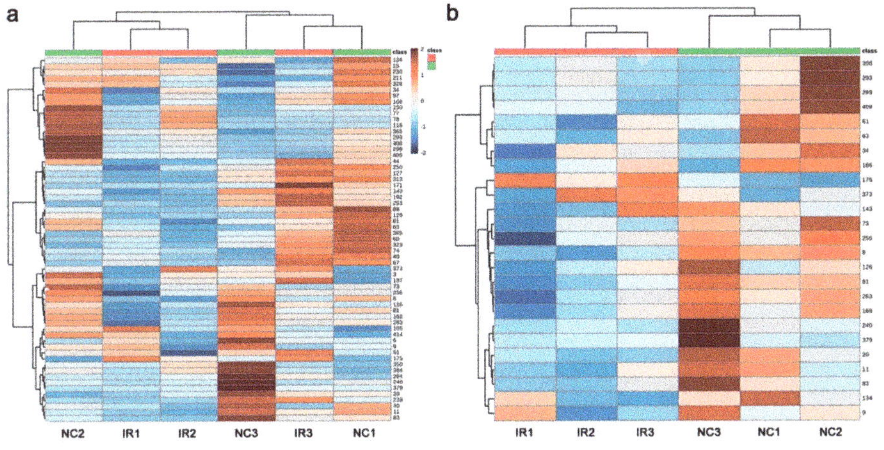

Figure 4. Heat maps from the targeted GC-MS analysis. At the top, the irradiated group was indicated by red horizontal bars, and the nonirradiated control group was indicated by green horizontal bars. (**a**) Heat map using all 61 peaks. (**b**) Heat map using the top 25 peaks.

To examine the differences in the peak area values between the irradiated and nonirradiated groups, we performed *t*-tests and fold change analyses. We detected four peaks with significant differences ($p < 0.05$) (Figure 5a). Similarly, we detected five peaks with $FC > 2.0$ or $FC < -2.0$ (Figure 5b). Notably, most peaks had negative *FC* values, indicating that they were mostly downregulated, which was consistent with the heat map results. It was notable that, in all five cases, the distribution ranges of the peak area values were much wider in the nonirradiated samples than in the irradiated samples, suggesting a stereotypical response among the three plants irrespective of the genetic differences. There were no peaks that satisfied both the *t*-test and fold change analysis conditions ($p < 0.05$ and $FC > 2.0$ or $FC < -2.0$) (Figure 5c).

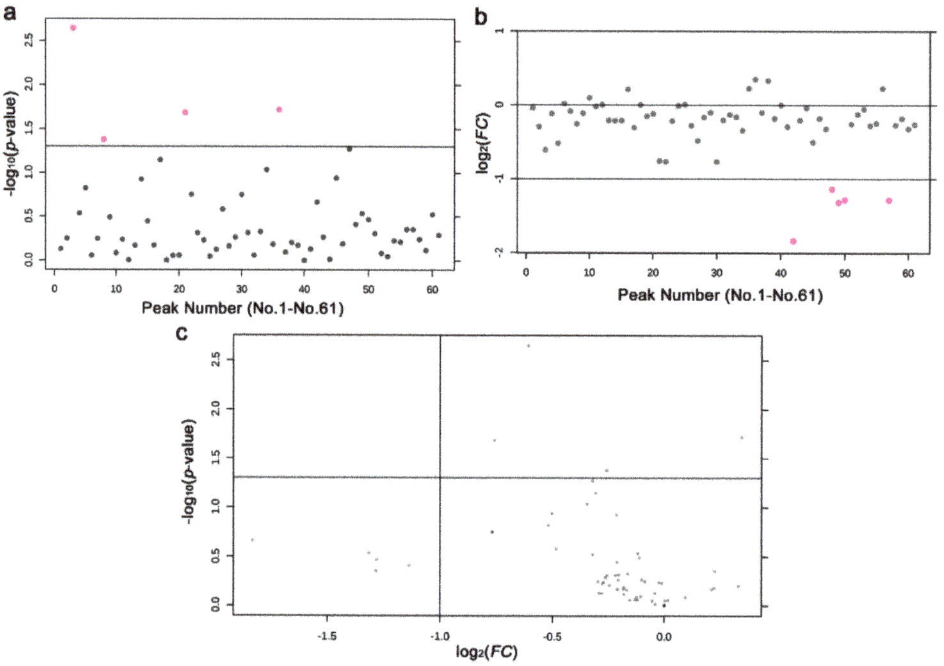

Figure 5. Results of the *t*-test and fold change analysis from the targeted GC-MS analysis between the irradiated and nonirradiated groups. (**a**) Distribution of the *p*-values (Student's *t*-test). The threshold was set at $p = 0.05$. The 4 pink dots indicate significant peaks. The peak numbers (No. 1–No. 61) here on the horizontal axis do not correspond to the original peak numbers shown in Figure 6, because the original peak numbers were assigned before elimination of the nonsense peaks by MetaboAnalyst (also in (**b**)). (**b**) Distribution of the *FC* values. The thresholds were set at $FC = 2.0$ and $FC = -2.0$. The 5 pink dots indicate peaks that were downregulated more than twofold. (**c**) Volcano plot. No peak satisfied both the *p*-value and *FC* value criteria.

The four peaks with $p < 0.05$ were annotated as caproic acid (No. 8, $p = 0.0022$), lauric acid (No. 175, $p = 0.019$), nonanoic acid (No. 81, $p = 0.021$), and 3-aminopropanoic acid (β-alanine) (No. 30, $p = 0.042$) (Figure 6a). Among them, only lauric acid was upregulated after irradiation, and the others were downregulated. Only caproic acid satisfied $p < 0.01$. The five peaks with $FC < 0.5$ (i.e., $FC < -2.0$) were annotated as azelaic acid, fructose, oleic acid, glucose, and fructose (two peaks with slightly different retention times were both annotated as fructose) (Figure 6b). These compounds were all downregulated after irradiation. The box plots indicated, in a cautionary manner, that their irradiated and nonirradiated distributions overlapped, except for in the case of azelaic acid. An outstanding single value for oleic acid in the nonirradiated group likely contributed greatly to $FC < 0.5$.

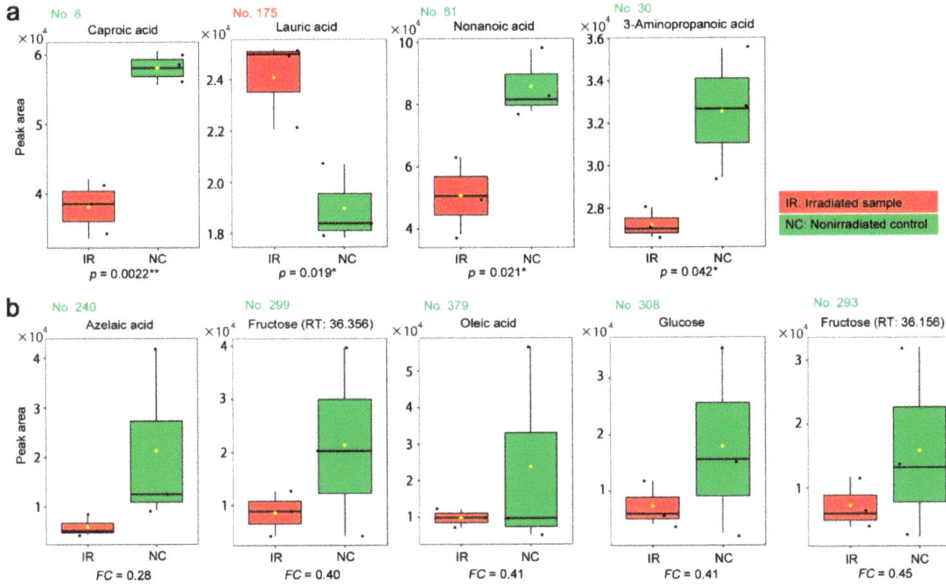

Figure 6. Box plots of the peak area (intensity) values from the targeted GC-MS analysis. The irradiated group (IR) is indicated by red boxes, and the nonirradiated control group (NC) is indicated by green boxes. The peak numbers are shown in red when upregulated, and they are shown in green when downregulated. (**a**) Peaks with significant differences between the irradiated and nonirradiated groups in the order of low-to-high p-values. There were 4 annotated peaks with $p < 0.05$ (Student's t-test). (**b**) Peaks identified by a fold change analysis between the irradiated and nonirradiated groups in the order of low-to-high FC values. There were 5 annotated peaks with $FC > 2.0$ or $FC < 0.5$. * $p < 0.05$, ** $p < 0.01$.

The *Oxalis* plant is known to contain a large amount of oxalic acid, as implied by the name. Oxalic acid in the leaves has been proposed to function as a feeding stimulant for larvae of the pale grass blue butterfly [62]. The abundant presence of oxalic acid in the leaves was verified as peak No. 15 in this analysis. Its peak levels were not significantly different between the irradiated and nonirradiated groups ($p = 0.87$; Student's t-test).

3.3. GC-MS: Nontargeted Method

In the nontargeted GC-MS method, 456 peaks were originally detected, including those detected only in a single sample. The shortest retention was 7.052 min, and the longest retention time was 63.095 min. Among them, 306 peaks were considered valid by MetaboAnalyst. We then performed PCA. PC1 and PC2 explained only 31.5% and 26.1% of the variance, respectively (Figure 7a). Notably, the irradiated and nonirradiated groups clustered individually, keeping the relative positions of the three samples intact. A likely interpretation of these data is that the nonirradiated samples in the negative PC2 area shifted up on the PC2 axis upon irradiation to be placed in the positive PC2 area as the irradiated samples. In other words, irradiation may be a major contributor to PC2, suggesting that a systematic change to the plant metabolites might have been caused by irradiation exposure. The loading plot showed that the peaks were scattered throughout, and some were clustered at the high PC1 region but not at the high PC2 region (Figure 7b), making a biological interpretation of the loading plot from the viewpoint of irradiation difficult.

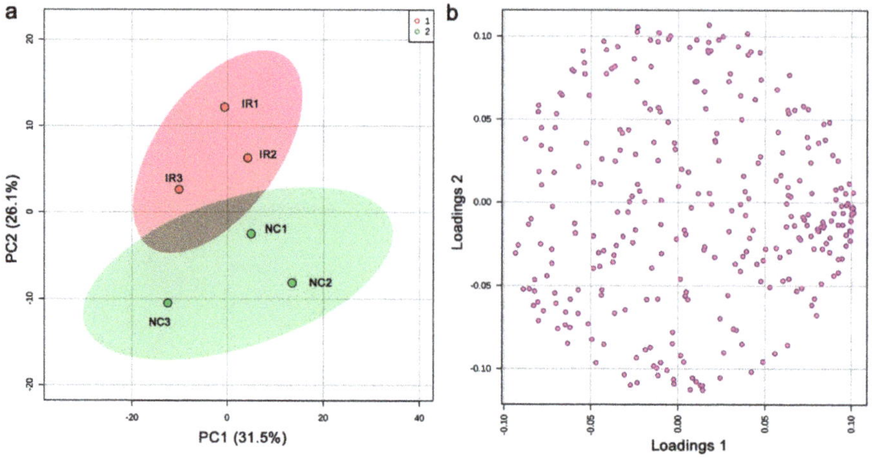

Figure 7. PCA of the nontargeted GC-MS analysis. (**a**) Score plot; 95% confidence ranges are colored. IR, irradiated; NC, nonirradiated. (**b**) Loading plot; 306 peaks are located in a 2D plane.

The heat map with all 306 peaks was too complex to decipher any legitimate pattern (Figure 8a), but a heat map with the top 25 peaks indicated differences in the metabolite levels between the irradiated and nonirradiated groups (Figure 8b). The three irradiated samples had a similar pattern to one another, and the three nonirradiated control samples also had a similar pattern to one another, which was different from that of the irradiated group. In other words, a systematic response to the irradiation treatment was observed in all the samples, which was consistent with the PCA results. As in the targeted analysis above, the nonirradiated group appeared to have more metabolites present at higher levels than the irradiated group, suggesting that irradiation might have caused an overall metabolic slowdown in this plant, at least as determined by the nontargeted GC-MS method.

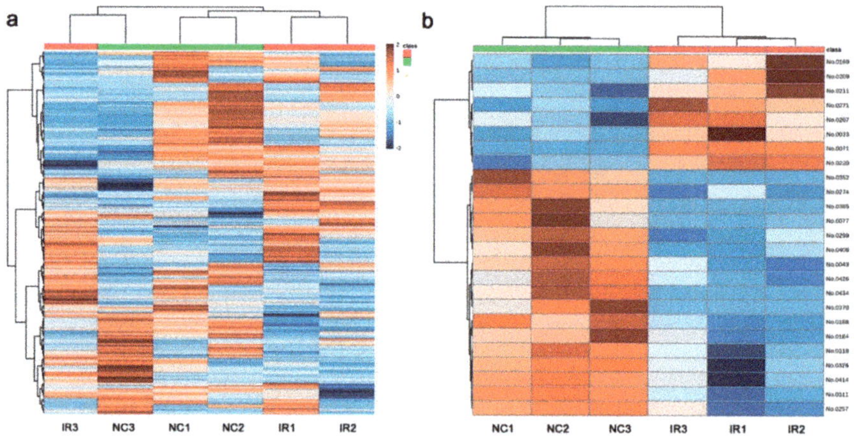

Figure 8. Heat maps from the nontargeted GC-MS analysis. At the top, the irradiated group was indicated by red horizontal bars, and the nonirradiated control group was indicated by green horizontal bars. (**a**) Heat map using all 306 peaks. (**b**) Heat map using the top 25 peaks.

We detected 28 significantly different peaks between the irradiated and nonirradiated groups using a *t*-test ($p < 0.05$) (Figure 9a); seven peaks had $p < 0.01$, and one had $p < 0.001$. Similarly, we performed a fold change analysis, in which 16 peaks were detected as $FC > 2.0$ and 27 peaks as $FC < -2.0$ (Figure 9b). It appears that more samples were downregulated (negative *FC* values) than upregulated (positive *FC* values), which was consistent with the results of the heat map above. Among them, seven peaks satisfied both conditions ($p < 0.05$ and $FC > 2.0$ or $FC < -2.0$) (Figure 9c).

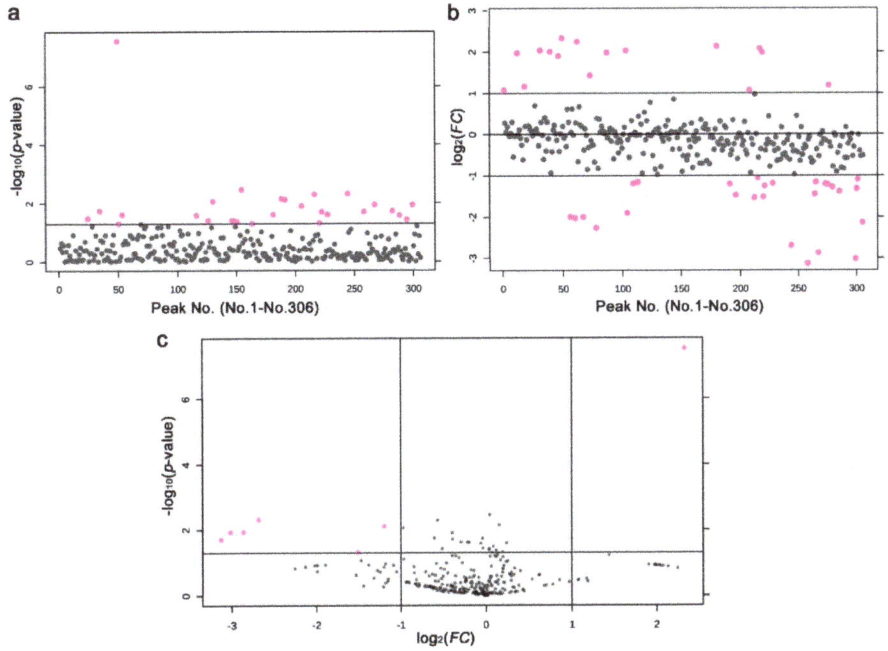

Figure 9. Results of the *t*-test and fold change analysis of the nontargeted GC-MS analysis between the irradiated and nonirradiated groups. (**a**) Distribution of the *p*-values (Student's *t*-test). The threshold was set at $p = 0.05$. The 28 pink dots indicate significant peaks. The peak numbers (No. 1–No. 306) here on the horizontal axis do not correspond to the original peak numbers shown in Figure 10, because the original peak numbers were assigned before elimination of the nonsense peaks by MetaboAnalyst (also in (**b**)). (**b**) Distribution of the *FC* values. The thresholds were set at $FC = 2.0$ and $FC = -2.0$. The 16 and 27 pink dots indicate peak upregulation and downregulation by more than twofold, respectively. (**c**) Volcano plot. The 7 pink dots indicate peaks that satisfy both the *p*-value and *FC* value criteria.

Among the peaks with $p < 0.05$, 10 peaks were upregulated upon irradiation, whereas 18 peaks were downregulated (Figure 10a). Only three peaks were annotated as follows: citric acid (No. 318, $p = 0.020$), nonanoic acid (No. 184, $p = 0.038$), and 3-aminopropanoic acid (No. 207, $p = 0.038$). The latter two were also found in the targeted method above. Citric acid and nonanoic acid were downregulated, whereas 3-aminopropanoic acid was upregulated in contrast to its downregulation in the targeted method, questioning the validity of this result.

Among the peaks with $FC > 2.0$ or $FC < -2.0$, 16 peaks were detected as upregulated and 27 as downregulated. However, among them, only one peak was annotated as oleic acid (No. 395) (Figure 10b). Oleic acid was also detected in the targeted method with $FC < -2.0$ (i.e., $FC < 0.5$), but its box plot in both methods suggested that this difference was dependent on a single outstanding value (Figures 6b and 10b), questioning the validity of this result.

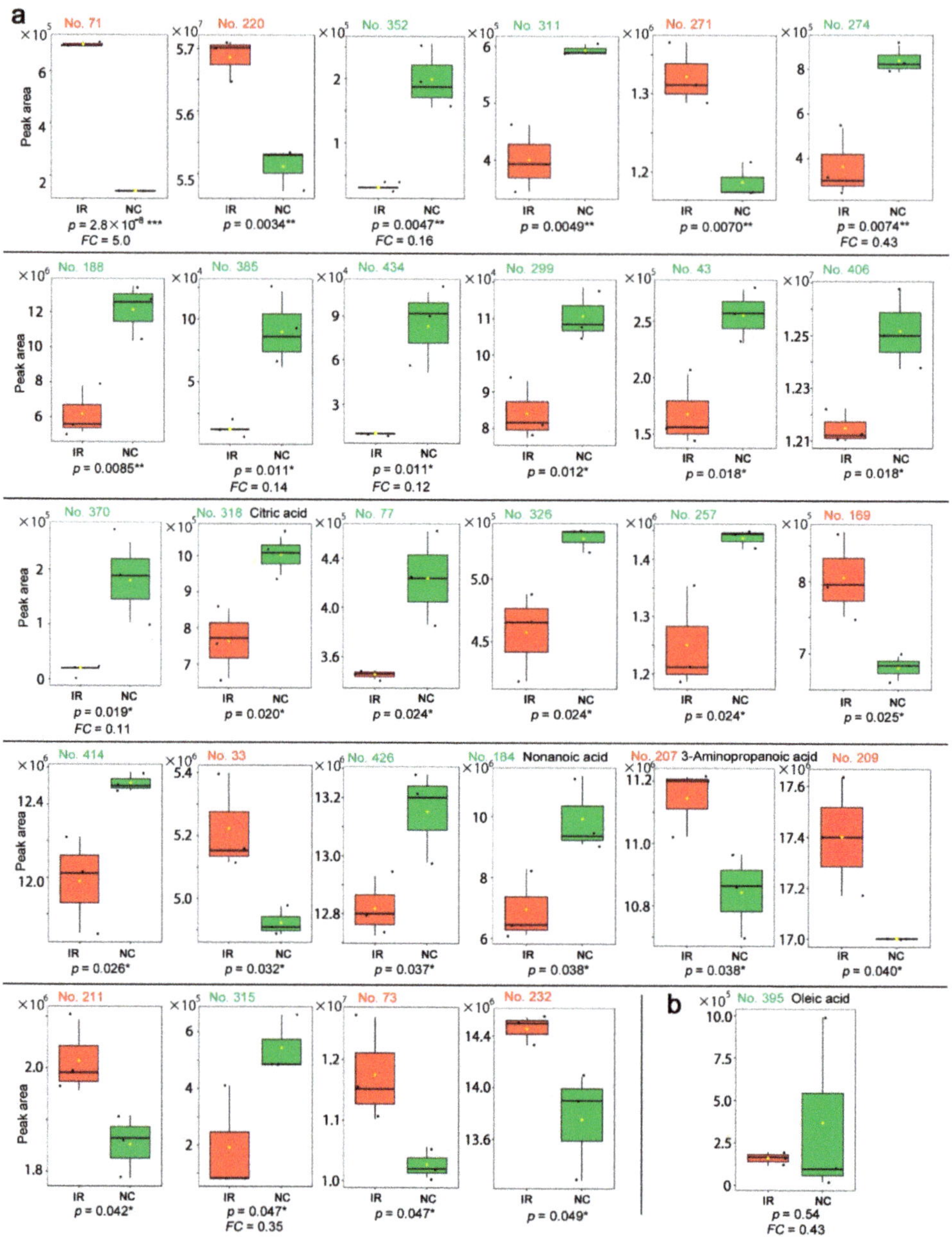

Figure 10. Box plots of the peaks in the nontargeted GC-MS. The irradiated group (IR) is indicated by red boxes, and the nonirradiated control group (NC) is indicated by green boxes. The peak numbers are shown in red when upregulated, and they are shown in green when downregulated. (**a**) Peaks with significant differences between the irradiated and nonirradiated groups in the order of low-to-high *p*-values. There were 28 peaks with $p < 0.05$ (Student's *t*-test), and among them, only 3 peaks were annotated. The *FC* values are shown when $FC > 2.0$ or $FC < 0.5$. (**b**) Peak No. 395 identified by the fold change analysis. This peak was annotated as oleic acid. * $p < 0.05$, ** $p < 0.01$, *** $p < 0.001$.

Oxalic acid was detected in a large amount as peak No. 76 from the nontargeted method, and its peak levels were not significantly different between the irradiated and nonirradiated groups ($p = 0.67$; Student's t-test).

3.4. LC-MS

From the LC-MS analysis, 9554 peaks were originally detected, including those detected in only a single sample; the shortest retention time was 3.001 min ($C_9H_{16}O_5N_2$), and the longest retention time was 24.996 min ($C_{43}H_{66}O_{14}$). Among them, 5418 peaks were considered valid by MetaboAnalyst. Based on the peak area (intensity) data, the PCA was performed (Figure 11a). PC1 and PC2 explained 47.8% and 22.3% of the variance, respectively. Irradiated and nonirradiated samples from the same plant individual were located close to each other, suggesting that the coordinates of the samples were determined mainly by genetic background and marginally by radiation response. The loading plot showed that there were many peaks in the negative area of the horizontal axis (Figure 11b), but the biological interpretation was difficult.

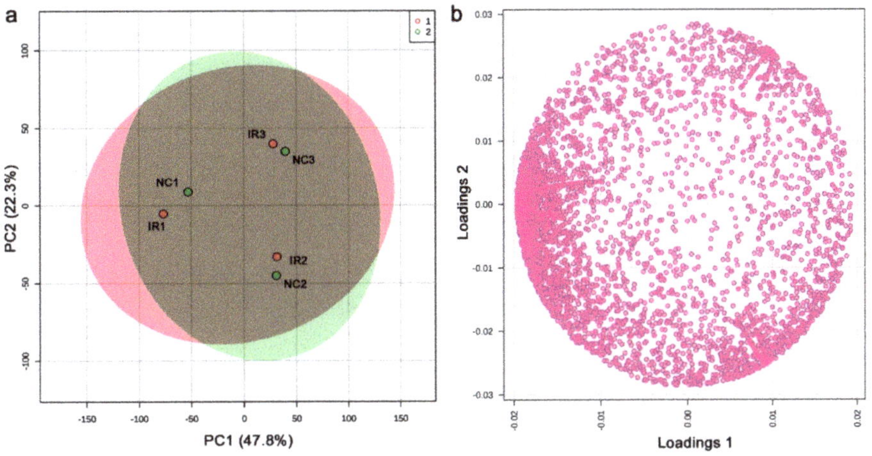

Figure 11. PCA results from the LC-MS analysis. (**a**) Score plot; 95% confidence ranges are colored. IR, irradiated; NC, nonirradiated. (**b**) Loading plot; 5418 peaks are located in a 2D plane.

A heatmap of all 5418 peaks also demonstrated that a clonal pair of irradiated and nonirradiated samples of the same plant exhibited similar patterns (Figure 12a), suggesting that genetic differences contributed more substantially to their metabolite differences. However, when the top 25 peaks were examined, the irradiated and nonirradiated groups showed different patterns, and the irradiated group appeared to have more samples with higher metabolite levels than the nonirradiated group (Figure 12b). This result was different from that of GC-MS.

We detected 36 significantly different peaks between the irradiated and nonirradiated samples using t-tests ($p < 0.05$) (Figure 13a). Similarly, we also performed a fold change analysis (Figure 13b). We found 902 upregulated and 547 downregulated peaks. Since there were too many peaks to investigate, they were not pursued further. Notably, 25 peaks satisfied both criteria ($p < 0.05$ and $FC > 2.0$ or $FC < -2.0$) (Figure 13c).

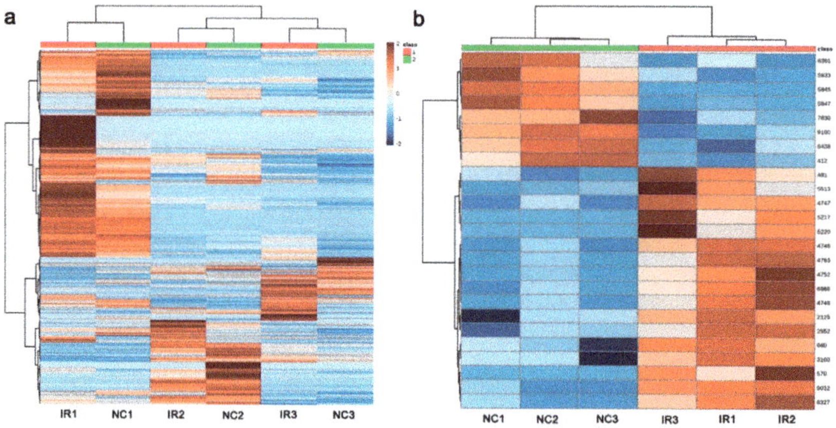

Figure 12. Heat maps from the LC-MS analysis. At the top, the irradiated group was indicated by red horizontal bars, and the nonirradiated control group was indicated by green horizontal bars. (**a**) Heat map using all 5418 peaks. (**b**) Heatmap using the top 25 peaks.

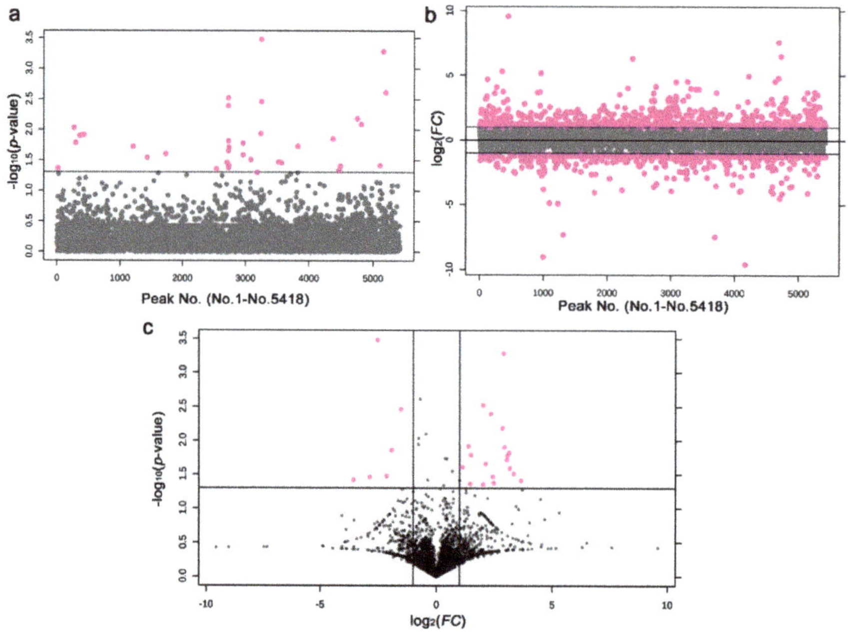

Figure 13. Results of the *t*-test and fold change analysis after the LC-MS analysis between the irradiated and nonirradiated groups. (**a**) Distribution of the *p*-values (Student's *t*-test). The threshold was set at $p = 0.05$. The 36 pink dots indicate significant peaks. The peak numbers (No. 1–No. 5418) here on the horizontal axis do not correspond to the original peak numbers shown in Figure 14, because the original peak numbers were assigned before elimination of the nonsense peaks by MetaboAnalyst (also in (**b**)). (**b**) Distribution of the *FC* values. The thresholds were set at $FC = 2.0$ and $FC = -2.0$. The 902 and 547 pink dots indicate peaks that were upregulated and downregulated more than twofold, respectively. (**c**) Volcano plot. The 25 pink dots indicate peaks that satisfy both the *p*-value and *FC* value criteria.

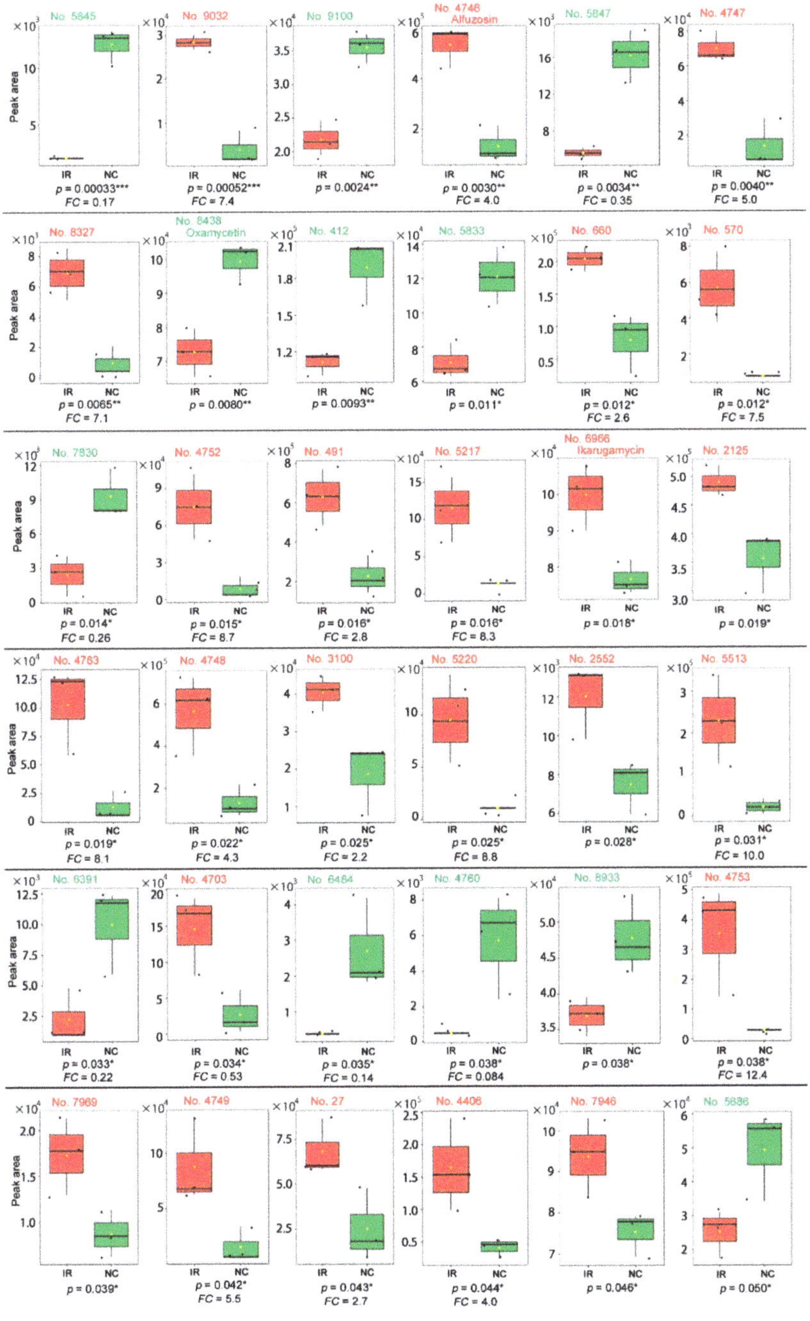

Figure 14. Box plots of the 36 peaks detected by the LC-MS analysis. The irradiated group (IR) is indicated by red boxes, and the nonirradiated control group (NC) is indicated by green boxes. The peak numbers are shown in red when upregulated, and they are shown in green when downregulated. Peaks with significant differences between the irradiated and nonirradiated groups ($p < 0.05$, Student's *t*-test) are shown in the order of low-to-high *p*-values. Among them, only 3 peaks are singularly annotated. The *FC* values are shown when $FC > 2.0$ or $FC < 0.5$. * $p < 0.05$, ** $p < 0.01$, *** $p < 0.001$.

Among the peaks with $p < 0.05$, 24 peaks were upregulated after irradiation, whereas 12 peaks were downregulated (Figure 14). Most peaks had no compound name annotations, whereas some were annotated as multiple compounds. Only three peaks were annotated as a single compound as follows: alfuzosin (No. 4746, $p = 0.0033$), oxamicetin (oxamycetin) (No. 8438, $p = 0.0080$), and ikarugamycin (No. 6966, $p = 0.018$).

3.5. Functional Categorization of the LC-MS Annotated Peaks

There were 12 peaks with multiple annotations and three peaks with a single annotation (Table 1). We checked their MS/MS fragment spectrograms, but similar ones were not found in the searched databases (Supplementary Table S2; MS/MS spectrograms can be seen by double-clicking the peak number in the MS2 tab in this Microsoft Excel file). These 15 peaks were classified into four functional categories (Table 2). The first functional category was "antioxidation", in which plant-derived antioxidants were included. These compounds were all upregulated upon irradiation. No. 660 ($p = 0.012$) was identified as either L-L-homoglutathione or S-methylglutathione. No. 2125 ($p = 0.019$), and No. 4406 ($p = 0.044$) had multiple candidates, but they were flavonoids, which is a group of phytochemical antioxidants [63].

The second functional category was the "stress response", in which plant-derived compounds for the stress resistance were included. All of these compounds, except one, were upregulated upon irradiation. No. 4746 ($p = 0.0030$) was alfuzosin, which is known to have pharmacological activity in animals [64–67]. No. 4763 ($p = 0.019$) and No. 2552 ($p = 0.028$) had multiple candidates, partly due to their small masses, but the candidates for No. 4763 and No. 2552 included dihydrobenzofuran (coumaran) and 2-acetylfuran, respectively. The former is well-known as a natural biopesticide [68]. No. 7969 ($p = 0.039$) had multiple candidates, but they were all alkaloids. No. 4753 ($p = 0.038$) was identified as either 7-chlorodeutziol or myobontioside A. These compounds belonged to a group of iridoids, which are plant chemicals that probably ward off insects and other animals [69,70]. No. 7830 ($p = 0.014$) had five (virtually three due to redundancy) candidate compounds that were all plant-derived secondary metabolites. One of them was a limonoid, a possible antifeedant [71], and another was a taxoid [72]. However, No. 7830 was downregulated.

The third functional category was "non-plant derivatives". All of these compounds, except one, were downregulated upon irradiation, probably because of the sterilization effect of irradiation. This plant sample might have been contaminated by microorganisms despite the washing process of the collected leaves. No. 8438 ($p = 0.0080$) and No. 6966 ($p = 0.018$) were identified as oxamicetin (oxamycetin) and ikarugamycin, respectively, and are known to be antifungal antibiotics produced by a group of soil bacteria, *Streptomyces* [73–75]. No. 412 ($p = 0.0093$) was identified as either leinamycin or uracil, and if the former is correct, this peak is another antibiotic produced by the same soil bacterial genus *Streptomyces* [76]. No. 5833 ($p = 0.011$) had three candidate compounds, all of which were mycotoxins from fungi [77,78]. The upregulation of No. 6966, annotated as ikarugamycin, was enigmatic, but a group of soil bacteria might have responded to the irradiation treatment.

The fourth functional category was simply "unknown". No. 4703 ($p = 0.034$) and No. 8933 ($p = 0.038$), included in this category, had various candidate compounds.

Table 1. Summary of the annotated LC-MS peaks.

No.	Formula	Exact Mass	Up/Down	Annotation (Compound Name)
4746	$C_{19}H_{27}O_4N_5$	389.206	UP	Alfuzosin
8438	$C_{29}H_{42}O_{10}N_6$	634.296	DOWN	Oxamicetin; Oxamycetin (produced by *Streptomyces inusitatus*)
412	$C_4H_4O_2N_2$	112.027	DOWN	Acetylenedicarboxamide; Acetylene diamide; Aquamycin; Cellocidin; Lenamycin; NSC 38643; NSC 65381 (produced by *Streptomyces* sp.); Uracil
5833	$C_{17}H_{22}O_7$	338.137	DOWN	Acetyldeoxynivalenol (a mycotoxin from *Fusarium graminearum*); O-Acetylcyclocalopin A (produced by the mushroom *Boletus calopus*); 15-Acetyl-4-deoxynivalenol (a mycotoxin from *Fusarium graminearum*)
660	$C_{11}H_{19}O_6N_3S$	321.099	UP	L-L-Homoglutathione (an antioxidant); S-Methylglutathione (an antioxidant)
7830	$C_{37}H_{46}O_{12}$	682.299	DOWN	Plant-derived pharmacological compounds (*1)
6966	$C_{29}H_{38}O_4N_2$	478.283	UP	Ikarugamycin (produced by *Streptomyces phaeochromogenes*)
2125	$C_{15}H_{18}O_{11}$	374.085	UP	Flavonoids (*2)
4763	C_8H_8O	120.058	UP	Various compounds (*3) including Dihydrobenzofuran in plants such as *Lantana camara* as a natural biopesticide
2552	$C_6H_6O_2$	110.037	UP	Various compounds (*4) including 2-Acetylfuran in plants
4703	$C_{10}H_{14}O_3$	182.094	UP	Various compounds (*5)
8933	$C_{13}H_{20}O$	192.151	DOWN	Various compounds (*6)
4753	$C_{15}H_{23}O_9Cl$	382.103	UP	7-Chlorodeutziol; Myobontioside A (plant derivatives)
7969	$C_{21}H_{23}O_6N$	385.153	UP	Colchicine-related alkaloids (*7)
4406	$C_{32}H_{36}O_{18}$	708.190	UP	Flavonoids (*8)

*1: 2,7-Dideacetyltaxuspine X; (+)-2,7-Dideacetyltaxuspine X; Swietephragmin C (a limonoid from an African medicinal plant); Taxezopidine K; (+)-Taxezopidine K (isolated and purified from the seeds of the Japanese yew *Taxus cuspidata* and an inhibitor of Ca^{2+}-induced depolymerization of microtubules). *2: 6-[5-(2-Carboxy-2-hydroxyethyl)-2-hydroxyphenoxy]-3,4,5-trihydroxyoxane-2-carboxylic acid; 6-{[3-(3,4-Dihydroxyphenyl)-2-hydroxypropanoyl]oxy}-3,4,5-trihydroxyoxane-2-carboxylic acid; 6-[5-(2-Carboxyethyl)-2,3-dihydroxyphenoxy]-3,4,5-trihydroxyoxane-2-carboxylic acid; 3,4,5-Trihydroxy-6-{[3-(3,4,5-trihydroxyphenyl)propanoyl]oxy}oxane-2-carboxylic acid; 6-[4-(2-Carboxyethyl)-2,6-dihydroxyphenoxy]-3,4,5-trihydroxyoxane-2-carboxylic acid; 6-[4-(2-Carboxy-2-hydroxyethyl)-2-hydroxyphenoxy]-3,4,5-trihydroxyoxane-2-carboxylic acid. *3: Phenylacetaldehyde; Styrene oxide; 3-Ethenylphenol; *p*-Methylbenzaldehyde; Dihydrobenzofuran (or Coumaran, an acetylcholinesterase (AChE) inhibitor produced by the leaves of *Lantana camara* as a biopesticide); Lentialexin (produced by a mixed culture of *Lentinus edodes* (Shiitake mushrooms) and *Trichoderma polysporum*); 2-Methylbenzaldehyde; Acetophenone; 3-Methylbenzaldehyde; 4-Vinylphenol. *4: Resorcinol; Pyrocatechol; 2-Acetylfuran (found in coffee, passion fruit, and others); 5-Methyl-2-furancarboxaldehyde; (*E,E*)-2,4-Hexadienedial; Hydroquinone. *5: 4-Ethyl-2,6-dimethoxyphenol; Eupatriol; Bombardolide B; Furfuryl pentanoate; Peperinic acid; *threo*-Anethole glycol; Amyl 2-furoate; Verimol J; Stagonolide E; (-)-Stagonolide E; Isoamyl 2-furoate; Fistupyrone; Annularin D; (+)-Annularin D; Furfuryl isovalerate; 2-Ethoxy-4-(methoxymethyl)phenol; Modiolide B; Bombardolide D; (-)-Bombardolide D; Multiflotriol; Crocusatin I; Sapinofuranone B; 4-Hydroxy-6-isopropyl-3,5-dimethyl-2H-pyran-2-one; GSIR-1; 2-(2-Hydroxy-4-methylphenyl)-1,3-propanediol; 9,10-Dihydroxythymol; 4-(Ethoxymethyl)-2-methoxyphenol; Dihydroconiferyl alcohol. *6: 2,5-Diisopropyl-3-methylphenol; Edulan I; 2,3-Diisopropyl-5-methylphenol; alpha-Damascone; Cycloionone; 2,4-Diisopropyl-3-methylphenol; Vitispirane; Isospirene; γ-Ionone; 4-(2,6,6-Trimethyl-1,3-cyclohexadien-1-yl)-2-butanone; 2,5-Diisopropyl-3-methylphenol; Pseudoionone; 2,6-Diisopropyl-3-methylphenol; (*E*)-5,8-Megastigmadien-4-one; δ-Damascone; 4-(2,6,6-Trimethylcyclohex-1-enyl)but-2-en-4-one; Phenethyl isoamyl ether; (2*E*,4*Z*,7*Z*)-2,4,7-Tridecatrienal; 2,4-Diidopropyl-5-methylphenol; 4-(4-Methyl-3-pentenyl)-3-cyclohexene-1-carboxaldehyde. *7: *N*-Formyl-*N*-deacetylcolchicine (an alkaloid from *Gloriosa superba*); Romucosine H (an alkaloid from *Annona cherimola*); *N*-Methylporphyroxine (a rhoeadine alkaloid); Papaverrubine F (an alkaloid); 3-Demethylcolchicine; 2-Demethylcolchicine; (-)-2-Demethylcolchicine; Glaucamine (an alkaloid); Polycarpine (a marine alkaloid). *8: Patuletin 3-rhamnoside-7-(3‴,4‴-diacetylrhamnoside); Patuletin 3-(4″-acetylrhamnoside)-7-(3‴-acetylrhamnoside); Patuletin 3,7-bis(3-acetylrhamnoside); Kaempferide 3-rhamnoside-7-(6″-succinylglucoside); Patuletin 3-(4″-acetylrhamnoside)-7-(2‴-acetylrhamnoside).

Table 2. Functional categories of the annotated LC-MS peaks.

Category No.	Possible Function	Peak No. (Up/Down)	UP/DOWN (Collective)
1	Antioxidation	660 (UP), 2125 (UP), 4406 (UP)	UP
2	Stress response	4746 (UP), 4763 (UP), 2552 (UP), 7969 (UP), 4753 (UP), 7830 (DOWN)	UP (or DOWN)
3	Non-plant derivatives	8438 (DOWN), 6966 (UP), 412 (DOWN), 5833 (DOWN)	DOWN (or UP)
4	Unknown	4703 (UP), 8933 (DOWN)	UP orDOWN

3.6. LC-MS/MS Spectrograms of Peak No. 4746 and Alfuzosin

As mentioned above, peak No. 4746 was singularly annotated as alfuzosin. Alfuzosin is a synthetic quinazoline derivative with an exact mass of 389.206. Since alfuzosin is not a plant-derived drug, peak No. 4746 should not be identical to alfuzosin. Peak No. 4746 was estimated to have a very similar mass; its *m/z* value for $[M+H]^+$ was 390.214. However, automated annotation does not guarantee chemical identification. The difference between the theoretical and experimental exact mass values (dPPM value) was calculated to be −3.37 ppm, which was acceptable for annotation but relatively large in absolute values among the annotated peaks. Here, we examined the LC-MS/MS spectrograms of peak No. 4746 and compared them to those of alfuzosin recorded in the Human Metabolome Database (HMDB) [60,61] (Figure 15).

Ten experimental $[M+H]^+$ spectrograms of alfuzosin revealed that the major fragments of alfuzosin were located at 235, 156, and 71 (*m/z*), and further fragmentation produced peaks with 231, 219, 147, 105, 78, and 63 (*m/z*) (Figure 15a–j). In contrast, the major fragments of No. 4746 in the present analysis were located at 279, 227, 167, 149, 121, 71, 65, and 57 (*m/z*); among which, the fragment at 149 (*m/z*) had the highest intensity (Figure 15k). Between the experimental records and the present analysis, only the fragment at 71 (*m/z*) was identical. The fragment at 149 (*m/z*) in the present analysis was close to the fragment at 147 (*m/z*) in the database, but their relative intensity values were different. Furthermore, one of the fragments from peak No. 4746 was predicted to be $C_6H_{10}O_5$. This means that No. 4746 had a hexose moiety, which was not found in alfuzosin. Taken together, it is likely that peak No. 4746, annotated as alfuzosin, was not actually alfuzosin itself but, instead, an unknown structural isomer of alfuzosin or an unknown compound similar to alfuzosin.

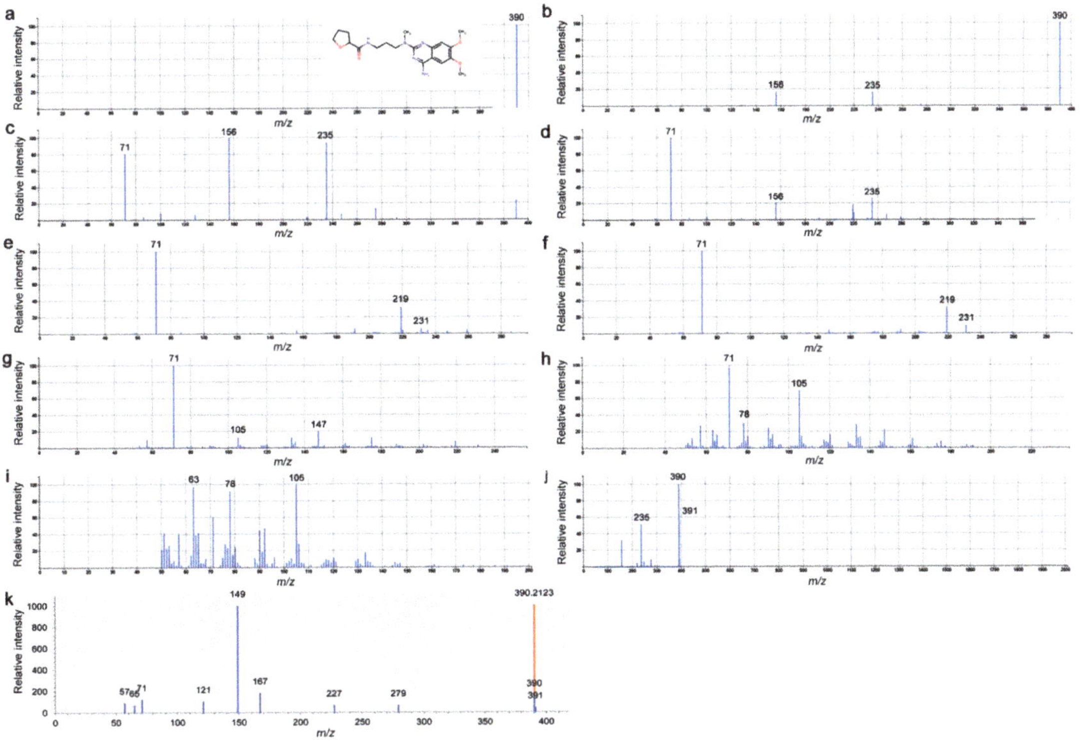

Figure 15. LC-MS/MS spectrograms of alfuzosin and peak No. 4746 (from the present study) after positive ionization ($[M+H]^+$). (**a**–**j**) Ten experimental HMDB records of alfuzosin. The highest intensity was set as 100, and the relative intensity values of the fragments are shown. The top 3 fragments are labeled according to their *m/z* values in each panel. The inset in (**a**) is the structure of alfuzosin, in which oxygen and nitrogen atoms and bonds are shown in red and purple, respectively. The splash keys for the identification of these spectrograms from (**a**) to (**j**) (after "splash10-") are as follows: 0006-0009000000-58fd75abdf178ea1b331, 0006-0119000000-90a7799e494035d224e7, 0abi-5791000000-648d5529faeac9e35dc0, 00di-9260000000-591f2bacf7d82aeba7e3, 00di-9250000000-d107a0e9675604325 bb1, 00di-9340000000-f708562f3998488d7fce, 00di-7910000000-3fe44ce2a42f727569e2, 05fr-8900000000-dca38c682ad0 b6c90b46, 0bvl-9300000000- bad5c36d48e0e93a82aa, and 0006-0249000000-93552ff8c72ff220f172. (**k**) Present study. The major fragments are labeled by *m/z* values. The full records of the ionized fragments $[M+H]^+$ (*m/z*) and their relative intensity values (in parenthesis; the highest intensity was set as 1000) are as follows: 390.212 (1,000) (shown in brown), 149.02 (999), 167.03 (179), 390.21 (152), 71.09 (120), 121.03 (100), 57.07 (87), 65.04 (66), 227.13 (62), 279.16 (62), 391.19 (38), 93.07 (36), 67.62 (32), 113.13 (17), 391.08 (17), 121.04 (17), 183.10 (16), 93.03 (16), 107.09 (16), 390.16 (14), 79.05 (13), 163.11 (13), 228.13 (13), 209.12 (13), 169.09 (12), 137.06 (12), 95.05 (11), 111.04 (11), 81.07 (11), 119.09 (10), 390.14 (10), 365.63 (10), 413.48 (10), 351.38 (10), 255.77 (9), 184.48 (9), 364.65 (9), 304.16 (9), 354.17 (8), 105.07 (8), 81.03 (8), 55.06 (8), 220.88 (8), 153.82 (7), 109.22 (7), 53.76 (7), and 91.48 (7). The fragment at 228.13 (*m/z*) indicates a loss of 162.08, suggesting the loss of a hexose moiety ($C_6H_{10}O_5$).

4. Discussion

4.1. Experimental System

In this study, we examined the metabolomic response of the creeping wood sorrel, the host plant of the pale grass blue butterfly, to low-dose radiation exposure. To reproduce the possible radiation environment of Fukushima, contaminated soil was collected from Fukushima, and plants collected from Okinawa, the least affected prefecture in Japan, were irradiated to exclude a history of previous exposure. We believe that this experimental system uniquely provided us with information on the acute perturbation of the plant in Fukushima immediately after the nuclear accident. Additionally, we may be able to obtain a hint at the ongoing chronic impacts of the Fukushima nuclear accident on the ecosystem. For accurate understanding of the chronic impacts, metabolomic analyses of field-collected plants from Fukushima should be performed. Having mentioned this point, we clarified that the major radionuclide in the soil used in the present study was anthropogenic ^{137}Cs, as expected.

The plant looked completely healthy after irradiation. In fact, the cumulative absorbed dose and dose rate in the creeping wood sorrel in this system were understood as "very low" compared with other low-dose (rate) experiments. For example, one experiment employed 185.4 mGy/h and 1.1 Gy for the low-dose exposure to mice [48], which is 5453-fold and 193-fold higher than in the present study, respectively. The external exposure experiment using the pale grass blue butterfly and its host plant in the original paper [19] was estimated to be 55 mSv and 125 mSv. These values are 9.6-fold and 22-fold higher than in the present study (5.7 mGy), assuming that Sv and Gy are interchangeable. Notable exceptions were in situ exposure experiments at Iitate Village, where the exposure level was 4 μSv/h for 3 days [15,17].

Considering the very low exposure level in the present study, it is not surprising that only a small number of significantly different peaks were detected by the GC-MS and LC-MS analyses. Rather, it may be surprising that the plant nonetheless responded metabolically to 5.7 mGy of irradiation treatment. We primarily used $p < 0.05$ as the criterion for differentially expressed peaks. This relatively generous criterion was unavoidable for this study to gain a substantial amount of information on the differentially expressed metabolites under very low exposure levels. Accordingly, at least some candidate peaks may be false positives. This problem may partially be resolved by increasing the number of samples. Another difficulty of this study was that the information in the databases on the secondary metabolites of this non-model organism appeared to be limited. Below, with these limitations in mind, we discuss the present results.

4.2. Interpretations of the GC-MS Results

In both the targeted and nontargeted GC-MS methods, the irradiated and nonirradiated groups showed different peak distribution patterns. It appears that, by GC-MS, the irradiated group showed the overall lower peak intensity values, which were clearly seen in the fold change distributions (Figures 5b and 9b). The PCA of the targeted method (Figure 3) showed that the irradiated group clustered in a much narrower range than the nonirradiated control group. The wider distribution of the nonirradiated group could be interpreted as the genetic difference, and the wider distribution was made narrower after irradiation, suggesting a stereotypical response. Similarly, the PCA of the nontargeted method (Figure 7) showed a systematic upward shift upon irradiation along PC2. In both cases, it is likely that the plant indeed responded to this level of irradiation systematically. In the nontargeted method, PC2 (26.1%) may represent the irradiation effects, and in contrast, PC1 (31.5%) may represent the possible genetic effects. The contribution of PC1 was larger than that of PC2, but they may be considered comparable. The peaks identified by the fold change analysis revealed that they were all downregulated, and their distribution ranges became much narrower in response to irradiation, being consistent with the PCA results. A rough interpretation could be that metabolomic changes of a group of plant chemicals in response to irradiation at the milligray level were as large as the genetic differences among the plant species in Okinawa.

We were able to examine box plots of each candidate peak from the GC-MS. From the targeted method, the peaks annotated as caproic acid, nonanoic acid, and 3-aminopropanoic acid were downregulated, whereas the peak annotated as lauric acid was upregulated (t-test). Caproic acid, also known as hexanoic acid, has been shown to be an inducer of plant defense mechanisms [79]. This function of caproic acid is different from, but similar to, that of azelaic acid [79], another candidate (fold change analysis). Moreover, azelaic acid is a derivative of oleic acid [79], which is yet another candidate (fold change analysis). Together, it can be speculated that their metabolic pathways and defense functions may be perturbed by irradiation. Nonanoic acid is known as a phytotoxic chemical that may function to eradicate other neighboring plants [80]. Lauric acid can function as an antifungal and possible antiviral agent [81]. It has been speculated that lauric acid is upregulated as a part of the plant defense mechanism under irradiation stress.

A carbohydrate pathway may be slowed down, since glucose and fructose may be downregulated (fold change analysis), as observed from the results of the targeted method. This is consistent with the finding that citric acid ($p = 0.020$; t-test) was also found to be downregulated by the nontargeted method, suggesting that the TCA cycle may be slowed down. Alternatively, a photosynthetic pathway that produces glucose and fructose may be slowed down. Nonanoic acid was found in not only the targeted method but, also, the nontargeted method (t-test). Oleic acid was also found in the nontargeted method (fold change analysis), although this result is not credible based on its box plot (Figure 10b) because of a single outstanding value. 3-Aminopropanoic acid, also known as β-alanine, is a compound related to the general stress response in plants [82] and was found by both the targeted and nontargeted methods (t-test) with different results; it was upregulated in the former and downregulated in the latter. This was the only case in which the targeted and nontargeted results contradicted each other. Since the targeted method is more reliable than the nontargeted method in terms of the alignment and annotation, the latter could simply be an annotation mistake. If so, the upregulation of 3-aminopropanoic acid could indicate the activation of a defense mechanism.

Notably, many peaks were not annotated in the nontargeted method, but some were highly significant. The most significant peak from the GC-MS analysis was No. 71 ($p = 2.8 \times 10^{-8}$), which showed a marked upregulation ($FC = 5.0$) (Figure 10a). We confirmed that oxalic acid was detected at a high level in both the irradiated and nonirradiated groups with no statistically significant differences in either the targeted or nontargeted methods. Even if the oxalic acid in the leaves is a feeding stimulant for larvae of the pale grass blue butterfly [62], irradiated leaves would not cause a feeding problem in terms of the level of oxalic acid.

4.3. Interpretations of the LC-MS Results

The LC–MS PCA results indicated that irradiated and nonirradiated samples of the same clone were located more closely together in the score plot than the irradiated or nonirradiated groups (i.e., the same treatment group). (Figure 11). This result was in contrast to the GC-MS results, in which a systematic change in the irradiated group along PC2 was observed. Furthermore, by LC-MS, the number of upregulated significant peaks was greater than the number of downregulated peaks (Figure 13b), which was also in contrast to the GC-MS results. An interpretation of these data is that the downregulation of a group of GC-MS-detected metabolites (including many primary metabolites) upon irradiation is a relatively stereotypical response in this plant, but the upregulation of a group of LC-MS-detected metabolites (including many secondary metabolites) upon irradiation is more dependent on the genetic background of the individual plant. In other words, the response profiles may be stereotyped but simultaneously allow individual variations. To examine by PCA if there are any systematic irradiation responses to the low-dose levels, GC-MS seems to be superior to LC-MS. On the other hand, to examine by t-test if there are any upregulated peaks, LC-MS seems to be more suitable than GC-MS.

A limited number of peaks were annotated, but the plants appeared to upregulate the production of antioxidants and certain chemicals important for stress resistance and defense, including potential natural insecticides (Table 2). We consider that these two functional categories represent reasonable mechanistic response profiles of this plant. How plants receive irradiation signals is not known, but reactive oxygen species (ROS) may be produced inside cells, which may activate a series of general stress responses. The antioxidants detected here may play a role in scavenging these ROS. Interestingly, it seems that not all of the potential defense metabolites were upregulated. As discussed before, most compounds for the defense mechanisms, such as caproic acid, were downregulated, whereas lauric acid was upregulated in GC-MS. Similarly, in LC-MS, most of the annotated peaks in the second functional category were upregulated, but one was downregulated, which was likely a limonoid. It appears that there are many defense pathways, and some pathways may be activated more often than others in response to low-level radiation stress. Similarly, only a limited number of potential antioxidants were upregulated in the present study.

Perhaps because of this stress response or because of the direct radiation sterilization, non-plant derivatives such as oxamicetin (p = 0.0080), which may be contaminants from soil microorganisms, tended to be downregulated in the irradiated group (Figure 14). Additionally, ikarugamycin (p = 0.018), which may also be a contaminant, was upregulated. These annotations may be due to coincidences in the exact mass values, but the positive involvement of soil microbes, including endophytic bacteria and fungi, during the radiation response of the plant cannot be ruled out.

Peak No. 4746 was singularly annotated as alfuzosin with a reasonably small p-value (p = 0.0030) and high FC value (FC = 4.0) (Figure 14). In reference to the HMDB LC-MS/MS spectrogram records, peak No. 4746 was not alfuzosin itself but, instead, an unknown isomer that may contain a hexose moiety (Figure 15). Thus, the alfuzosin annotation for peak No. 4746 may be considered just a coincidence. Nonetheless, a functional coincidence may follow if their masses and affinities detected by LC-MS were identical. It is noteworthy that alfuzosin is a pharmaceutical drug that blocks α_1-adrenergic receptors in humans [64–67]. Considering that biogenic amines such as tyramine and octopamine play important physiological roles in insects similar to adrenergic transmitters in vertebrates [83–86], it may be possible that this unknown alfuzosin isomer in the plant functions as an antagonist of biogenic amine receptors after ingestion by insects to cause pharmacological effects on insect feeding behaviors and the metabolism.

We also found other unannotated plant derivatives that were significantly different between the irradiated and nonirradiated groups. The lowest p-value was assigned to peak No. 5845 ($p = 3.3 \times 10^{-4}$) with downregulation (FC = 0.17) (Figure 14). The peak with the second-lowest p-value was No. 9032 ($p = 5.2 \times 10^{-4}$) with upregulation (FC = 7.4) (Figure 14). Unfortunately, their structural identities are unknown.

4.4. Ecological Field Effects

Here, we found metabolomic changes in the investigated plant in response to very low levels of radiation exposure. We think that the detected upregulated or downregulated metabolites with or without annotation together might have contributed to the high mortality and morphological abnormality rates of the pale grass blue butterfly reported previously. In other words, changing the levels of the leaf metabolites may be an important part of the mechanism of the ecological field effects that were caused by the Fukushima nuclear accident. It should also be tested whether the pale grass blue butterfly responds to this radiation source (i.e., contaminated soil from Fukushima) physiologically or morphologically when reared together with the host plant.

Additionally, other ecological field effect modes of action may also be valid and important [27,32]. For example, a sodium deficiency is probably an important mechanism of the ecological field effects [38]. Another possibility in the plant is a decrease in their levels of vitamins available for the larvae when irradiated. Although some vitamins were found in the list of annotated compounds in the present study, no statistically significant difference was observed between the irradiated and nonirradiated groups. Furthermore, the biological impacts of the Fukushima nuclear accident include transgenerational effects caused by the initial high-dose exposure [19,34,36].

5. Conclusions

This study revealed that the creeping wood sorrel *Oxalis corniculata* responded at the metabolome level to low-level radiation exposure from soil contaminated by the Fukushima nuclear accident. Upon irradiation, the plant may reduce the levels of some compounds in carbohydrate metabolic pathways and some stress-related or defense mechanisms while simultaneously increasing the levels of some secondary metabolites that function as antioxidants and stress-related or defense chemicals. Some plant chemicals, including the discovered alfuzosin isomer, might have contributed to the high rates of mortality and abnormality observed among the butterflies in the field. Taken together, the results of this study demonstrated the metabolomic response of this *Oxalis* plant to low-dose radiation exposure and implicated the potential ecological field effects of low-level radioactive contamination from the Fukushima nuclear accident.

Supplementary Materials: The following are available online at https://www.mdpi.com/article/10.3390/life11090990/s1: Supplementary Table S1: GC-MS_Result Kazusa DNA Institute. Supplementary Table S2: LC-MS_Result Kazusa DNA Institute. Supplementary Table S3: GC-MS target data file. Supplementary Table S4: GC-MS nontarget data file. Supplementary Table S5: LC-MS data file.

Author Contributions: Conceptualization, J.M.O., W.T. and K.S.; methodology, W.T. and K.S.; software, W.T.; validation, J.M.O.; formal analysis and investigation, K.S., W.T. and J.M.O.; resources, K.S. and W.T.; data curation, K.S. and J.M.O.; writing—original draft preparation, J.M.O.; writing—review and editing, J.M.O., K.S. and W.T.; visualization, K.S. and J.M.O.; and supervision, project administration, and funding acquisition, J.M.O. All authors have read and agreed to the published version of the manuscript.

Funding: This research was funded by the Asahi Glass Foundation (Tokyo). The APC was funded by the Asahi Glass Foundation (Tokyo).

Institutional Review Board Statement: Not applicable.

Informed Consent Statement: Not applicable.

Data Availability Statement: The data presented in this study and the source data are available in this article and in the Supplementary Materials.

Acknowledgments: We thank K. Yoshida (Minamisoma City) for assisting in the collection of the contaminated soil from Fukushima; Y. Iraha, Y. Yona, and A. Morita for their technical assistance using MetaboAnalyst; and other members of the BCPH Unit of Molecular Physiology for their suggestions and technical help. We also thank the staff of the Kazusa DNA Research Institute for the practical advice and technical assistance. We would like to dedicate this paper to J. Nohara for his anniversary in the era of COVID-19.

Conflicts of Interest: The authors declare no conflict of interest. The funders had no role in the design of the study; in the collection, analyses, or interpretation of the data; in the writing of the manuscript; or in the decision to publish the results.

References

1. Arapis, G.D.; Karandinos, M.G. Migration of ^{137}Cs in the soil of sloping semi-natural ecosystems in Northern Greece. *J. Environ. Radioact.* **2004**, *77*, 133–142. [CrossRef]
2. Tahir, S.N.A.; Jamil, K.; Zaidi, J.H.; Arif, M.; Ahmed, N. Activity concentration of ^{137}Cs in soil samples from Punjab province (Pakistan) and estimation of gamma-ray dose rate for external exposure. *Radiat. Prot. Dosim.* **2006**, *118*, 345–351. [CrossRef]
3. Ambrosino, F.; Stellato, L.; Sabbarese, C. A case study on possible radiological contamination in the Lo Uttara landfill site (Caserta, Italy). *J. Phys. Conf. Ser.* **2020**, *1548*, 012001. [CrossRef]
4. Endo, S.; Kimura, S.; Takatsuji, T.; Nanasawa, K.; Imanaka, T.; Shizuma, K. Measurement of soil contamination by radionuclides due to the Fukushima Dai-ichi Nuclear Power Plant accident and associated estimated cumulative external dose estimation. *J. Environ. Radioact.* **2021**, *111*, 18–27. [CrossRef] [PubMed]
5. Møller, A.P.; Hagiwara, A.; Matsui, S.; Kasahara, S.; Kawatsu, K.; Nishiumi, I.; Suzuki, H.; Mousseau, T.A. Abundance of birds in Fukushima as judges from Chernobyl. *Environ. Pollut.* **2012**, *164*, 36–39. [CrossRef]
6. Bonisoli-Alquati, A.; Koyama, K.; Tedeschi, D.J.; Kitamura, W.; Sukuzi, H.; Ostermiller, S.; Arai, E.; Møller, A.P.; Mousseau, T.A. Abundance and genetic damage of barn swallows from Fukushima. *Sci. Rep.* **2015**, *5*, 9432. [CrossRef]
7. Murase, K.; Murase, J.; Horie, R.; Endo, K. Effects of the Fukushima Daiichi nuclear accident on goshawk reproduction. *Sci. Rep.* **2015**, *5*, 9405. [CrossRef] [PubMed]
8. Hayama, S.; Tsuchiya, M.; Ochiaki, K.; Nakiri, S.; Nakanishi, S.; Ishii, N.; Kato, T.; Tanaka, A.; Konno, F.; Kawamoto, Y.; et al. Small head size and delayed body weight growth in wild Japanese monkey fetuses after the Fukushima Daiichi nuclear disaster. *Sci. Rep.* **2017**, *7*, 3528. [CrossRef] [PubMed]

9. Ochiai, K.; Hayama, S.; Nakiri, S.; Nakanishi, S.; Ishii, N.; Uno, T.; Kato, T.; Konno, F.; Kawamoto, Y.; Tsuchida, S.; et al. Low blood cell counts in wild Japanese monkeys after the Fukushima Daiichi nuclear disaster. *Sci. Rep.* **2014**, *4*, 5793. [CrossRef]
10. Urushihara, Y.; Suzuki, T.; Shimizu, Y.; Ohtaki, M.; Kuwahara, Y.; Suzuki, M.; Uno, T.; Fujita, S.; Saito, A.; Yamashiro, H.; et al. Haematological analysis of Japanese macaques (*Macaca fuscata*) in the area affected by the Fukushima Daiichi Nuclear Power Plant accident. *Sci. Rep.* **2018**, *8*, 16748. [CrossRef]
11. Horiguchi, T.; Yoshii, H.; Mizuno, S.; Shiraishi, H. Decline in intertidal biota after the 2011 Great East Japan Earthquake and Tsunami and the Fukushima nuclear disaster: Field observations. *Sci. Rep.* **2016**, *6*, 20416. [CrossRef]
12. Akimoto, S. Morphological abnormalities in gall-forming aphids in a radiation-contaminated area near Fukushima Daiichi: Selective impact of fallout? *Ecol. Evol.* **2014**, *4*, 355–369. [CrossRef]
13. Akimoto, S.I.; Li, Y.; Imanaka, T.; Sato, H.; Ishida, K. Effects of radiation from contaminated soil and moss in Fukushima on embryogenesis and egg hatching of the aphid *Prociphilus oriens*. *J. Hered.* **2018**, *109*, 199–205. [CrossRef] [PubMed]
14. Ohmori, Y.; Kajikawa, M.; Nishida, S.; Tanaka, N.; Kobayashi, N.I.; Tanoi, K.; Furukawa, J.; Fujiwara, T. The effect of fertilization on cesium concentration of rice grown in a paddy field in Fukushima Prefecture in 2011 and 2012. *J. Plant Res.* **2014**, *127*, 67–71. [CrossRef] [PubMed]
15. Hayashi, G.; Shibato, J.; Imanaka, T.; Cho, K.; Kubo, A.; Kikuchi, S.; Satoh, K.; Kimura, S.; Ozawa, S.; Fukutani, S.; et al. Unraveling low-level gamma radiation-responsive changes in expression of early and late genes in leaves of rice seedlings at Iitate Village, Fukushima. *J. Hered.* **2014**, *105*, 723–738. [CrossRef] [PubMed]
16. Watanabe, Y.; Ichikawa, S.; Kubota, M.; Hoshino, J.; Kubota, Y.; Maruyama, K.; Fuma, S.; Kawaguchi, I.; Yoschenko, V.I.; Yoshida, S. Morphological defects in native Japanese fir trees around the Fukushima Daiichi Nuclear Power Plant. *Sci. Rep.* **2015**, *5*, 13232. [CrossRef]
17. Yoschenko, V.; Nanba, K.; Yoshida, S.; Watanabe, Y.; Takase, T.; Sato, N.; Keitoku, K. Morphological abnormalities in Japanese red pine (*Pinus densiflora*) at the territories contaminated as a result of the accident at Fukushima Dai-ichi Nuclear Power Plant. *J. Environ. Radioact.* **2016**, *165*, 60–67. [CrossRef]
18. Rakwal, R.; Hayashi, G.; Shibato, J.; Deepak, S.A.; Gundimeda, S.; Simha, U.; Padmanaban, A.; Gupta, R.; Han, S.; Kim, S.T.; et al. Progress toward rice seed OMICS in low-level gamma radiation environment in Iitate Village, Fukushima. *J. Hered.* **2018**, *109*, 2089–2211. [CrossRef]
19. Hiyama, A.; Nohara, C.; Kinjo, S.; Taira, W.; Gima, S.; Tanahara, A.; Otaki, J.M. The biological impacts of the Fukushima nuclear accident on the pale grass blue butterfly. *Sci. Rep.* **2012**, *2*, 570. [CrossRef]
20. Hiyama, A.; Nohara, C.; Taira, W.; Kinjo, S.; Iwata, M.; Otaki, J.M. The Fukushima nuclear accident and the pale grass blue butterfly: Evaluating biological effects of long-term low-dose exposures. *BMC Evol. Biol.* **2013**, *13*, 168. [CrossRef]
21. Taira, W.; Nohara, C.; Hiyama, A.; Otaki, J.M. Fukushima's biological impacts: The case of the pale grass blue butterfly. *J. Hered.* **2014**, *105*, 710–722. [CrossRef]
22. Nohara, C.; Hiyama, A.; Taira, W.; Tanahara, A.; Otaki, J.M. The biological impacts of ingested radioactive materials on the pale grass blue butterfly. *Sci. Rep.* **2014**, *4*, 4946. [CrossRef]
23. Nohara, C.; Taira, W.; Hiyama, A.; Tanahara, A.; Takatsuji, T.; Otaki, J.M. Ingestion of radioactively contaminated diets for two generations in the pale grass blue butterfly. *BMC Evol. Biol.* **2014**, *14*, 193. [CrossRef]
24. Taira, W.; Hiyama, A.; Nohara, C.; Sakauchi, K.; Otaki, J.M. Ingestional and transgenerational effects of the Fukushima nuclear accident on the pale grass blue butterfly. *J. Radiat. Res.* **2015**, *56*, i2–i18. [CrossRef]
25. Hiyama, A.; Taira, W.; Nohara, C.; Iwasaki, M.; Kinjo, S.; Iwata, M.; Otaki, J.M. Spatiotemporal abnormality dynamics of the pale grass blue butterfly: Three years of monitoring (2011–2013) after the Fukushima nuclear accident. *BMC Evol. Biol.* **2015**, *15*, 15. [CrossRef]
26. Taira, W.; Iwasaki, M.; Otaki, J.M. Body size distributions of the pale grass blue butterfly in Japan: Size rules and the status of the Fukushima population. *Sci. Rep.* **2015**, *5*, 12351. [CrossRef] [PubMed]
27. Otaki, J.M. Fukushima's lessons from the blue butterfly: A risk assessment of the human living environment in the post-Fukushima era. *Integr. Environ. Assess. Manag.* **2016**, *12*, 667–672. [CrossRef] [PubMed]
28. Hiyama, A.; Taira, W.; Iwasaki, M.; Sakauchi, K.; Gurung, R.; Otaki, J.M. Geographical distribution of morphological abnormalities and wing color pattern modifications of the pale grass blue butterfly in northeastern Japan. *Entomol. Sci.* **2017**, *20*, 100–110. [CrossRef]
29. Hiyama, A.; Taira, W.; Iwasaki, M.; Sakauchi, K.; Iwata, M.; Otaki, J.M. Morphological abnormality rate of the pale grass blue butterfly *Zizeeria maha* (Lepidoptera: Lycaenidae) in southwestern Japan: A reference data set for environmental monitoring. *J. Asia Pac. Entomol.* **2017**, *20*, 1333–1339. [CrossRef]
30. Nohara, C.; Hiyama, A.; Taira, W.; Otaki, J.M. Robustness and radiation resistance of the pale grass blue butterfly from radioactively contaminated areas: A possible case of adaptive evolution. *J. Hered.* **2018**, *109*, 188–198. [CrossRef]
31. Otaki, J.M.; Taira, W. Current status of the blue butterfly in Fukushima research. *J. Hered.* **2018**, *109*, 178–187. [CrossRef] [PubMed]
32. Otaki, J.M. Understanding low-dose exposure and field effects to resolve the field-laboratory paradox: Multifaceted biological effects from the Fukushima nuclear accident. In *New Trends in Nuclear Science*; Awwad, N.S., AlFaify, S.A., Eds.; IntechOpen: London, UK, 2018; pp. 49–71. [CrossRef]
33. Gurung, R.D.; Taira, W.; Sakauchi, K.; Iwata, M.; Hiyama, A.; Otaki, J.M. Tolerance of high oral doses of nonradioactive and radioactive caesium chloride in the pale grass blue butterfly *Zizeeria maha*. *Insects* **2019**, *10*, 290. [CrossRef]

34. Hancock, S.; Vo, N.T.K.; Omar-Nazir, L.; Batlle, J.V.I.; Otaki, J.M.; Hiyama, A.; Byun, S.H.; Seymour, C.B.; Mothersill, C. Transgenerational effects of historic radiation dose in pale grass blue butterflies around Fukushima following the Fukushima Dai-ichi Nuclear Power Plant meltdown accident. *Environ. Res.* **2019**, *168*, 230–240. [CrossRef]
35. Sakauchi, K.; Taira, W.; Toki, M.; Iraha, Y.; Otaki, J.M. Overwintering states of the pale grass blue butterfly *Zizeeria maha* (Lepidoptera: Lycaenidae) at the time of the Fukushima nuclear accident in March 2011. *Insects* **2019**, *10*, 389. [CrossRef]
36. Sakauchi, K.; Taira, W.; Hiyama, A.; Imanaka, T.; Otaki, J.M. The pale grass blue butterfly in ex-evacuation zones 5.5 years after the Fukushima nuclear accident: Contributions of initial high-dose exposure to transgenerational effects. *J. Asia Pac. Entomol.* **2020**, *23*, 242–252. [CrossRef]
37. Otaki, J.M. The pale grass blue butterfly as an indicator for the biological effect of the Fukushima Daiichi Nuclear Power Plant accident. In *Low-Dose Radiation Effects on Animals and Ecosystems*; Fukumoto, M., Ed.; Springer: Singapore, 2020; pp. 239–247. [CrossRef]
38. Sakauchi, K.; Taira, W.; Toki, M.; Tsuhako, M.; Umetsu, K.; Otaki, J.M. Nutrient imbalance of the host plant for larvae of the pale grass blue butterfly may mediate the field effect of low-dose radiation exposure in Fukushima: Dose-dependent changes in the sodium content. *Insects* **2021**, *12*, 149. [CrossRef] [PubMed]
39. Otaki, J.M.; Hiyama, A.; Iwata, M.; Kudo, T. Phenotypic plasticity in the range-margin population of the lycaenid butterfly *Zizeeria maha*. *BMC Evol. Biol.* **2010**, *10*, 252. [CrossRef]
40. Hiyama, A.; Taira, W.; Sakauchi, K.; Otaki, J.M. Sampling efficiency of the pale grass blue butterfly *Zizeeria maha* (Lepidoptera: Lycaenidae): A versatile indicator species for environmental risk assessment in Japan. *J. Asia Pac. Entomol.* **2018**, *21*, 609–615. [CrossRef]
41. Hiyama, A.; Otaki, J.M. Dispersibility of the pale grass blue butterfly *Zizeeria maha* (Lepidoptera: Lycaenidae) revealed by one-individual tracking in the field: Quantitative comparisons between subspecies and between sexes. *Insects* **2020**, *11*, 122. [CrossRef]
42. Hiyama, A.; Iwata, M.; Otaki, J.M. Rearing the pale grass blue *Zizeeria maha* (Lepidoptera, Lycaenidae): Toward the establishment of a lycaenid model system for butterfly physiology and genetics. *Entomol. Sci.* **2010**, *13*, 293–302. [CrossRef]
43. Iwata, M.; Hiyama, A.; Otaki, J.M. System-dependent regulations of colour-pattern development: A mutagenesis study of the pale grass blue butterfly. *Sci. Rep.* **2013**, *3*, 2379. [CrossRef]
44. Iwata, M.; Taira, W.; Hiyama, A.; Otaki, J.M. The lycaenid central symmetry system: Color pattern analysis of the pale grass blue butterfly *Zizeeria maha*. *Zool. Sci.* **2015**, *32*, 233–239. [CrossRef]
45. Hirata, K.; Otaki, J.M. Real-time in vivo imaging of the developing pupal wing tissues in the pale grass blue butterfly *Zizeria maha*: Establishing the lycaenid system for multiscale bioimaging. *J. Imaging* **2019**, *5*, 42. [CrossRef]
46. Coy, S.L.; Cheema, A.K.; Tyburski, J.B.; Laiakis, E.C.; Collins, S.P.; Fornace Jr, A.J. Radiation metabolomics and its potential in biodosimetry. *Int. J. Radiat. Biol.* **2011**, *87*, 802–823. [CrossRef]
47. Goudarzi, M.; Weber, W.; Mak, T.D.; Chung, J.; Doyle-Eisele, M.; Melo, D.; Brenner, D.J.; Guilmette, R.A.; Fornace, A.J. Development of urinary biomarkers for internal exposure by cesium-137 using a metabolomics approach in mice. *Radiat. Res.* **2014**, *181*, 54–64. [CrossRef]
48. Goudarzi, M.; Mak, T.D.; Chen, C.; Smilenov, L.B.; Brenner, D.J.; Fornace, A.J. The effect of low dose rate on metabolomic response to radiation in mice. *Radiat. Environ. Biophys.* **2014**, *53*, 645–657. [CrossRef] [PubMed]
49. Menon, S.S.; Uppal, M.; Randhawa, S.; Cheema, M.S.; Aghdam, N.; Usala, R.L.; Ghosh, S.P.; Cheem, A.K.; Dritschilo, A. Radiation metabolomics: Current status and future directions. *Front. Oncol.* **2016**, *6*, 20. [CrossRef] [PubMed]
50. Pannkuk, E.L.; Fornace Jr, A.J.; Laiakis, E.C. Metabolomic applications in radiation biodosimetry: Exploring radiation effects through small molecules. *Int. J. Radiat. Biol.* **2017**, *93*, 1151–1176. [CrossRef] [PubMed]
51. Suzuki, B.; Tanaka, S.; Nishikawa, K.; Yoshida, C.; Yamada, T.; Abe, Y.; Fukuda, T.; Kobayashi, J.; Hayashi, G.; Suzuki, M.; et al. Transgenerational effects on calf spermatogenesis and metabolome associated with paternal exposure to the Fukushima Nuclear Power Plant accident. In *Low-Dose Radiation Effects on Animals and Ecosystems*; Fukumoto, M., Ed.; Springer: Singapore, 2020; pp. 125–138.
52. Sakurai, N.; Ara, T.; Enomoto, M.; Motegi, T.; Morishita, Y.; Kurabayashi, A.; Iijima, Y.; Ogata, Y.; Nakajima, D.; Suzuki, H.; et al. Tools and databases of the KOMICS web portal for preprocessing, mining, and dissemination of metabolomics data. *BioMed Res. Int.* **2014**, *2014*, 194812. [CrossRef]
53. Sakurai, N.; Shibata, D. Tools and databases for an integrated metabolite annotation environment for liquid chromatography-mass spectrometry-based untargeted metabolomics. *Carotenoid Sci.* **2017**, *22*, 16–22.
54. Sakurai, N.; Narise, T.; Sim, J.-S.; Lee, C.-M.; Ikeda, C.; Akimoto, N.; Kanaya, S. UC2 search: Using unique connectivity of uncharged compounds for metabolite annotation by database searching in mass-spectrometry-based metabolomics. *Bioinformatics* **2018**, *34*, 698–700. [CrossRef]
55. Afendi, F.M.; Okada, T.; Yamazaki, M.; Hirai-Morita, A.; Nakamura, Y.; Nakamura, K.; Ikeda, S.; Takahashi, H.; Altaf-Ul-Amin, M.; Darusman, L.K.; et al. KNApSAcK family databases: Integrated metabolite-plant species databases for multifaceted plant research. *Plant Cell Physiol.* **2021**, *53*, e1. [CrossRef] [PubMed]
56. Sakurai, N.; Ara, T.; Kanaya, S.; Nakamura, Y.; Iijima, Y.; Enomoto, M.; Motegi, T.; Aoki, K.; Suzuki, H.; Shibata, D. An application of a relational database system for high-throughput prediction of elemental compositions from accurate mass values. *Bioinformatics* **2013**, *29*, 290–291. [CrossRef]

57. Xia, J.; Wishart, D.S. Web-based inference of biological patterns, functions and pathways from metabolomic data using MetaboAnalyst. *Nat. Protocol* **2011**, *6*, 743–760. [CrossRef]
58. Xia, J.; Sinelnikov, I.; Han, B.; Wishart, D.S. MetaboAnalyst 3.0—making metabolomics more meaningful. *Nucl. Acids Res.* **2015**, *43*, W251–W257. [CrossRef] [PubMed]
59. Pang, Z.; Chong, J.; Zhou, G.; de Lima Morais, D.A.; Chang, L.; Barrette, M.; Gauthier, C.; Jacques, P.É.; Li, S.; Xia, J. MetaboAnalyst 5.0: Narrowing the gap between raw spectra and functional insights. *Nucl. Acids Res.* **2021**, *49*, W388–W396. [CrossRef] [PubMed]
60. Wishart, D.S.; Tzur, D.; Knox, C.; Eisner, R.; Guo, A.C.; Young, N.; Cheng, D.; Jewell, K.; Arndt, D.; Sawhney, S.; et al. HMDB: The Human Metabolome Database. *Nucl. Acid Res.* **2007**, *35*, D521–D526. [CrossRef] [PubMed]
61. Wishart, D.S.; Feunang, Y.D.; Marcu, A.; Guo, A.C.; Liang, K.; Vázquez-Fresno, R.; Sajed, T.; Johnson, D.; Li, C.; Karu, N.; et al. HMDB 4.0: The human metabolome database for 2018. *Nucl. Acids Res.* **2018**, *46*, D608–D617. [CrossRef] [PubMed]
62. Yamaguchi, M.; Matsuyama, S.; Yamaji, K. Oxalic acid as a larval feeding stimulant for the pale grass blue butterfly *Zizeeria maha* (Lepidoptera: Lycaenidae). *Appl. Entomol. Zool.* **2016**, *51*, 91–98. [CrossRef]
63. Sakihama, Y.; Yamasaki, H. Phytochemical antioxidants: Past, present and future. In *Antioxidants—Benefits, Sources, Mechanisms of Action*; Waisundara, V., Ed.; IntechOpen: London, UK, 2021. [CrossRef]
64. Langer, S.Z. History and nomenclature of α_1-adrenoceptors. *Eur. Urol.* **1999**, *36*, 2–6. [CrossRef]
65. MacDonald, R.; Wilt, T.J. Alfuzosin for treatment of lower urinary tract symptoms compatible with benign prostatic hyperplasia: A systematic review of efficacy and adverse effects. *Urology* **2005**, *66*, 780–788. [CrossRef]
66. McKeage, K.; Plosker, G.L. Alfuzosin: A review of the therapeutic use of the prolonged-release formulation given once daily in the management of benign prostatic hyperplasia. *Drugs* **2002**, *62*, 633–653. [CrossRef] [PubMed]
67. Wilde, M.I.; Fitton, A.; McTavish, D. Alfuzosin. A review of its pharmacodynamic and pharmacokinetic properties, and therapeutic potential in benign prostatic hyperplasia. *Drugs* **1993**, *45*, 410–429. [CrossRef]
68. Rajashekar, Y.; Raghavendra, A.; Bakthavatsalam, N. *Acetylcholinesterase* inhibition by biofumigant (coumaran) from leaves of *Lantana camara* in stored grain and household insect pests. *BioMed Res. Int.* **2014**, *2014*, 187019. [CrossRef]
69. Engvild, K. Chlorine-containing natural compounds in higher plants. *Phytochemistry* **1986**, *25*, 781–791. [CrossRef]
70. Musa, A. Phytochemistry, pharmacological potency, and potential toxicity of Mycoporum spp. *Rec. Nat. Prod.* **2021**, *15*, 148–168. [CrossRef]
71. Sun, Y.-P.; Jin, W.-F.; Wang, Y.-Y.; Wang, G.; Morris-Natschke, S.; Liu, J.-S.; Wang, G.-K.; Lee, K.-H. Chemical structures and biological activities of limonoids from the genus *Swietenia* (Meliaceae). *Molecules* **2018**, *23*, 1588. [CrossRef] [PubMed]
72. Shigemori, H.; Sakurai, C.A.; Hosoyama, H.; Kobayashi, A.; Kajiyama, S.; Kobayashi, J. Taxezopidines J, K, and L, new toxoids from Taxus cuspidate inhibiting Ca^{2+}-induced depolymerization of microtubules. *Tetrahedron* **1999**, *55*, 2553–2558. [CrossRef]
73. Tomita, K.; Uenoyama, Y.; Fujisawa, K.; Kawaguchi, H. Oxamicetin, a new antibiotic of bacterial origin. III. Taxonomy of the oxamicetin-producing organism. *J. Antibiotics* **1973**, *26*, 765–770. [CrossRef] [PubMed]
74. Hasegawa, T.; Yamano, T.; Yoneda, M. *Streptomyces inusitatus* sp. nov. *Int. J. Syst. Bacteriol.* **1978**, *28*, 407–410. [CrossRef]
75. Jomon, K.; Kuroda, Y.; Ajisaka, M.; Sakai, H. A new antibiotic, ikarugamycin. *J. Antibiot.* **1972**, *25*, 271–280. [CrossRef]
76. Sekizawa, Y. A new antitumour substance lenamycin isolated and properties. *J. Biochem.* **1958**, *45*, 159–162. [CrossRef]
77. Mirocha, C.J.; Abbas, H.K.; Windels, C.E.; Xie, W. Variation in deoxynivalenol, 15-acetyldeoxynivalenol, 3-acetyldeoxynivalenol, and zearalenone production by *Fusarium graminearum* isolates. *Appl. Environ. Microbiol.* **1989**, *55*, 1315–1316. [CrossRef] [PubMed]
78. Hellwig, V.; Dasenbrock, J.; Gräf, C.; Kahner, L.; Schumann, S.; Steglich, W. Calopins and cyclocalopins—Bitter principles from *Boletus calopus* and related mushrooms. *Eur. J. Org. Chem.* **2002**, *2002*, 2895–2904. [CrossRef]
79. Djami-Tchatchou, A.T.; Ncube, E.N.; Steenkamp, P.A.; Dubery, I.A. Similar, but different: Structurally related azelaic acid and hexanoic acid trigger differential metabolomic and transcriptomic responses in tobacco cells. *BMC Plant Biol.* **2017**, *17*, 227. [CrossRef] [PubMed]
80. Dombrowski, J.; Martin, R.C. Green leaf volatiles, fire and nonanoic acid activate MAPkinases in the model grass species *Lolium temulentum*. *BMC Res. Notes* **2014**, *7*, 807. [CrossRef]
81. Walters, D.R.; Walker, R.L.; Walker, K.C. Lauric acid exhibits antifungal activity against plant pathogenic fungi. *J. Phytopathol.* **2003**, *151*, 228–230. [CrossRef]
82. Parthasarathy, A.; Savka, M.A.; Hudson, A.O. The synthesis and role of β-alanine in plants. *Front. Plant Sci.* **2019**, *10*, 921. [CrossRef]
83. Roeder, T.; Seifert, M.; Kähler, C.; Gewecke, M. Tyramine and octopamine: Antagonistic modulators of behavior and metabolism. *Arch. Insect Biochem. Physiol.* **2003**, *54*, 1–13. [CrossRef]
84. Roeder, T. Tyramine and octopamine: Ruling behavior and metabolism. *Annu. Rev. Entomol.* **2005**, *50*, 447–477. [CrossRef]
85. Selcho, M.; Pauls, D. Linking physiological processes and feeding behaviors by octopamine. *Curr. Opin. Insect Sci.* **2019**, *36*, 125–130. [CrossRef] [PubMed]
86. Roeder, T. The control of metabolic traits by octopamine and tyramine in invertebrates. *J. Exp. Biol.* **2020**, *223*, jeb194282. [CrossRef] [PubMed]

life

MDPI

Article

Efficacy of a Graphene Oxide/Chitosan Sponge for Removal of Radioactive Iodine-131 from Aqueous Solutions

Tanate Suksompong [1], **Sirikanjana Thongmee** [1] **and Wanwisa Sudprasert** [2,*]

[1] Department of Physics, Faculty of Science, Kasetsart University, Bangkok 10900, Thailand; tanate.su@ku.th (T.S.); fscisjn@ku.ac.th (S.T.)
[2] Department of Applied Radiation and Isotopes, Faculty of Science, Kasetsart University, Bangkok 10900, Thailand
* Correspondence: fsciwasu@ku.ac.th; Tel.: +668-19892902

Abstract: Iodine-131 is increasingly used for diagnostic and therapeutic applications. The excretion of radioactive iodine is primarily through the urine. The safe disposal of radioactive waste is an important component of overall hospital waste management. This study investigated the feasibility of using graphene oxide/chitosan (GO/CS) sponges as an adsorbent for the removal of iodine-131 from aqueous solutions. The adsorption efficiency was investigated using iodine-131 radioisotopes to confirm the results in conjunction with stable isotopes. The results revealed that the synthetic structure consists of randomly connected GO sheets without overlapping layers. The equilibrium adsorption data fitted well with the Langmuir model. The separation factor (R_L) value was in the range of 0–1, confirming the favorable uptake of the iodide on the GO/CS sponge. The maximum adsorption capacity of iodine-131 by GO/CS sponges was 0.263 MBq/mg. The highest removal efficiency was 92.6% at pH 7.2 ± 0.2. Due to its attractive characteristics, including its low cost, the ease of obtaining it, and its eco-friendly properties, the developed GO/CS sponge could be used as an alternative adsorbent for removing radioiodine from wastewater.

Keywords: graphene oxide/chitosan sponge; adsorbent; adsorption isotherm; iodine-131; radioactive waste management

Citation: Suksompong, T.; Thongmee, S.; Sudprasert, W. Efficacy of a Graphene Oxide/Chitosan Sponge for Removal of Radioactive Iodine-131 from Aqueous Solutions. *Life* **2021**, *11*, 721. https://doi.org/10.3390/life11070721

Academic Editors: Fabrizio Ambrosino and Supitcha Chanyotha

Received: 22 May 2021
Accepted: 19 July 2021
Published: 20 July 2021

1. Introduction

Iodine is one of the best-known elements because it is an essential mineral for the body. Iodine, which enters the body with food, binds to the amino acid tyrosine and forms a hormone called thyroxin in the follicular cells of the thyroid gland. On the other hand, if the body is exposed to radioactive iodine, such as iodine-131, it can be dangerous. A considerable body of evidence from research has shown the serious effects of iodine-131 exposure on the risk of thyroid cancer [1,2]. Recent research by Zupunski et al. [3] found a correlation between subjects exposed to iodine-131 from Chernobyl's fallout during childhood at age ≤18 years and thyroid cancer risk. This is of great concern, as iodine-131 is routinely administered as a radiopharmaceutical in nuclear medicine in the form of sodium iodide (NaI) for diagnosis and treatment of thyroid disease [4]. Approximately 80% of the administered iodine-131 is excreted in the urine [5]; therefore, it has potential to be discharged to sewage wastewater if not carefully controlled. Elevated concentrations of iodine-131 have been observed in municipal wastewater treatment plants following the application of therapeutic doses [6]. Rose and Swanson [7] found iodine-131 in sewage sludge from three water pollution control plants, ranging from 0.027 ± 0.002 to 148 ± 4 Bq/g dry weight.

The harmful effects of iodine-131 are gaining more attention around the world as an increase in iodine-131 contamination has been detected in the environment. Recently, iodine-131 has been detected in river water and aquatic organisms in South Korea. This indicates that radioactive iodine-131 can enter the environment through wastewater treatment plants through increased medical use [8]. Additionally, in 2020, Mosos et al. [9] reported the results of iodine-131 measurements at a wastewater treatment plant in Bogota, Colombia, and found the highest activity in raw water in the morning each day. The concentration of iodine-131 exceeds the reference value for drinking water and is close to the discharge limit in water bodies in Columbia. This indicates the risk of iodine-131 contamination of the environment from medical activities. The United States Environmental Protection Agency (USEPA) has established drinking water standards that specify no more than 4 millirems of beta radiation per year, which is equivalent to iodine-131 activity of not more than 3.0 pCi/L or 0.111 Bq/L [10].

The World Health Organization (WHO) has issued guidance on the level of iodine-131 that can be released into water bodies through sewage effluent: not more than 10 Bq/L [11]. In general, a waste collection system must be designed to store radioactive waste over a period of time. However, as the number of patients increases, many older facilities have limited space for waste storage. Therefore, they are unable to develop new systems, or these systems require large investments. With respect to this, if iodine-131-adsorbing materials can be developed, they will reduce the radioactivity of the radioactive waste released into the sewage waste drainage system. The use of absorbents can significantly reduce the volume of radioactive waste destined for temporary decay storage. This development is very beneficial for the safety of nearby personnel and the environment. In this research, we synthesized GO/CS sponges, a material with the potential to absorb or separate iodine-131 from water, which could be used in the future for radioactive waste management in hospitals.

2. Materials and Methods

2.1. Materials

Chitosan was obtained from Hunan Insen Biotech Co., Ltd in China (molecular weight: 10,000–20,000 Daltons). Graphite flakes were procured from Jiangsu XfNano materials Tech Co., Ltd, China, with a mean size of 100 mesh and a purity of 99.5% (model number XF051). Sulfuric acid (H_2SO_4, 95.0–98.0%) was obtained from Ajax Finechem (Australia); potassium permanganate ($KMnO_4$, ≥99.0%), phosphoric acid (H_3PO_4, ≥99.0%), sodium nitrate ($NaNO_3$, ≥99.0%), sodium hydroxide (NaOH, ≥99.0%), sodium iodide (NaI, ≥99.0%), and ethanol (C_2H_5OH, ≥99.9%) were obtained from Merck (Kenilworth, NJ, USA). Hydrochloric acid (HCl, ≥99.0%) and hydrogen peroxide (H_2O_2, 30%) were obtained from Sigma Aldrich (St. Louis, MO, USA).

2.2. Synthesis of Graphene Oxide Suspension

GO was prepared using a modified Hummer's method [12]. Briefly, graphite flakes (3.0 g) were mixed with $KMnO_4$ (9.0 g); then, the mixture of H_2SO_4 (360 mL) and H_3PO_4 (40 mL) was slowly added under continuous stirring in a water bath and stirred for 15 min. Next, 3.0 g of $NaNO_3$ was added to the solution, followed by 400 mL of deionized water. This mixture was magnetically stirred for 4 h. After that, 30 mL of H_2O_2 was added to the solution. The filtrate obtained was washed with 200 mL of HCl (10%), then 200 mL of ethanol was added and the suspension was stirred at room temperature for 60 h. The GO was washed with 5% HCl and DI water until a neutral pH was achieved. Subsequently, the suspension was kept without disturbance around 3–4 h until the GO settled on the bottom of the beaker and the volume of the suspension had decreased to 250 mL. The final concentration of the GO suspension was 12.0 mg/mL.

2.3. Preparation of GO/CS Sponges

CS was prepared by dissolving 6.0 g of chitosan in 294 mL of acetic acid (2.0%). The GO suspension and the CS solution were mixed at a ratio of 2:1. The obtained solution of GO/CS was sonicated for an additional 1 h and aged for 12 h to form a homogeneous GO/CS suspension [13]. The final concentration of the stock GO/CS suspension was 0.01467 mg/μL. Preparation of the GO/CS sponges was started by pouring the GO/CS suspension into a Teflon cup before freezing it at −20 °C, followed by freeze-drying (vacuum of 0.1 mbar for 24 h), then baking it at 60 °C for 48 h to eliminate moisture.

2.4. Characterization of GO/CS Sponges

The GO/CS sponge samples were characterized by different techniques. Scanning electron microscopy (SEM; Quanta model 450) was used to observe the surface morphology of the GO and the GO/CS suspension. Raman spectra were obtained with a DXR Raman microscope (aperture: 50 micrometer slit; laser spot size: approximately 3.1 micrometers; power: 10 mW) with a 532 nm laser. The functional groups on the adsorbent samples were investigated with a Fourier-transform infrared spectrometer (FT-IR; Perkin Elmer, Spectrum model).

2.5. Adsorption Experiments

The adsorption experiments were carried out by using the stable isotope of iodine-127 and the radioactive isotope of iodine-131. As the stable and radioactive iodine isotopes have similar chemical properties, the stable isotope was mainly used to perform all the adsorption experiments except for the adsorption ability test in order to avoid radiation exposure and radioactive contamination of the devices and instruments. In addition, the adsorption efficiency of GO/CS for removing iodine-131 was easily determined by radioactivity analysis using gamma spectrometry, rather than indirect analysis by measuring the iodine concentration with other instruments.

The adsorption capacity of the GO/CS sponge was determined using absorption techniques with a UV–vis spectrophotometer (SHIMADZU UV-2600). The iodide adsorption spectra were observed at 226 nm [14]. The calibration curve for the relationship between the concentration of the NaI solution (0.1 to 4.0 mg/L) and its absorbance was established. The adsorption capacity at various iodine concentration was determined with 2.0 mg of GO/CS sponges mixed in a 50 mL NaI solution at concentrations of 0.1–4 mg/L. Next, the mixture was agitated in a shaker at room temperature for 24 h. The GO/CS sponges were separated from the solution using glass microfiber filters with a 0.2 μm pore size. The concentration of iodide in the remaining solution was measured with a UV–vis spectrophotometer, and the equilibrium adsorption amounts, q_e (mg/g), were calculated via Equation (1) [15].

$$q_e = \frac{(C_i - C_t)V}{m} \quad (1)$$

where C_i (mg/L) and C_t (mg/L) are the initial sodium iodide concentration and the concentration at any time t (min), respectively; V (mL) is the volume of the sodium iodide solution; and m (g) is the mass of the CS/GO sponges.

The effects of the parameters, including adsorbent dosage, contact temperature, contact time, and pH, on the removal of iodide were studied. To explore the effects of adsorbent dosage, the GO/CS sponges (1.5–6.0 mg) were mixed in a 50 mL NaI solution (1.5 mg/L) for 2 h. The mixture was adjusted to pH 7.2 ± 0.2. Afterwards, the GO/CS sponges were filtered, and the residual concentration of the iodide in the supernatant was estimated spectrophotometrically using the established calibration curve. The adsorption capacity at equilibrium (q_e) was calculated from Equation (1). The adsorption capacity test at different temperatures was conducted by mixing 2.0 mg of GO/CS sponges with a 50 mL NaI solution (1.5 mg/L) at pH 7.2 ± 0.2 for 2 h. The mixing temperature was varied from 5 to 45 °C. The q_e at each temperature was determined as previously described. The effect of contact time on the adsorption capacity was studied by setting up the experiment as above and then varying the contact time from 5 min to 24 h at room temperature. The influence of the solution's pH on the adsorption capacity was evaluated by adjusting the pH of solution with HCl/NaOH solution to pH values of 4 to 9.

The adsorption isotherm experiment was conducted by adding 2.0 mg of CS/GO sponges into a 50 mL NaI solution at different concentrations (0.5, 1.0, 1.5, 2.0, and 4.0 mg/L). The adsorption data were fitted with the Langmuir and Freundlich models [16,17] as presented in Equations (2) and (3), respectively.

$$\text{Langmuir model}: \frac{C_e}{q_e} = \frac{1}{K_L\, q_{max}} + \frac{C_e}{q_{max}} \tag{2}$$

where C_e (mg/L) is the concentration of NaI at equilibrium, q_e (mg/g) is the amount of adsorbed sodium iodide on the surface of the GO/CS sponges at equilibrium, K_L (L/mg) is the Langmuir constant related to the adsorption capacity, and q_{max} (mg/g) is the maximum adsorption capacity of the GO/CS sponges.

$$\text{Freundlich model}: \log q_e = \log K_F + \left(\frac{1}{n}\right) logC_e \tag{3}$$

where K_F (mg/g) and $1/n$ are Freundlich constants representing the coefficient and intensity of adsorption, respectively.

The removal efficiency of iodine-131 by GO/CS was determined. A stock solution of iodine-131 in NaI with a radioactive concentration of 11.84 MBq/mL was obtained from the Thailand Institute of Nuclear Technology (Public Organization). It was diluted with deionized water until a final concentration of 140.78 kBq/mL was reached. The adsorption experiments were carried out in 50 mL plastic tubes to prevent the adsorption of radionuclides onto the glass wall. The experiment was conducted to explore the optimum equilibrium time by varying the incubation period from 0.5 to 48 h. The 30 mL iodine-131 solutions with 4.223 MBq of activity were added into each tube, which contained GO/CS sponges at the same concentration in each tube. The mixtures were then stirred for 10 min and allowed to equilibrate for 0.5 to 48 h at room temperature. The GO/CS suspension was filtered out by glass microfiber filters with a pore size of 0.2 µm. Subsequently, the radioactivity of the iodine-131 in the solution was measured with a gamma spectrometer (Canberra, model DSA 1000) equipped with a coaxial HPGe detector (model GR2519) 51.7 mm in diameter and 58.5 mm in length. This system delivered a resolution of 1.82 keV for 1332.5 keV gamma rays from cobalt-60, with an efficiency of 0.7011% for 364 keV gamma rays of iodine-131 calculated using the standard source (Eckert & Ziegler source No. 1868-30) with energies in the range of 88–1836 keV. Gamma spectrum analysis was performed using Genie 2000 software. The radioactivity of iodine-131 [18], given in Becquerel (Bq), was calculated using Equation (4):

$$\mathrm{A} = \frac{NC_e}{t\varepsilon pV} \tag{4}$$

where A is the activity concentration of iodine-131 (Bq/L), N is the number of counts in the photo-peak at an energy level of 364 keV, C_e is the true coincidence summing correction factor, t is the measurement time (s), ε is the detector's efficiency, p is the probability of disintegration of the radionuclide, and V is the volume of the solution (L).

The removal efficiency of iodine-131 by GO/CS was determined by Equation (5):

$$\text{Removal efficiency } (\%) = \left[\frac{A_0 - A_1}{A_0}\right] \times 100 \tag{5}$$

where A_0 is the initial radioactivity of iodine-131 before GO/CS adsorption and A_1 is the radioactivity of the resulting supernatant.

3. Results and Discussion

3.1. Characterization of GO/CS Sponges

GO/CS sponges with a proportion of 2:1 (v/v) were generated by a freeze-drying process at −20 °C. The shape was similar to the Teflon cup. Their physical characteristics were solid and light with a three-dimensional structure, as shown in Figure 1.

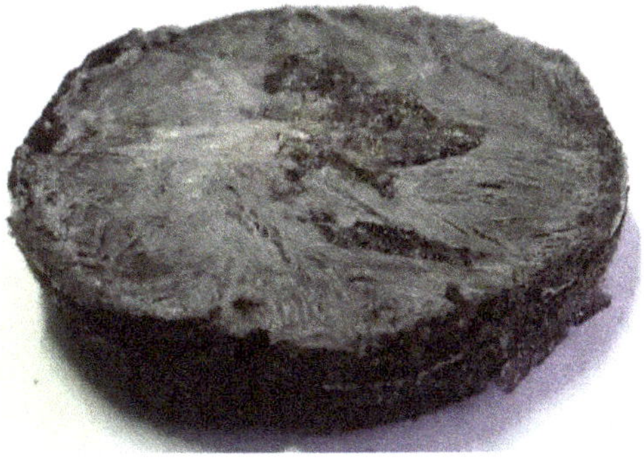

Figure 1. A GO/CS sponge after the freeze-drying process.

SEM images were used to determine the morphology of GO and GO/CS sponges. Figure 2a shows the GO, revealing that the surface of GO is generally smooth, whereas the edges of the GO/CS sponge are raised and bent, as seen in Figure 2b. This may be due to the different structure and hardness between CS and GO. It was proposed that the adequate dispersion of GO in CS was achieved through the interactions between the amino groups (-NH_2) in CS and the carboxylic groups (-COOH) in GO [19,20]. Moreover, Figure 2c shows that the GO sheets lined up randomly and the GO sheets connected together without any overlay, resulting in gaps. In Figure 2d, the gaps inside the GO/CS sponge's structure beneath the surface were caused by air that replaced the liquid during the freeze-drying process. This demonstrates that the internal structure of the GO/CS sponge was full of a GO sheet network.

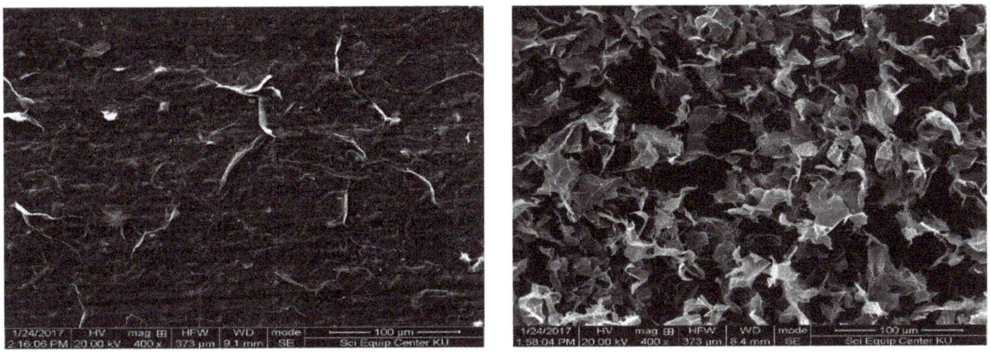

(a) (b)

Figure 2. *Cont.*

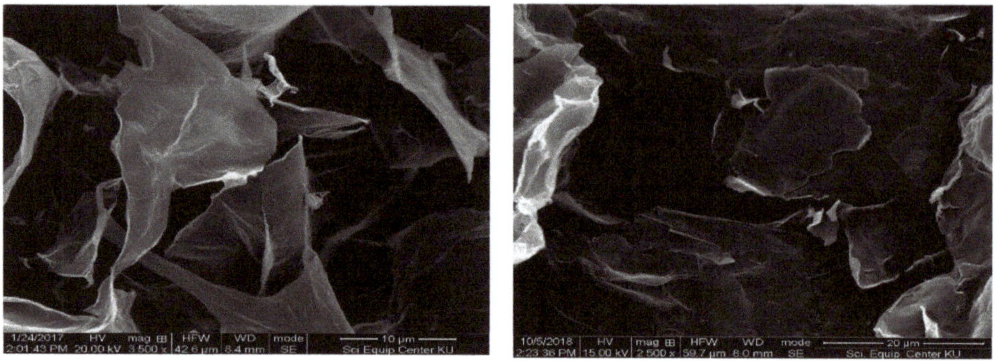

(c) (d)

Figure 2. SEM images of (**a**) GO and (**b**–**d**) GO/CS sponge at different magnifications.

To observe the ordered and disordered crystal structures of GO and GO/CS, Raman spectroscopy was used. Figure 3 shows the Raman spectra of GO and GO/CS sponges. We can see that the results of these two samples are similar. The Raman spectrum of GO shows two characteristic peaks at 1340 cm^{-1} (D-band, C-C) and 1588 cm^{-1} (G-band, C=C). The D-band is caused by the breathing modes of carbon atoms (six-atom rings) and the G-band corresponds to the high-frequency E_{2g} mode of sp^2-hybridized carbon atoms [21]. The D-band to G-band intensity ratio (I_D/I_G) for GO was 0.8032. For the GO/CS suspension, shifts at the D-bands (20 cm^{-1}) and G-bands (2 cm^{-1}) were observed compared with GO. The size of the I_D/I_G ratio of GO/CS (0.933) was slightly greater than the I_D/I_G ratio of GO (0.8032). This result suggests that the structure of the GO/CS suspensions was similar to that of ordered carbon nano-sheets [22].

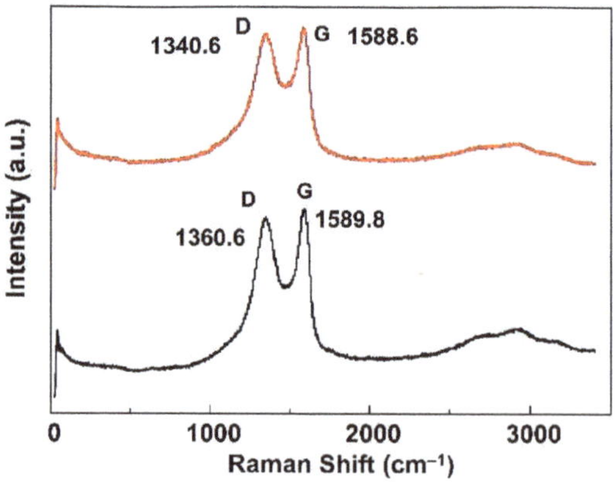

Figure 3. Raman spectra of GO and GO/CS sponges.

The functional groups of GO, CS, and GO/CS were investigated by FT-IR, as shown in Figure 4. The GO spectrum showed adsorption peaks that were assigned to the functional group of O-H bonds at 3416 cm^{-1}, the C=O stretches of the carboxylic group (1729 cm^{-1}), the stretching vibrations of aromatics (C=C) at 1624 cm^{-1}, epoxy C−O−C stretches at 1217 cm^{-1}, and alkoxy C−O bonds at 1054 cm^{-1}. Our results were similar to the results reported by Liu et al. [23]. The CS spectrum showed the main adsorption peak of the NH_2 group of CS at 1574 cm^{-1}. In the case of the GO/CS suspension, the FT-IR spectrum showed a new adsorption peak at 1638 cm^{-1} related to amide (N-H), whereas the C=O at

1729 cm^{-1} disappeared. This might have been caused by the NH_2 of CS reacting with the C=O of GO to form the functional group of N-H in the GO/CS suspension. Moreover, the C−O in the spectrum of GO/CS at 1073 cm^{-1} was less intense than that in the spectrum of GO at 1054 cm^{-1}. This is because the CS interacted with the OH groups of GO [24]. A diagram of the mechanism of the GO/CS suspension from the reaction between GO and CS is displayed in Figure 5. This reaction scheme explains that, when GO is mixed with the CS solution, electrostatic interactions are formed between the C=O group of GO and the N-H groups of CS, resulting in the stable GO/CS composite with improved elasticity [24].

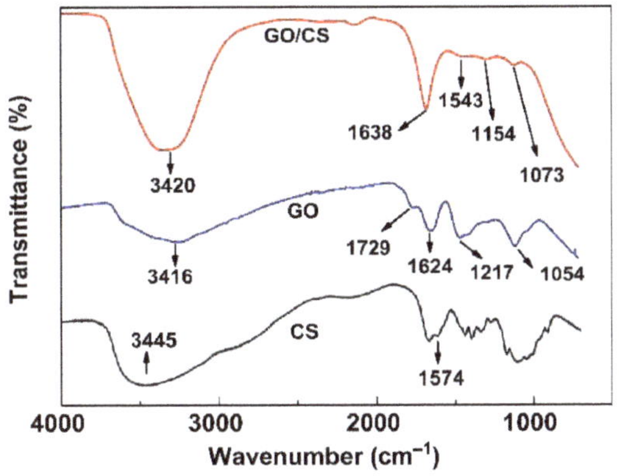

Figure 4. FT-IR spectra of GO, CS, and GO/CS.

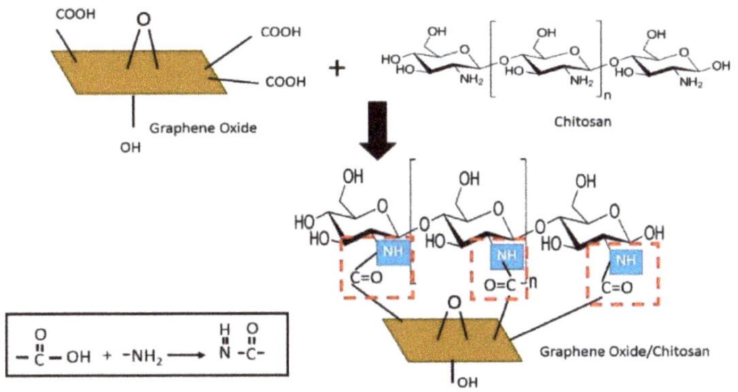

Figure 5. Proposed mechanism of the reaction between GO and CS to form the GO/CS suspension.

3.2. Adsorption Capacity

A series of experiments were conducted to determine the effect of adsorbent concentration, temperature, contact time, and pH on the adsorption rate. UV–vis spectroscopy was used to observe the concentration of NaI remaining in the supernatant after adsorption by the GO/CS sponges. The UV-vis absorption spectra of different NaI concentrations are shown in Figure 6a. The maximum absorption was found at a wavelength of 226 nm [25]. A calibration curve was established to determine the concentration of NaI derived from the absorbance by using the regression equation y = 0.7834x + 0.1373, as shown in Figure 6b.

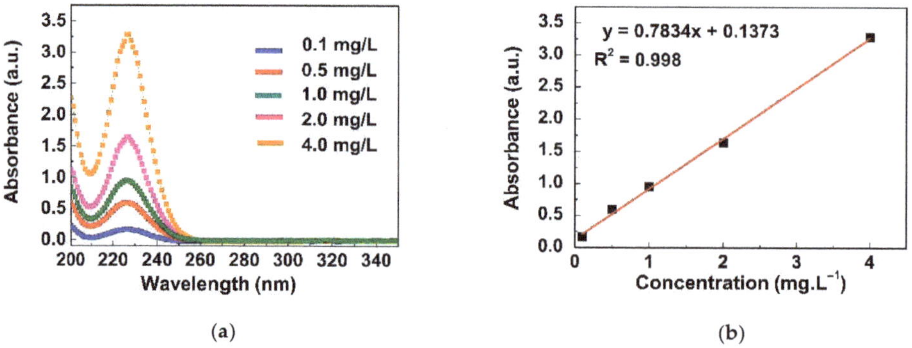

Figure 6. (**a**) UV–visible absorption spectra of NaI at concentrations of 0.1–4.0 mg/L at 25 °C and pH 7.2 ± 0.2; (**b**) calibration curve used for determination of the NaI concentration.

Figure 7 shows the adsorption capacity results for 2.0 mg of the GO/CS sponges mixed with the NaI solution at concentrations of 0.1–4 mg/L. The adsorption capacity at equilibrium (q_e) did not change in the 30 mg/g range, and the highest q_e was found at 30.52 mg/g with a concentration of 1.5 mg/L. This was probably due to the surface of the GO/CS sponge being limited and thus insufficient to adsorb NaI from the solution.

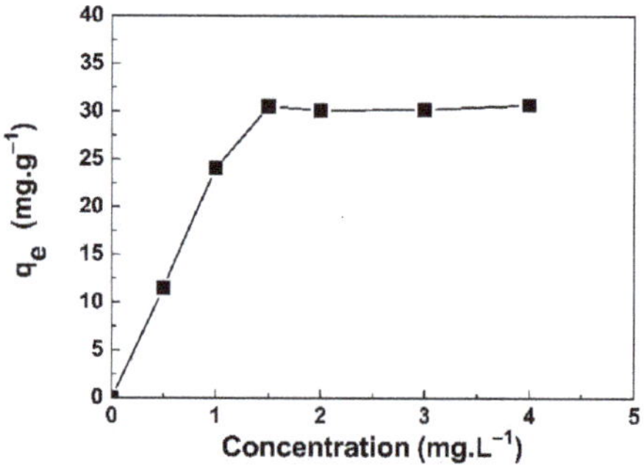

Figure 7. The NaI adsorption capacity (q_e) of 2.0 mg of GO/CS sponges with a contact time of 24 h at a pH of 7.2 ± 0.2 and room temperature.

The effect of adsorbent dosage on adsorption capacity was investigated in the range of 1.5–6.0 mg of GO/CS sponges. The q_e and removal percentage of NaI were plotted against dosage, as illustrated in Figure 8a,b, respectively. The q_e increased as the adsorbent dose increased from 1.5 to 2.0 mg, then decreased from 2.0 mg onward. The highest q_e was found at 30.17 mg/g, whereas the lowest q_e was found at 11.9 mg/g. The removal percentage increased rapidly from 45 to 95% at 1.5–2.5 mg, and it tended to be stable at approximately 95%.

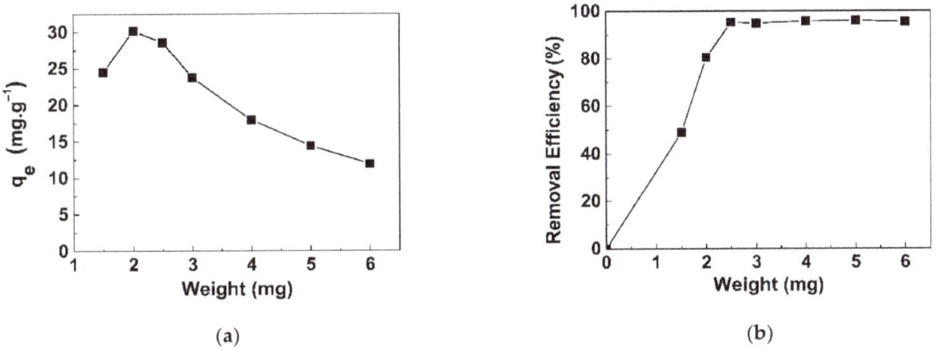

Figure 8. Effect of adsorbent dosage on NaI adsorption onto GO/CS sponges: (**a**) q_e and (**b**) removal percentage at a concentration of 1.5 mg/L at 25 °C, pH 7.2 ± 0.2.

The effect of temperature on adsorption capacity was investigated from 5 to 50 °C. The adsorption capacity increased with the increase in temperature from 5 to 35 °C then decreased until the temperature was 50 °C, as shown in Figure 9a. The highest removal was found at 95% (Figure 9b), which was equivalent to 35.3 mg/g at 35 °C. This observation might be explained by the iodide ions (I^-) moving and reacting with hydrogen when the temperature increased from 5 °C to 35 °C; after that, the adsorption decreased because the physical adsorption mechanism between the -NH_3^+ and I^- functional groups on the GO/CS sponge was an exothermic process. The higher the temperature, the lower the adsorption capacity (30.1 mg/g). These results, according to studies by Besemer et al. [26], suggest that iodide ions react with hydrogen on the surface of an adsorbent, resulting in two types of hydrogen bonds, namely O···H–O–H···I and I^-···H–O–H···I^-. For implementation, it might be more practical to use room temperature (25 °C) rather than 35 °C. This is because of the small difference in adsorption capacity found between these two temperatures.

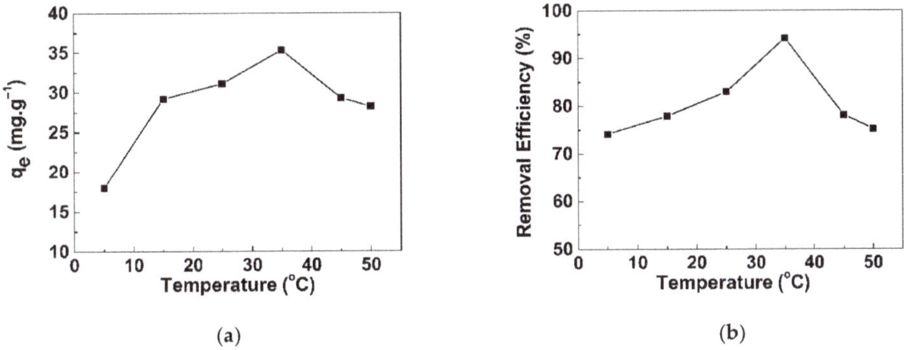

Figure 9. Effect of temperature on NaI adsorption onto GO/CS sponges: (**a**) q_e and (**b**) percentage removal at a concentration of 1.5 mg/L at pH 7.2 ± 0.2.

The effects of contact time from 5 to 180 min and from 5 min to 24 h on the adsorption capacity are shown in Figure 10a,b, respectively. The q_e rapidly increased during the first 30 min as the adsorbent still had a large surface, then it slowly increased until it became stable when the adsorption reached the saturation point. From 30 min to 24 h, the q_e was relatively stable in the range of 30.8–31.8 mg/g.

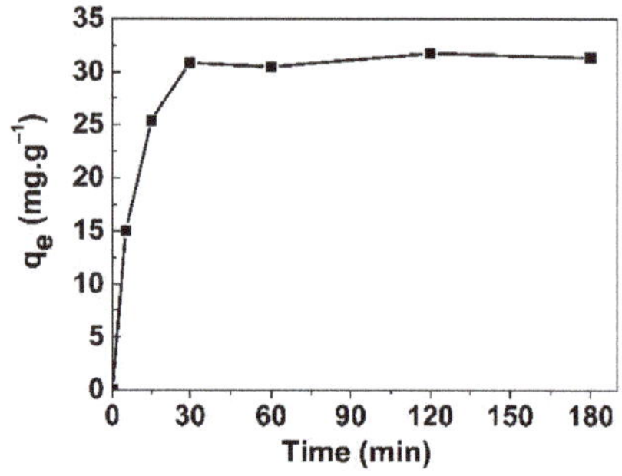

(a)

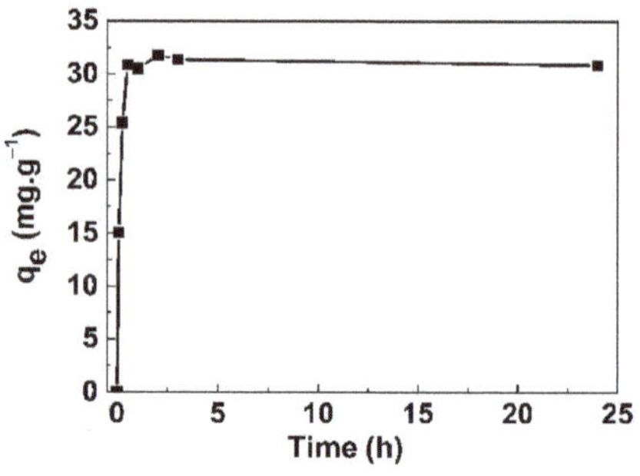

(b)

Figure 10. Effect of contact time on NaI adsorption onto GO/CS sponges: (**a**) 5 min to 180 min and (**b**) 5 min to 24 h.

To explore the effects of pH on the adsorption capacity, the pH of the mixing solution was varied from 3 to 12. Figure 11 shows the adsorption capacity of the GO/CS sponges for iodide ions at pH 3 to pH 12. The results show that the lowest adsorption was found at pH 3 and adsorption increased until pH 8. After that, the adsorption continuously decreased from pH 9 to pH 12. This may be because the capabilities of H^+ and I^- can be combined in the form of hydrogen iodide (HI). Normally, at a low pH [27], the -NH_2 group of CS will change to $-NH_3{}^+$, resulting in its binding to I^-. However, when GO was mixed with CS, the -NH_2 groups in CS bound with GO instead of I^-, and only GO could adsorb I^-, resulting in less adsorption at low pH. The maximum adsorption of I^- onto the GO/CS sponges was found at pH 7.2 (q_e = 32.1 mg/g), where I^- could create a bond between the positively charged surface of GO without the effect of H^+ and OH^-. When the pH value increased, the surface charge of GO was fundamentally negative; this may have led to the decrease in the adsorption capacity of GO/CS sponges [28].

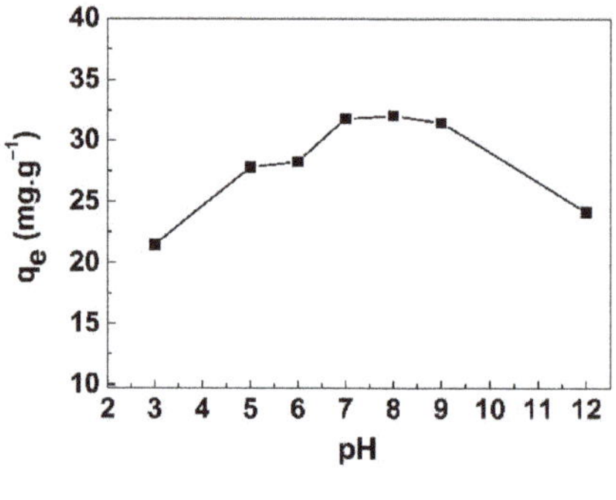

Figure 11. Iodide ion adsorption capacity of GO/CS sponges at pH 3 to pH 12.

Adsorption isotherms are very useful for demonstrating the amount of material adsorbed per unit of mass of the adsorbent as a function of the equilibrium concentration of the adsorbate. The Langmuir and Freundlich adsorption isotherms for iodide removal from aqueous solutions by the GO/CS sponges are shown in Figure 12a,b, respectively. In the case of the Langmuir isotherm model, the linear plot between $1/q_e$ and $1/C_e$ resulted in a slope of $1/(q_{max}{\cdot}K_L)$ and an intercept of $1/q_{max}$. For the Freundlich model, a graph between log C_e and log q_e resulted in a slope of $1/n$ and an intercept of log K_F [29]. The Langmuir and Freundlich isotherm parameters calculated from the slopes and intercepts of the linear plots are presented in Table 1. The maximum adsorption capacity was found to be 30.5 mg/g, and the Langmuir constant (K_L) was 0.0182 L/mg. In order to describe the affinity between the adsorbent and adsorbate, the Langmuir isotherm can be expressed by a dimensionless constant called the separation factor (R_L) [30], as shown in Equation (6):

$$R_L = \frac{1}{1 + K_L C_o} \tag{6}$$

where C_o is the initial concentration of iodide ions and K_L is the Langmuir constant. The R_L value indicates the shape of the isotherm if it is consistent with the adsorption. A value of $R_L > 1$ indicates unfavorable uptake, $R_L = 1$ indicates linear uptake, $0 < R_L < 1$ indicates favorable uptake, and $R_L = 0$ indicates irreversible uptake [31]. The R_L value (0.6295) observed in our study is in the range of 0–1, confirming the favorable uptake of the iodide by the GO/CS sponges. Furthermore, the correlation coefficients (R^2) of the Langmuir and Freundlich isotherms were found to be 0.9986 and 0.7645, respectively, demonstrating that the Langmuir isotherm completely conformed to the experimental data. This finding confirmed that the GO/CS sponges' structure was uniform monolayer GO sheets.

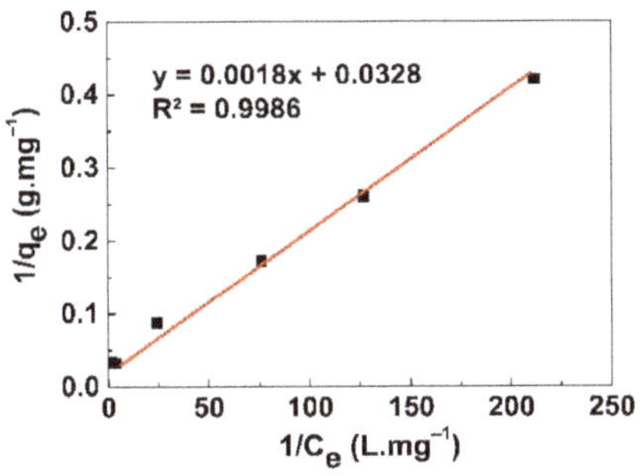

(a)

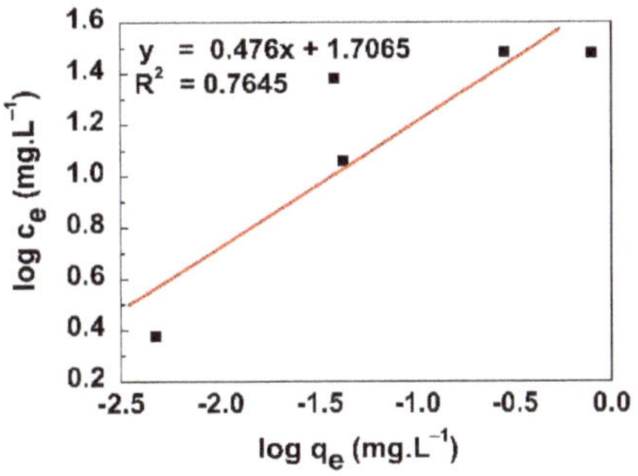

(b)

Figure 12. (**a**) Langmuir and (**b**) Freundlich adsorption isotherms for iodide removal at various concentrations.

Table 1. Langmuir and Freundlich isotherm parameters for adsorption of iodide ions from a solution by GO/CS sponges.

Langmuir Isotherm Model				Freundlich Isotherm Model		
q_{max} (mg/g)	K_L (L/mg)	R^2	R_L	K_F (mg/g)	1/n	R^2
30.4878	0.0182	0.9986	0.6295	50.8745	0.4760	0.7645

In light of the Langmuir isotherm parameters derived from our experiments, they can be used to determine the adsorption capacity of GO/CS sponges to remove radioactive iodine-131 from wastewater. As iodine-131 is classified as a carrier-free radioisotope, its specific activity can be determined by the equation $A = \lambda N$, where A is radioactivity (Bq), λ is the decay constant (where $\lambda = 0.693/T_{1/2}$), and N is the number of radioactive atoms [32]. The maximum adsorption for iodine-131 calculated from the Langmuir isotherm parameters is 1.1927×10^8 MBq/g.

In order to determine the efficiency of the synthesized GO/CS sponges for the removal of radioactive iodine from wastewater, adsorption experiments were conducted by varying the incubation time and the weight of the GO/CS sponges. The effects of the incubation period on adsorption capacity are shown in Figure 13. It is apparent that adsorption was rapid in the first 1 h but then slowed until reaching equilibrium. In the initial stage, the high removal rate was probably due to the rapid contact of sodium iodide molecules with the active sites on the external surfaces of the GO/CS sponge adsorbent. The subsequently constant rate might be attributed to the diminishing availability of the remaining active sites. However, a 24 h incubation period was chosen in our study to guarantee sufficient adsorption.

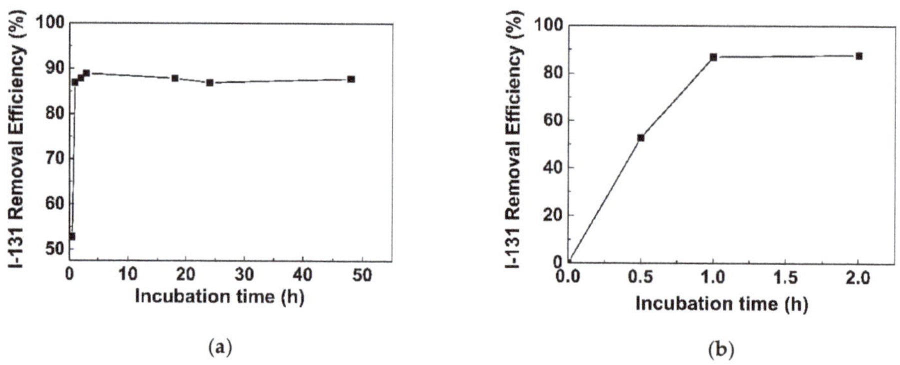

Figure 13. The removal efficiency of I-131 by GO/CS at different incubation times: (**a**) 0.5 h to 48 h and (**b**) 0.5 h to 2 h.

Figure 14 shows the removal efficiency of iodine-131 by GO/CS sponges calculated by Equations (4) and (5). The removal efficiency increased with an increase in the weight of the GO/CS sponges from 1.0 to 2.0 mg. After that, the removal efficiency was relatively constant. At a weight of 1.5 mg, the removal efficiency was 68.9%. The highest removal efficiency was found at 92.6% for the weight of 4.0 mg, which is equivalent to 0.263 MBq iodine-131 per milligram of adsorbent at pH 7.2 $\pm$ 0.2 for 24 h of adsorption. Based on the release of iodine-131 demonstrated by Larsen et al. [33], a treatment dose of ~3.8 GBq (102.6 mCi) of iodine-131 in a patient at a local hospital resulted in iodine-131 releases of approximately 3.2 GBq (87 mCi) entering the sewer system over a 40 h period. In comparison with those observations, the synthesized GO/CS demonstrated reasonable adsorbing capacity, i.e., 12.167 g of GO/CS was needed to remove 3.2 GBq. Since the cost of GO/CS is relatively cheap compared with various commercially available adsorbents, it would be possible to replace the costly adsorbents with the GO/CS sponges developed here.

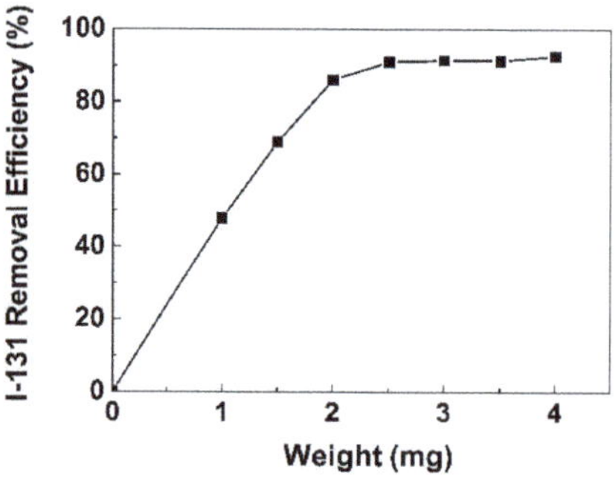

Figure 14. The removal efficiency of iodine-131 by GO/CS sponges at 24 h.

When the GO/CS sponges were compared with conventional adsorbents such as activated carbon, although GO/CS sponges are more difficult to synthesize than activated carbon, our study showed that the synthesized GO/CS sponges were quite effective. Iodine can be absorbed to the maximum adsorption capacity in as short as 1 h, whereas activated carbon has a very slow adsorption time of 48 h [34]. The distinguishing feature of the GO/CS sponges is that they are easily shaped; therefore, they are convenient for filtering or trapping iodine at specific points such as in hospital sewer systems. Another interesting point of the GO/CS sponges is their handling after iodine adsorption in the solution. They can be easily separated from the solution by means of a clamping device. Because of the physical nature of the GO/CS sponges, in which the graphene oxide sheets are connected to chitosan, they are not easily broken. Unlike activated carbon, which is a fine powder mixed with a solution, filtration is required to separate them. In addition, our study showed that the developed GO/CS sponges adsorbed iodine-131 at a high efficiency of 90% within 1 h without external energy use. This finding is interesting because radiation containers can be placed under radiation-shielding materials. This supports the best practices of working with radiation: ALARA ("As Low as Reasonably Achievable"), which means that all reasonable effort is needed to keep radiation exposure below the radiation dose limit as much as possible.

Among the many advantages of the GO/CS sponges, they can be shaped in a wide variety of containers used in the freeze-drying process and the sponges do not dissolve into the water; thus, the synthesized sorbent can be used in practical applications. For example, they may be added to a column to adsorb iodine-131 from pouring wastes, which can significantly reduce the amount of liquid radioactive waste that must be stored until being drained. These advantages not only reduce the cost of the waste storage system but also reduce storage space, as the secondary wastes arising from sorbents are several times smaller than liquid wastes. Due to the adsorption efficiency of the sorbent (12.167 g sorbent required to remove 3.12 GBq), the amount of solid secondary waste generated would be extremely low and can be stored in a small lead-shielded container until decay. We expect to explore the renewable performance of adsorbents further. This would greatly minimize the secondary waste generated. In addition, the sorbent will be particularly useful for increasing the number of patients receiving radioiodine therapy, as it can speed up the treatment time of contaminated liquid waste. Most importantly, the waste management system to be built for a new hospital does not have to be at a large scale that would normally be required for basic radioactive waste treatment principles: (1) concentrate and contain (2) delay and decay, and (3) dilute and disperse.

4. Conclusions

We successfully synthesized graphene oxide/chitosan (GO/CS) sponges as an adsorbent for the removal of iodine-131 from aqueous solutions. The adsorption isotherm experiment was consistent with Langmuir's isotherm model, which indicates that the synthetic structure consisted of randomly connected GO sheets without overlapping layers. Our study demonstrated that the synthesized GO/CS sponges could effectively adsorb radioactive iodine-131 in aqueous solution at pH 7.2 ± 0.2. The maximum adsorption capacity of iodine-131 by GO/CS was 0.263 MBq/mg. Based on the information of iodine-131 released from patients undergoing therapeutic application and entering a municipal sewer, the amount of GO/CS needed for radioactive iodide treatment is feasible. Due to its attractive characteristics, including the low cost, the ease of obtaining it, and its eco-friendly properties, the GO/CS developed here could be used as an alternative adsorbent for removing radioiodine from wastewater. It may be applied to adsorb liquid iodine-131 wastes from wards before releasing them into the sewage waste drainage system.

Author Contributions: Conceptualization, T.S., S.T. and W.S.; methodology, T.S. and W.S.; software, T.S.; validation, T.S., S.T. and W.S.; formal analysis, T.S.; investigation, T.S. and W.S.; resources, T.S.; data curation, T.S.; writing—original draft preparation, T.S. and W.S.; writing—review and editing, S.T. and W.S.; visualization, T.S.; supervision, S.T. and W.S.; project administration, S.T. and W.S.; funding acquisition, T.S. and W.S. All authors have read and agreed to the published version of the manuscript.

Funding: This work was partially funded by the Faculty of Science, Kasetsart University, Bangkok, Thailand.

Institutional Review Board Statement: Not applicable.

Informed Consent Statement: Not applicable.

Data Availability Statement: Not applicable.

Acknowledgments: The nuclear instruments were supplied by the Department of Applied Radiation and Isotope, Faculty of Science, Kasetsart University, Thailand. The sodium iodide ($Na^{131}I$) solutions were provided by the Thailand Institute of Nuclear Technology (Public Organization), Thailand.

Conflicts of Interest: The authors declare no conflict of interest.

References

1. Yamashita, S.; Suzuki, S. Risk of thyroid cancer after the Fukushima nuclear power plant accident. *Respir. Investig.* **2013**, *51*, 128–133. [CrossRef]
2. Tronko, M.D.; Howe, G.R.; Bogdanova, T.I.; Bouville, A.C.; Epstein, O.V.; Brill, A.B.; Likhtarev, I.A.; Fink, D.J.; Markov, V.V.; Greenebaum, E.; et al. A cohort study of thyroid cancer and other thyroid diseases after the chornobyl accident: Thyroid cancer in Ukraine detected during first screening. *J. Natl. Cancer Inst.* **2006**, *98*, 897–903. [CrossRef]
3. Zupunski, L.; Ostroumova, E.; Drozdovitch, V.; Veyalkin, I.; Ivanov, V.; Yamashita, S.; Cardis, E.; Kesminiene, A. Thyroid cancer after exposure to radioiodine in childhood and adolescence: ^{131}I-related risk and the role of Selected host and environmental factors. *Cancers* **2019**, *11*, 1481. [CrossRef]
4. Sondorp, L.H.J.; Ogundipe, V.M.L.; Groen, A.H.; Kelder, W.; Kemper, A.; Links, T.P.; Coppes, R.P.; Kruijff, S. Patient-derived papillary thyroid cancer organoids for radioactive iodine refractory screening. *Cancers* **2020**, *12*, 3212. [CrossRef] [PubMed]
5. Haghighatafshar, M.; Banani, A.; Zeinali-Rafsanjani, B.; Etemadi, Z.; Ghaedian, T. Impact of the amount of liquid intake on the dose rate of patients treated with radioiodine. *Indian J. Nucl. Med.* **2018**, *33*, 10–13. [PubMed]
6. Esparza, D.; Valiente, M.; Borràs, A.; Villar, M.; Leal, L.O.; Vega, F.; Cerdà, V.; Ferrer, L. Fast-response flow-based method for evaluating ^{131}I from biological and hospital waste samples exploiting liquid scintillation detection. *Talanta* **2020**, *206*, 120224. [CrossRef]
7. Rose, P.S.; Swanson, R.L. Iodine-131 in sewage sludge from a small water pollution control plant serving a thyroid cancer treatment facility. *Health Phys.* **2013**, *105*, 115–120. [CrossRef] [PubMed]
8. Lee, U.; Kim, M.J.; Kim, H.R. Radioactive iodine analysis in environmental samples around nuclear facilities and sewage treatment plants. *Nucl. Eng. Technol.* **2018**, *50*, 1355–1363. [CrossRef]
9. Mosos, F.; Velásquez, A.M.; Mora, E.T.; Tello, C.D. Determination of ^{131}I activity concentration and rate in main inflows and outflows of Salitre wastewater treatment plant (WWTP), Bogota. *J. Environ. Radioact.* **2020**, *225*, 106425. [CrossRef]
10. U.S. Environmental Protection Agency (EPA). *National Primary Drinking Water Standards, December 1999*; EPA: Washington, DC, USA, 2002. Available online: https://nepis.epa.gov/Exe/ZyPURL.cgi?Dockey=200024LJ.txt (accessed on 10 March 2021).
11. World Health Organization (WHO). *Guidelines for Drinking-Water Quality: Fourth Edition Incorporating the First Addendum*; WHO: Geneva, Switzerland, 2017; pp. 210–211.
12. Chen, J.; Yao, B.; Li, C.; Shi, G. An improved hummers method for eco-friendly synthesis of graphene oxide. *Carbon* **2013**, *64*, 225–229. [CrossRef]
13. Yadav, M.; Rhee, K.Y.; Park, S.J.; Hui, D. Mechanical properties of Fe_3O_4/GO/chitosan composites. *Compos. B. Eng.* **2014**, *66*, 89–96. [CrossRef]
14. Kazantseva, N.N.; Ernepesova, A.; Khodjamamedov, A.; Geldyev, O.A.; Krumgalz, B.S. Spectrophotometric analysis of iodide oxidation by chlorine in highly mineralized solutions. *Anal. Chim. Acta.* **2002**, *456*, 105–119. [CrossRef]
15. Vijayakumar, G.; Tamilarasan, R.; Dharmendirakumar, M. Adsorption, kinetic, equilibrium and thermodynamic studies on the removal of basic dye rhodamine-B from aqueous solution by the use of natural adsorbent perlite. *J. Mater. Environ. Sci.* **2012**, *3*, 157–170. [CrossRef]
16. Martín, F.S.; Kracht, W.; Vargas, T. Attachment of *Acidithiobacillus ferrooxidans* to pyrite in fresh and saline water and fitting to Langmuir and Freundlich isotherms. *Biotechnol. Lett.* **2020**, *42*, 957–964. [CrossRef]
17. Boparai, H.K.; Joseph, M.; O'Carroll, D.M. Kinetics and thermodynamics of cadmium ion removal by adsorption onto nano zerovalent iron particles. *J. Hazard. Mater.* **2011**, *186*, 458–465. [CrossRef]
18. He, L.C.; Diao, L.J.; Sun, B.H.; Zhua, L.H.; Zhao, J.W.; Wang, M.; Wang, K. Summing coincidence correction for γ-ray measurements using the HPGe detector with a low background shielding system. *Nucl. Instrum. Methods Phys. Res. A* **2018**, *880*, 22–27. [CrossRef]
19. Zhang, L.; Wang, Y.; Yang, X. Structure and properties of graphene oxide-chitosan composite membrane prepared by the plasma acid solvent. *J. Fiber Sci. Technol.* **2018**, *74*, 143–149. [CrossRef]
20. Gong, Y.; Yu, Y.; Kang, H.; Chen, X.; Liu, H.; Zhang, Y.; Sun, Y.; Song, H. Synthesis and characterization of graphene oxide/chitosan composite aerogels with high mechanical performance. *Polymers* **2019**, *11*, 777. [CrossRef]
21. Kim, S.G.; Park, O.K.; Lee, J.H.; Ku, B.C. Layer-by-layer assembled graphene oxide films and barrier properties of thermally reduced graphene oxide membranes. *Carbon Lett.* **2013**, *14*, 247–250. [CrossRef]
22. Yadav, M.; Ahmad, S. Montmorillonite/graphene oxide/chitosan composite: Synthesis, characterization and properties. *Int. J. Biol. Macromol.* **2015**, *79*, 923–933. [CrossRef]
23. Liu, S.; Ouyang, J.; Luo, J.; Sun, L.; Huang, G.; Ma, J. Removal of uranium (VI) from aqueous solution using graphene oxide functionalized with diethylenetriaminepentaacetic phenylenediamine. *J. Nucl. Sci. Technol.* **2018**, *55*, 781–791. [CrossRef]
24. Emadi, F.; Amini, A.; Gholami, A.; Ghasemi, Y. Functionalized graphene oxide with chitosan for protein nanocarriers to protect against enzymatic cleavage and retain collagenase activity. *Sci. Rep.* **2017**, *7*, 42258. [CrossRef]

25. Inglezakis, V.J.; Satayeva, A.; Yagofarova, A.; Tauanov, Z.; Meiramkulova, K.; Farrando-Pérez, J.; Bear, J.C. Surface interactions and mechanisms study on the removal of iodide from water by use of natural zeolite-based silver nanocomposites. *Nanomaterials* **2020**, *10*, 1156. [CrossRef]
26. Besemer, M.; Bloemenkamp, R.; Ariese, F.; Manen, H.J. Identification of multiple water–iodide species in concentrated NaI solutions based on the Raman bending vibration of water. *J. Phys. Chem. A* **2016**, *120*, 709–714. [CrossRef] [PubMed]
27. Kamal, M.A.; Bibi, S.; Bokhari, S.W.; Siddique, A.H.; Yasin, T. Synthesis and adsorptive characteristics of novel chitosan/graphene oxide nanocomposite for dye uptake. *React. Funct. Polym.* **2017**, *110*, 21–29. [CrossRef]
28. Guerrero-Fajardo, C.A.; Giraldo, L.; Moreno-Pirajan, J.C. Preparation and characterization of graphene oxide for Pb(II) and Zn(II) ions adsorption from aqueous solution: Experimental, thermodynamic and kinetic study. *Nanomaterials* **2020**, *10*, 1022. [CrossRef] [PubMed]
29. Ali, I.H.; Al Mesfer, M.K.; Khan, M.I.; Danish, M.; Alghamdi, M.M. Exploring adsorption process of lead (II) and chromium (VI) ions from aqueous solutions on acid activated carbon prepared from *Juniperus procera* leaves. *Processes* **2019**, *7*, 217. [CrossRef]
30. Ayawei, N.; Angaye, S.S.; Wankasi, D.; Dikio, E.D. Synthesis, characterization and application of Mg/Al layered double hydroxide for the degradation of congo red in aqueous solution. *Open J. Phys. Chem.* **2015**, *5*, 56–70. [CrossRef]
31. Oluwafemi, O.; Ojo, A.A.; Ezekiel, A.; Adebayo, O.L. Adsorptive removal of anionic dye from aqueous solutions by mixture of kaolin and bentonite clay: Characteristics, isotherm, kinetic and thermodynamic studies. *Iran. J. Energy Environ.* **2015**, *6*, 147–153. [CrossRef]
32. El-gammal, W.; Abu-Zied, H.; Nada, A.; Aki, Z.F. Age determination of Pu-bearing samples using gamma spectrometry for safeguards and nuclear forensics applications. *IOSR-JAP* **2020**, *12*, 8–13. [CrossRef]
33. Larsen, I.L.; Stetar, E.A.; Giles, B.G.; Garrison, B. Concentrations of iodine-131 released from a hospital into a municipal sewer. *RSO Mag.* **2001**, *6*, 13–18.
34. Bara, Y.; Nakao, T.; Inoue, K.; Nakamori, I. Adsorption mechanism of iodine on activated carbon. *Sep. Sci. Technol.* **1985**, *20*, 21–31. [CrossRef]

life

MDPI

Article

Radiological Risk to Human and Non-Human Biota Due to Radioactivity in Coastal Sand and Marine Sediments, Gulf of Oman

Ibrahim I. Suliman [1,*] and Khalid Alsafi [2]

[1] Department of Physics, College of Science, Imam Mohammad Ibn Saud Islamic University (IMSIU), Riyadh 11642, Saudi Arabia
[2] Department of Radiology, Medical Physics Unit, King Abdulaziz University, P.O. Box 80215, Jeddah 21589, Saudi Arabia; kalsafi@kau.edu.sa
* Correspondence: iiidris@imamu.edu.sa; Tel.: +966-5389-18127

Abstract: Natural and ^{137}Cs radioactivity in coastal marine sediment samples was measured using gamma spectrometry. Samples were collected at 16 locations from four beaches along the coastal area of Muscat City, Gulf of Oman. Radioactivity in beach sand was used to estimate the radiological risk parameters to humans, whereas the radioactivity in marine sediments was used to assess the radiological risk parameters to non-human biota, using the ERICA Tool. The average radioactivity concentrations (Bqkg^{-1}) of ^{226}Ra, ^{232}Th, ^{40}K, ^{210}Pb and ^{137}Cs in sediments (sand) were as follows: 16.2 (16.3), 34.5(27.8), 54.7 (45.6), 46.8 (44.9) and 0.08 (0.10), respectively. In sand samples, the estimated average indoor (D_{in}) and outdoor (D_{out}) air absorbed dose rates due to natural radioactivity were 49.26 and 27.4 and the total effective dose (AED_{Total}; μSvy^{-1}) ranged from 150.2 to 498.9 (average: 275.2). The measured radioactivity resulted in an excess lifetime cancer risk (ELCR) in the range of 58–203 (average: 111) in and an average gonadal dose (AGD; μGy.y^{-1}) ranged from 97.3 to 329.5 (average: 181.1). Total dose rate per marine organism ranged from 0.035 μGy h^{-1} (in zooplankton) to 0.564 μGy h^{-1} (in phytoplankton). The results showed marine sediments as an important source of radiation exposure to biota in the aquatic environment. Regular monitoring of radioactivity levels is vital for radiation risk confinement. The results provide an important radiological risk profile parameter to which future radioactivity levels in marine environments can be compared.

Keywords: radioactivity; gamma spectrometry; absorbed dose rates; radiation risk; aquatic environment; non-human biota

Citation: Suliman, I.I.; Alsafi, K. Radiological Risk to Human and Non-Human Biota Due to Radioactivity in Coastal Sand and Marine Sediments, Gulf of Oman. *Life* **2021**, *11*, 549. https://doi.org/10.3390/life11060549

Academic Editors: Fabrizio Ambrosino and Supitcha Chanyotha

Received: 3 May 2021
Accepted: 7 June 2021
Published: 11 June 2021

1. Introduction

Radioactivity naturally exists in the environment in different conditions such as soil, underground water, marine sediment, and biota. Radioactivity enters the marine environment through different pathways, including via river and rainwater transport into the sea; however, this is often due to nuclear waste disposal, which is discharged from nuclear power plants as well as from medical, industrial, research, and educational uses of radionuclides [1–3]. Sources of marine radioactivity are numerous: uranium isotopes are present in large amounts in seas and oceans; thorium in water is hydrolysed and attached to particle surfaces, and is thus not as soluble in water; and ^{226}Ra and ^{40}K are highly soluble in water. On the other hand, ^{210}Pb enters the atmosphere via ^{222}Rn diffusion and in rainfall. ^{137}Cs in the environment poses radiation protection concerns given its high yield, long half-life, and significant uptake and retention in biological organisms. The principal sources of ^{137}Cs released in the environment have included atmospheric nuclear weapons testing and releases during nuclear reactor accidents [4].

Regarding the radiation risk to non-human biota, the concerning biological effects include those that could lead to changes in population size or structure. Among these endpoints are early mortality, some forms of morbidity, impairment of reproductive capacity, and the induction of chromosomal damage. Therefore, it is necessary to estimate the doses received and then compare such data with the nearest relevant data for reference organisms to evaluate the likely radiation effects for such organisms in an environmental context [5].

Assessing radiation exposure among humans requires a better understanding of the radionuclide's behaviour in pertinent environments [6–8]. Thus, the primary aims of nearly all marine radioactivity studies have been to form a scientific foundation upon which to determine the radiological risk of radioisotopes in marine environments. This is an enormously important issue that is in alignment with the present radiation protection standards [6,7]. Considering the importance of the subject, several radioactivity studies were performed in the marine environments of Gulf countries [9–13]. The results revealed a high degree of variability in radioactivity levels and emphasised the importance of such studies from a radiation protection standpoint. In Oman, not much work has been done to explore environmental and marine radioactivity. In fact, the only study we found was carried out by Salih, who studies radioactivity levels in marine environments [14]. In these studies, the radiological risk to non-human biota was not covered.

Thus, we sought to assess radiation exposure, radiological hazards, and attributed cancer risk from naturally occurring radioactivity found in marine sediments along the coastal area of the Gulf of Oman. This is important, as oceans and seas are directly impacted and ultimately serve as a sink of radioactivity and other contaminants, as they link diverse geographical areas with one another, representing a major source of marine pollution.

2. Materials and Methods

2.1. Study Location

The study was performed in the Muscat principality in Oman (23.5859° N, 58.4059° E) which had a population of approximately 1.28 million in 2015. Figure 1 shows the map of Oman, which highlights Muscat city and the four sample locations. These are Manuma Beach (A), Seeb Beach (B), Aziba Beach (C), and Qurum Beach (D), which are the four major beaches. These beaches are important sightseeing destinations and are among the most abundantly frequented in the city, especially in summer. Thus, it is essential that radioactivity levels are studied to determine the extent of the associated radiological hazards.

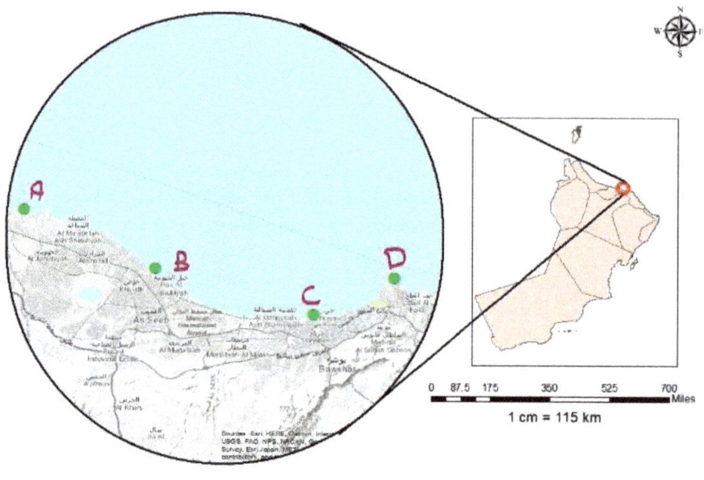

Figure 1. Map of Oman showing Muscat city, Gulf of Oman and the four sample locations: Manuma Beach (**A**); Seeb Beach (**B**); Aziba Beach (**C**); and Qurum Beach (**D**).

For the coastal sands, each sample was taken at a depth of 5 cm at an average interval of 200 m between two locations. Samples of marine sediment were taken from the area covered by sea water at about 20 m from the beach to provide a fair representation of the area's geological and sediment characteristics, which are the greatest determinants of the types of radionuclides present. The samples were taken to the laboratory at the Medical Physics Department, Sultan Qaboos University, where they were dried for 24 h in an oven set at 80 °C. To better estimate the specific activity of radium, samples were tightly sealed in Marinelli beakers and left for 4 weeks to achieve equilibrium.

2.2. Radioactivity Measurements

Spectrometric measurements were performed using a p-type high purity germanium (HPGe) detector, with a relative efficiency of 40% to that of NaI spectrometry (ORTEC, Oak Ridge, TN, USA). Gamma Vision-32 software was used for spectrum analysis (ORTEC, Oak Ridge, TN, USA). Energy and efficiency calibrations were performed before the measurements were taken using a standard mixture of sources from the International Atomic Energy Agency (IAEA). Background measurements were performed without a sample in place, and these measurements were subsequently subtracted from the measured activity concentrations.

The ^{226}Ra activity was estimated from the ^{214}Pb and ^{214}Bi radionuclide activities measured directly determined from their gamma-ray energy lines 351.92 keV and 609.31 keV, respectively. The activity of ^{232}Th was estimated from the ^{212}Bi, ^{212}Pb, and ^{228}Ac radionuclide activities measured directly from their gamma-ray energy lines 727.17, 238.63, and 911.60 keV, respectively. The activity concentrations of ^{40}K, ^{210}Pb, and ^{137}Cs were measured directly using their gamma ray lines 1460.81, 46.5, and 662 keV, respectively [4]. Using these parameters, the specific activity (A) of a given radionuclide in the sample was determined as follows:

$$A = \frac{N}{PE \cdot \varepsilon \cdot Tc \cdot M \cdot k} \tag{1}$$

where M is the mass of the sample in kg, N is the sample net area in the peak range, PE is the gamma emission probability, Tc is the counting time, and ε is the photo peak efficiency [7]. k is the product of all correction factors ($k = k_1 \cdot k_2 \cdot k_3 \cdot k_4 \cdot k_5$); where k_1, k_2, k_3, k_4, and k_5 are correction factors to account for the radionuclide decay, the nuclide decay during counting, self-attenuation, pulses loss due to random summing, and the coincidence, respectively [15,16].

$$\frac{u(A)}{A} = \sqrt{\left(\frac{u(N)}{N}\right)^2 + \left(\frac{u(P_E)}{P_E}\right)^2 + \left(\frac{u(\varepsilon)}{\varepsilon}\right)^2 + \left(\frac{u(T_C)}{T_C}\right)^2 + \left(\frac{u(M)}{M}\right)^2 + \left(\frac{u(k)}{k}\right)^2} \tag{2}$$

where $\frac{u(N)}{N}$, $\frac{u(P_E)}{P_E}$, $\frac{u(\varepsilon)}{\varepsilon}$, $\frac{u(T_C)}{T_C}$, $\frac{u(M)}{M}$ and $\frac{u(k)}{k}$ are the relative uncertainties of the counting rate, gamma emission probability, photo peak efficiency, counting time, sample mass and correction factors, respectively. The standard uncertainty in the correction factors is determined as: ($\frac{u(k)}{k} = \sqrt{\left(\frac{u(k_1)}{k_1}\right)^2 + \left(\frac{u(k_2)}{k_2}\right)^2 + \left(\frac{u(k_3)}{k_3}\right)^2 + \left(\frac{u(k_4)}{k_4}\right)^2 + \left(\frac{u(k_5)}{k_5}\right)^2}$).

The standard uncertainties in activity measurements for ^{40}K, ^{210}Pb and ^{137}Cs radionuclides were used for the determination of the expanded uncertainty. For ^{226}Ra and ^{232}Th which are determined from other radionuclides, the combined uncertainty was determined as the square root of the quadratic sum of the relative standard uncertainties of respective radionuclides as shown in Equations (3) and (4) [15,16].

$$\frac{u(A_{Ra-226})}{A_{Ra-226}} = \sqrt{\left(\frac{u(A_{Pb-214})}{A_{Pb-214}}\right)^2 + \left(\frac{u(A_{Bi-214})}{A_{Bi-214}}\right)^2} \tag{3}$$

$$\frac{u(A_{Ra-226})}{A_{Ra-226}} = \sqrt{\left(\frac{u(A_{Pb-212})}{A_{Pb-212}}\right)^2 + \left(\frac{u(A_{Bi-212})}{A_{Bi-212}}\right)^2 + \left(\frac{u(A_{Ac-228})}{A_{Ac-228}}\right)^2} \tag{4}$$

Standard uncertainty in activity concentrations, as shown in Equation (2), was determined using software. The overall uncertainties in the measurement results were quoted as expanded uncertainty at 95% confidence level with coverage factor (k = 2) [15].

3. Results and Discussion

3.1. Radioactivity Contents in Marine Sediment

This study presents an effort to assess the magnitude of environmental and artificial radionuclides in marine environments. The specific activity (Bqkg^{-1}) of natural radionuclides ^{226}Ra, ^{232}Th, ^{40}K, and ^{210}Pb in coastal marine sands and sediments in the Gulf of Oman are presented in Table 1.

Table 1. Radioactivity concentrations (Bqkg^{-1}) of ^{226}Ra, ^{232}Th, ^{40}K, ^{210}Pb, and ^{137}Cs in coastal marine sands and sediments.

Sample Code	Weight (kg)	Activity Concentrations (Bqkg^{-1})				
		^{226}Ra	^{232}Th	^{40}K	^{210}Pb	^{137}Cs
			Beach Sand			
S01	1421	21.5 ± 1.4	28.0 ± 2.8	78.7 ± 4.8	**	0.11
S02	1371	24.8 ± 1.2	54.9 ± 3.1	74.4 ± 3.2	42.7 ± 19.4	0.05
S05	1592	14.3 ± 1.0	29.3 ± 2.2	29.5 ± 2.0	(125.5 ± 12.2)	0.04
S06	1568	14.3 ± 1.0	10.4 ± 1.1	30.6 ± 2.1	24.7 ± 11.4	0.07
S09	1501	13.6 ± 0.9	43.3 ± 3.3	56.9 ± 3.7	67.4 ± 13.5	0.19
S10	1414	9.3 ± 0.7	30.0 ± 2.6	29.0 ± 2.1	**	0.13
S13	1038	15.6 ± 1.0	13.2 ± 1.3	32.0 ± 2.1	**	0.07
S14	1376	17.0 ± 1.1	13.6 ± 1.3	34.0 ± 2.3	**	0.15
Average		16.30	27.84	45.64	44.9	0.10
			Marine sediments			
S03	1363	21.0 ±1.0	50.4 ± 3.0	93.9 ± 3.8	44.9 ± 1.9	0.05
S04	1425	19.4 ± 1.4	47.3 ± 3.8	93.4 ± 5.7	**	0.07
S07	1532	12.8 ± 0.8	31.6 ± 1.2	39.6 ± 2.3	(158.9 ± 12.9)	0.07
S08	1750	11.5 ± 0.8	31.0 ± 2.4	33.5 ± 2.6	42.9 ± 12.5	0.09
S11	1496	13.8 ± 0.8	22.4 ± 2.2	30.8 ± 2.8	53.0 ± 11.4	0.13
S12	1421	13.2 ± 1.0	28.0 ± 1.8	42.9 ± 2.0	**	0.08
S15	1332	17.7 ± 1.4	32.6 ± 2.6	42.2 ± 3.9	65.1 ± 14.2	0.09
S16	1423	20.4 ± 1.2	32.5 ± 1.6	61.3 ± 2.7	28.1 ± 13.3	0.08
Average		16.2	34.5	54.7	46.8	0.08

The results in the brackets ( ) are outliers and are excluded from the average; ** indicate that the activity is less than the minimum detectable activity (MDA).

As shown, the radioactivity (Bqkg^{-1}) of ^{226}Ra, ^{232}Th, ^{40}K and ^{210}Pb ranges were 9.324.8 (average: 16.3), 10.4–54.9 (average: 27.8), 29.0–78.7 (average: 45.6), and 24.7–67.4 (average: 44.9) Bqkg^{-1}, respectively, in coastal sands, and from 11.0–21.0 (average: 16.2), 22.0–50.4 (average: 34.5), 30.8–93.4 (average: 54.7), and 28.1–65.1Bqkg^{-1} (average: 46.8), respectively, in marine sediments. Figures 2 and 3 show the boxplot distribution of the radioactivity distribution of ^{226}Ra, ^{232}Th, ^{40}K, and ^{210}Pb radionuclides in sand and sediment, respectively. A large variability in activity concentrations is shown among radionuclides, reflecting the geological and morphological characteristics of the collected sediments, as well as their respective radionuclide contents. A high degree of variability in the measured radioactivity was shown in the studied samples, as these samples reflect the geological characteristics of their sites of origin. Usually, the radioactivity of ^{238}U and ^{232}Th is linked with heavy minerals, while that of ^{40}K is associated with clay minerals.

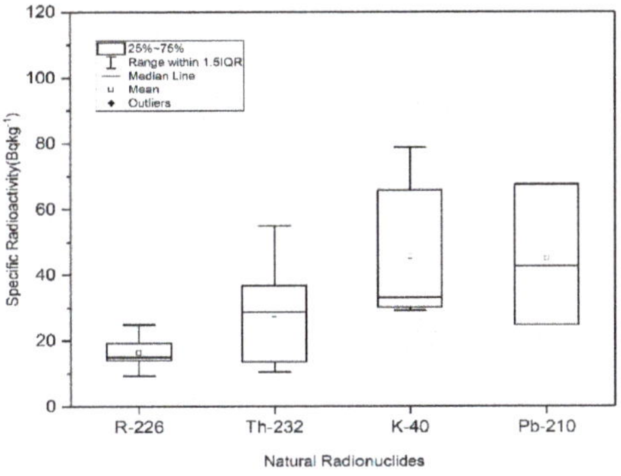

Figure 2. Boxplot of distributions of the radioactivity concentrations for ^{226}Ra, ^{232}Th, ^{40}K, and ^{210}Pb (natural radionuclides) in marine coastal sands.

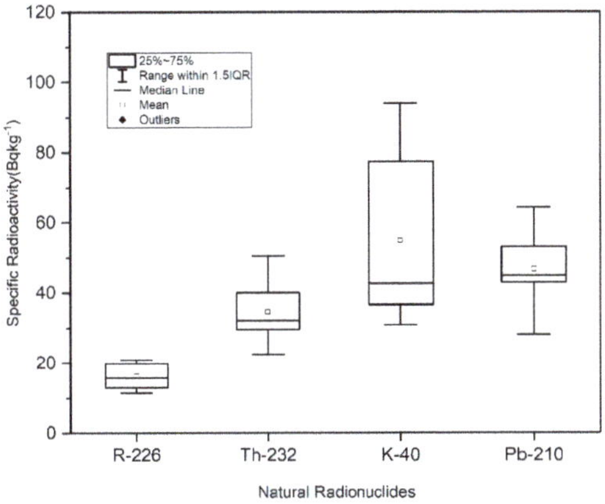

Figure 3. Boxplot illustrating the distributions of radioactivity concentrations for ^{226}Ra, ^{232}Th, ^{40}K, and ^{210}Pb natural radionuclides in marine sediments.

The radioisotope of ^{210}Pb revealed relatively high activity concentration (especially in the S05 and S07 sampling locations), considering a potential different origin than that of the local mineralogy, which could be due to submarine groundwater discharge sources in these areas.

In Table 2, a comparison is given of the average (range) radioactivity concentrations (Bqkg^{-1}) obtained in this work versus in the literature. According to the IAEA, when the activity of the ^{238}U or ^{232}Th decay series is $\leq$1000 Bqkg^{-1}and that of ^{40}K is $\leq$10,000 Bqkg^{-1}, the radioactive material may not be regarded as naturally occurring and is thus exempt from regulations [14]. The measured ^{210}Pb in marine sediment originated from their parents ^{222}Rn and ^{226}Ra, which depend on several natural and environmental processes [17]. The radioactivity of ^{210}Pb in the present work is comparable to values reported in the literature [9–13,17–19].

Table 2. Comparison of average (range) radioactivity concentrations ($Bqkg^{-1}$) obtained in this work versus those in the literature.

Location	^{226}Ra	^{232}Th	^{40}K	^{210}Pb	^{137}Cs	References
World	35	30	400	**	**	[2]
Qatar	4.2–19.5	1.0–6.0	11–188	**	0.18–0.66	[9]
Kuwait	17.3–20.5	15–16.4	353–445	23.6–44.3	1.0–3.1	[10]
Iran	11.8–22.7	10.7–25	223–535	**	0.14–2.8	[11]
Saudi Arabia	4.4–19.3	5.3–58.9	324.6–1133	**	0.6–8.7	[12]
Kuwait	18.6–21.4	14.0–17.1	351.2–404.0	**	1.5–2.9	[13]
Greece	18–86	20–31	368–610	47–105	0.7–3.8	[17]
China	13.7–52.	26.1–71.9	392–898	**	**	[19]
Egypt	38.51	**	33.35	659.18	**	[20]
Oman	16.2 (16.3)	34.5 (27.8)	54.7 (45.6)	46.8 (44.9)	0.1 (0.1)	This stud

** indicate that the activity is less than the minimum detectable activity (MDA).

Figure 4 shows a bar chart illustrating the distribution of activity concentrations of ^{137}Cs among different samples. As shown, the radioactivity concentration of the artificial radionuclide ^{137}Cs varied from 0.04–0.19 $Bqkg^{-1}$ (average: 0.09). The ^{137}Cs radioactivity levels in the current study were very low in most samples, suggestive of a low level of contamination. ^{137}Cs in marine and other environments may have radiological impacts given its long half live, high yield, and high uptake and retention in biological systems. The results of our study were compared with the results of similar studies reported in different countries around the world (Table 2). As shown, the radioactivity levels in our study are comparable to those reported in Kuwait, Qatar, Saudi Arabia, and Greece, and can be explained by the fact that these studies were carried out in adjacent marine environments in which these radionuclides could be easily transported.

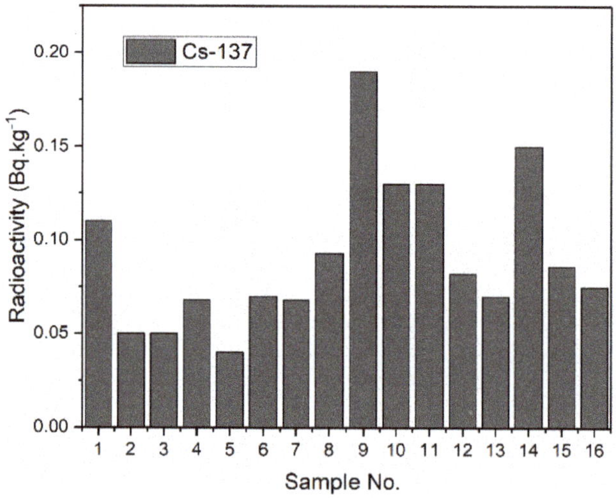

Figure 4. Bar charts of the radioactivity concentration of the artificial radionuclide, ^{137}Cs.

Correlations between the activity concentrations of the radionuclides are presented in Table 3. The results are graphically depicted in Figure 5, showing the correlations between the activity concentrations of ^{226}Ra and ^{232}Th. Figure 6 shows the correlations between the activity concentrations of the naturally occurring radionuclides ^{226}Ra and ^{40}K are illustrated using corresponding colour codes.

Table 3. Correlation between activity concentrations of natural radionuclides.

Radionuclide	Statistics	^{226}Ra	^{232}Th	^{40}K
^{226}Ra	Correlation coefficient	1	0.47	0.77
	p-value	-	0.07	<0.001
^{232}Th	Correlation coefficient	0.47	1	0.75
	p-value	0.07	-	<0.001
^{40}K	Correlation coefficient	0.77	0.75	-
	p-value	<0.001	<0.001	<0.001

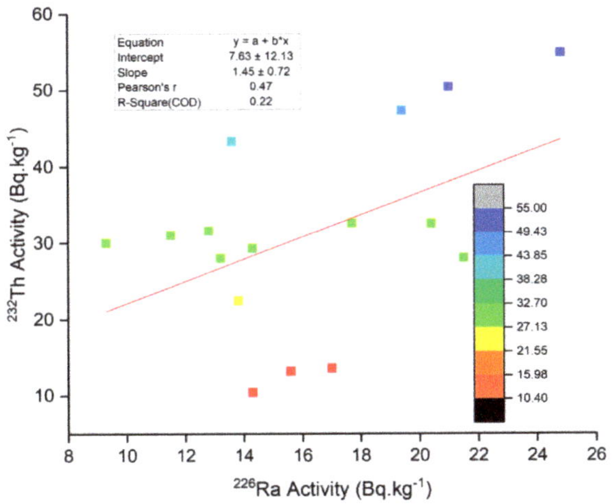

Figure 5. Correlation between the activity concentrations of ^{226}Ra and ^{232}Th (natural radionuclides); the colour codes correspond with ^{232}Th activity.

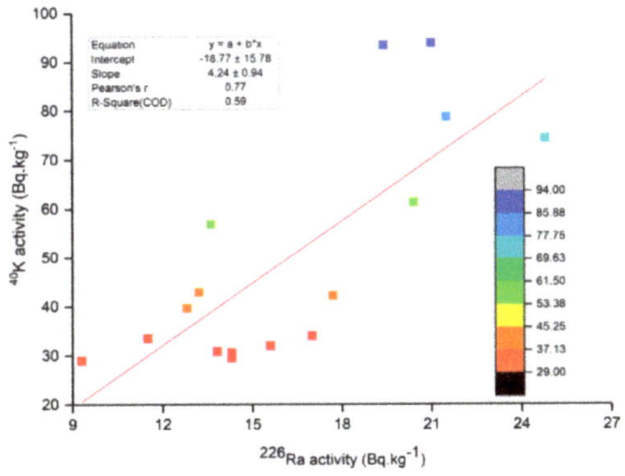

Figure 6. Correlation between the activity concentrations of ^{226}Ra and ^{40}K (natural radionuclides); the colour codes correspond with ^{40}K activity.

The correlation between ^{226}Ra and ^{232}Th was not significant ($r = 0.47$, $p > 0.05$), whereas a highly significant correlation was observed between ^{226}Ra and ^{40}K ($r = 0.77$, $p < 0.001$) and between ^{232}Th and ^{40}K ($r = 0.75$, $p < 0.001$). The observed correlations could be attributed to the origin of these radionuclides.

3.2. Assessing Radiological Hazards

3.2.1. Radium-Equivalent Activity

Radium-equivalent activity (Ra_{eq}) is a single parameter that represents the collective risk of ^{226}Ra, ^{232}Th, and ^{40}K radioactivity [21,22]. This parameter can be used to assess whether external doses to the public exceed the recommended annual dose limit of 1 mSv. The Ra_{eq} is determined using Equation (5):

$$Ra_{eq} = A_{Ra} + 1.43A_{th} + 0.077A_{K} \quad (5)$$

where A_{Ra}, A_{Th}, and A_k are the specific activities (Bq kg^{-1}) of ^{226}Ra, ^{232}Th, and ^{40}K, respectively, in the studied samples. Table 4 shows the radium equivalent activity (Ra_{eq}), absorbed dose rates, effective rates, and external hazard index associated with the radioactivity in sand. As presented, the Ra_{eq} ranged from 31.5 to 109.0 $Bqkg^{-1}$ (average: 59.6), with values < 370 $Bqkg^{-1}$ representing the recommended limit for radiological risk control [2].

Table 4. Radium-equivalent activity, absorbed dose rates, effective rates, excessive cancer risk, annual gonadal dose and external hazard index (H_{ex}) associated with the radioactivity in coastal sand.

Sample Code	R_{aq} ($Bqkg^{-1}$)	Dose Rate ($nGy.h^{-1}$)		AED_{Total} (μSvy^{-1})	ELCR per 10^{-6}	AGD $\mu Gy.y^{-1}$	H_{ex}
		D_{in}	D_{out}				
S01	67.6 ± 5.7	56.9	31.1	317.2	126	208.2	0.18
S02	109.0 ± 4.6	89.2	50.2	498.9	203	329.5	0.30
S05	58.5 ± 3.1	47.7	26.8	267.1	108	175.9	0.16
S06	31.5 ± 2.6	27.0	14.3	150.2	58	97.3	0.09
S09	79.5 ± 5.0	64.7	36.9	362.6	149	240.9	0.22
S10	54.4 ± 3.4	43.9	25.1	246.0	101	163.2	0.15
S13	36.9 ± 2.7	31.4	16.8	174.8	68	113.4	0.10
S14	39.1 ± 2.9	33.3	17.7	185.2	72	120.1	0.11
Average	59.56	49.26	27.4	275.2	111	181.1	0.16

Raq is the radium equivalent activity, D_{in} and D_{out} are the indoor and outdoor air absorbed, respectively. AED_{Total} is the total effective doses due to internal and external radiation exposure. ECR, excessive cancer risk. AGD ($\mu Gy.y^{-1}$), annual gonadal dose and Hex is the external hazard index.

3.2.2. External Hazard Index (H_{ex})

The external radiation exposure due to natural radioactivity is defined in terms of the external hazard index (H_{ex}), calculated as follows [2,23]:

$$H_{ex} = \left(\frac{A_{Ra}}{370} + \frac{A_{Th}}{259} + \frac{A_K}{4810}\right) \leq 1 \quad (6)$$

where A_{Ra}, A_{Th}, and A_k are the specific activities (Bq kg^{-1}) of ^{226}Ra, ^{232}Th, and ^{40}K, respectively in the studied samples. To comply with the requirements of the 1 mSv annual dose limit for the public, H_{ex} should be <1, as shown above [2]. As seen in Table 4, the H_{ex} values ranged from 0.09 to 0.30. These results ensure that the public's exposure to the environmental radioactivity of ^{226}Ra, ^{232}Th, and ^{40}K radionuclides in coastal sand remain within acceptable limits.

3.3. External Absorbed Dose Rates

The naturally occurring radioactivity in the environment is a major source of external exposure to the world's population. The indoor (D_{in}) and outdoor (D_{out}) external gamma doses due to the presence of ^{226}Ra, ^{232}Th, and ^{40}K in coastal sand 1 m above the ground surface can be calculated as follows [2,23]:

$$D_{out}\left(nGyh^{-1}\right) = 0.427A_{Ra} + 0.662A_{Th} + 0.043A_{K} \quad (7)$$

$$D_{in}\left(nGyh^{-1}\right) = 0.92A_{Ra} + 1.1A_{Th} + 0.081A_{K} \quad (8)$$

where A_{Ra}, A_{Th}, and A_k are the specific activities (Bq kg^{-1}) of ^{226}Ra, ^{232}Th, and ^{40}K, respectively, in the studied samples. As shown in Table 4, the D_{in} values (nGy.h^{-1}) ranged from 27.0 to 89.2 (average: 49.26), whereas D_{out} values (nGy.h^{-1}) ranged from 14.3 to 50.2 (average: 27.4). The current average dose figures fell below the global average value (55 nGy.h^{-1}) for areas that were deemed to have normal levels of natural background radiation. Our results are lower than the doses reported by in Pakistan (87.47 nGy.h^{-1}) [24]. According to UNSCEAR reports, a conversion coefficient, absorbed dose to effective dose received by adults of 0.7 Sv/Gy, and an outdoor occupancy factor of 0.2 were used [2]. Thus, the annual effective radioactivity dose in coastal sand can be estimated according to the following equations:

$$AED_{out}\left(Svy^{-1}\right) = D_{out}\left(nGyh^{-1}\right) \times 8760hy^{-1} \times 0.2 \times 0.7SvGy^{-1} \times 10^{-3} \tag{9}$$

$$AED_{in}\left(Svy^{-1}\right) = D_{in}\left(nGyh^{-1}\right) \times 8760hy^{-1} \times 0.8 \times 0.7SvGy^{-1} \times 10^{-3} \tag{10}$$

$$AED_{total}\left(Svy^{-1}\right) = AED_{out} + AED_{in} \tag{11}$$

The total effective dose and AED_{Total} (μSvy^{-1}) ranged from 150.2 to 498.9 (average: 275.2) (Table 4). The global average annual effective dose from natural radionuclides (i.e., the sum of effective doses from both indoor and outdoor occupations) is 0.48 mSvy^{-1}. The results for individual countries generally fall within the range of 0.3–0.6 mSv. The effective dose value obtained in this study is almost half of the value reported in Pakistan (0.92 mSvy^{-1}) [24]. The effective dose is an important dosimetric quantity that allows different ionising radiation exposure categories to be compared and can be used to obtain broad estimates of radiation-attributed cancer incidents.

3.4. Excess Lifetime Cancer Risk (ELCR)

Low doses of ionising radiation, such as those encountered in response to natural radioactivity, are known to cause stochastic effects in the form of cancer. The probability with which these risks occur increases with increasing doses. The International Commission on Radiological Protection (ICRP) has estimated the number of fatalities per 1 Sv effective dose to be 0.05; this is known as the fatal risk factor. The ELCR can thus be determined using Equation (12) [2,23]:

$$ELCR = AED_{total}\left(Svy^{-1}\right) \times LF \times RF\left(Sv^{-1}\right) \tag{12}$$

where $AED_{total}\left(Svy^{-1}\right)$ is the annual effective dose calculated from indoor and outdoor exposure, LE is life expectancy (66 years), and RF is fatal risk factor per Sievert, which is 0.05 Sv^{-1}, as per ICRP Report [6]. The average number of ECR per million population ranged from 58–203 (average: 111) due to radioactivity from sand (Table 4).

3.5. Annual Gonadal Dose Equivalent (AGDE)

The AGDE was computed from activity using Equation (13) [2,23]:

$$AGDE\left(mSv.y^{-1}\right) = 3.09A_{Ra} + 4.18A_{Th} + 0.314A_K \tag{13}$$

where A_{Ra}, A_{Th}, and A_k are the specific activities (Bq kg^{-1}) of ^{226}Ra, ^{232}Th, and ^{40}K, respectively, in the studied samples. The average AGD (μGy.y^{-1}) ranged from 97.3 to 329.5 (average: 181.1) in coastal sand (Table 4). The global value is about 300 μGy.y^{-1} according to UNSCEAR reports [2].

3.6. Radiological Risk to Non-Human Biota

We have used the ERICA Tool software (Environmental Risk from Ionising Contaminants: Assessment and Management) to estimate the radiological risk parameters to non-human biota in marine environments [25]. ERICA Tool is a dosimetric model that enables calculations of internal and external absorbed dose rates to non-human biota covering a wide range of body masses and habitats for all radionuclides of interest. In addition, the software estimates the activity concentrations in biota; total absorbed dose rates, and risk quotients from the media (sediment) activity concentrations.

Table 5 shows the activity concentration in reference organisms in the marine environment determined using ERICA Tool. Total absorbed dose rate per organism as well as risk coefficients to non-human biota are presented in Table 6.

Table 5. Activity concentration in organism [Bq kg^{-1} f.w.].

Isotope	Activity in Sediment (Bqkg^{-1} d.w.)	Activity Concentration in Organism (Bq kg^{-1} f.w.)					
		Benthic Fish	Macroalgae	Mollusc-Bivalve	Pelagic Fish	Phytoplankton	Zooplankton
Ra-226	16.2	0.43	0.27	0.20	0.43	3.48	0.25
Th-232	34.3	0.01	0.020	0.01	0.01	3.15	0.031
Pb-210	46.80	5.80	0.18	1.11	5.80	84.3	2.99
Cs-137	0.08	0.0006	0.0007	0.0004	0.0006	0.0001	0.0010

Table 6. Total dose rate per organism and risk coefficients due to radioactivity in marine sediments calculated using the ERICA Tool.

Organism	Background Dose Rates	Screening Value [µGy h^{-1}]	Total Dose Rate per Organism [µGy h^{-1}]	Risk Quotient
Benthic fish	0.58	10	0.067	0.007
Macroalgae	0.87	10	0.048	0.005
Mollusc-bivalve	2.0	10	0.036	0.004
Pelagic fish	0.42	10	0.059	0.006
Phytoplankton	0.38	10	0.564	0.056
Zooplankton	0.94	10	0.035	0.003

As shown in Table 5, the highest radioactivity was evident in phytoplankton, followed by benthic fish. The levels of ^{210}Pb were significantly high in phytoplankton compared to those of the sediments indicating high ^{210}Pb bioaccumulation in phytoplankton as suggested in the literature [26,27].

Table 6 presents the total absorbed dose rate to marine organisms and risk quotients due to radioactivity in marine sediments calculated using the ERICA Tool.

As shown, excluding phytoplankton, the estimated total dose rate per organism was below the background dose rates (Table 6). However, the total dose rate for phytoplankton exceeds the background dose rate by 48 %, which is due to the radioactivity bioaccumulation in phytoplankton. Thus, the total dose rate and risk quotients are comparable to those presented by Botwe et al. [28] in Ghana.

4. Conclusions

To recapitulate, radioactivity levels were determined for common natural and anthropogenic radionuclides in costal sand and marine sediments. The results show varying levels of natural radioactivity that were comparable to those reported in similar studies. A significant correlation was shown for ^{232}Th and ^{40}K, and for ^{226}Ra and ^{232}Th; these relationships could be attributed to the origin of these radionuclides. The radioactivity levels in sediments are a source of radiation exposure for marine organisms. Regular monitoring of radioactivity levels is vital for radiation risk confinement. The results provide important baseline data to which future radioactivity levels in marine environments can be compared. Considering the fact that oceans and seas form the ultimate sink of contaminants, including radioactivity, future research initiatives that study radioactivity levels in marine environments and assess associated radiological hazards to the population are of utmost importance in order to ensure protection of the marine environment. Such a project should also consider investigating radioactivity from artificial radionuclides.

Author Contributions: Conceptualization and Methodology, I.I.S.; Validation and Formal Analysis, I.I.S. and K.A.; Investigation, K.A.; Resources, I.I.S.; Writing—Original Draft Preparation, K.A.; Writing—Review and Editing, I.I.S. and K.A.; Supervision, I.I.S.; Funding Acquisition, N/A. Both authors have read and agreed to the published version of the manuscript.

Funding: This research received no external funding.

Institutional Review Board Statement: Not applicable.

Informed Consent Statement: Not applicable.

Data Availability Statement: The data presented in this study are available on request from the corresponding author. The data are not publicly available due to privacy reasons.

Conflicts of Interest: The authors declare no conflict of interest.

References

1. Yii, M.W.; Zaharudin, A.; Abdul-Kadir, I. Distribution of naturally occurring radionuclides activity concentration in East Malaysian marine sediment. *Appl. Radiat. Isot.* **2009**, *67*, 630–635. [CrossRef] [PubMed]
2. United Nations Scientific Committee on the Effects of Atomic Radiation (UNSCEAR). *Sources, Effects and Risks of Ionization Radiation*; Report to the General Assembly, with Scientific Annexes B: Exposures from Natural Radiation Sources; UNSCEAR: New York, NY, USA, 1993.
3. Pálsson, S.E.; Skuterud, L.; Fesenko, S.; Golikov, V. Radionuclide transfer in arctic ecosystems. In *Quantification of Radionuclide Transfers in Terrestrial and Freshwater Environments for Radiological Assessments*; IAEATECDOC-1616; IAEA: Vienna, Austria, 2009; pp. 381–396.
4. International Atomic Energy Agency (IAEA). *Extent of Environmental Contamination by Naturally Occurring Radioactive Material (NORM) and Technological Options for Mitigation*; IAEA Technical Reports Series No. 419; IAEA: Vienna, Austria, 2003.
5. Valentin, J. *Environmental Protection: The Concept and Use of Reference Animals and Plants. Annals of the ICRP*; ICRP Publication 108; ICRP: Ottawa, ON, Canada, 2008.
6. ICRP. *The 2007 Recommendations of the International Commission on Radiological Protection (ICRP)*; ICRP Publication 103; Ann. ICRP 37; Pergamon Press: Oxford, UK, 2007.
7. IAEA. *Radiation Protection and Safety of Radiation Sources: International Basic Safety Standards*; International Atomic Energy Agency (IAEA): Vienna, Austria, 2014.
8. Adreani, T.E.; Mattar, E.; Alsafi, K.; Sulieman, A.; Suliman, I.I. Natural radioactivity and radiological risk parameters in local and imported building materials used in Sudan. *Appl. Ecol. Environ. Res.* **2020**, *18*, 7563–7572. [CrossRef]
9. Al-Qaradawi, I.; Abdel-Moati, M.; Al-Yafei, M.A.A.; Al-Ansari, E.; Al-Maslamani, I.; Holm, E.; Al-Shaikh, I.; Mauring, A.; Pinto, P.V.; Abdulmalik, D.; et al. Radioactivity levels in the marine environment along the Exclusive Economic Zone (EEZ) of Qatar. *Mar. Pollut. Bull.* **2015**, *90*, 323–329. [CrossRef] [PubMed]
10. Uddin, S.; Aba, A.; Fowler, S.W.; Behbehani, M.; Ismaeel, A.; Al-Shammari, H.; Alboloushi, A.; Mietelski, J.W.; Al-Ghadban, A.; Al-Ghunaim, A.; et al. Radioactivity in the Kuwait marine environment—Baseline measurements and review. *Mar. Pollut. Bull.* **2015**, *100*, 651–661. [CrossRef]
11. Zare, M.R.; Mostajaboddavati, M.; Kamali, M.; Abdi, M.R.; Mortazavi, M.S. 235U, 238U, 232Th, 40K and 137Cs activity concentrations in marine sediments along the northern coast of Oman Sea using high-resolution gamma-ray spectrometry. *Mar. Pollut. Bull.* **2012**, *64*, 1956–1961. [CrossRef] [PubMed]
12. Al-Trabulsy, H.A.; Khater, A.E.M.; Habbani, F.I. Radioactivity levels and radiological hazard indices at the Saudi coastline of the Gulf of Aqaba. *Radiat. Phys. Chem.* **2011**, *80*, 343–348. [CrossRef]
13. Al-Zamel, A.Z.; Bou-Rabee, F.; Olszewski, M.; Bem, H. Natural radionuclides and 137Cs activity concentration in the bottom sediment cores from Kuwait Bay. *J. Radioanal. Nucl. Chem.* **2005**, *266*, 269–276. [CrossRef]
14. Saleh, I.H. Radioactivity of ^{238}U, ^{232}Th, ^{40} K, and ^{137}Cs and assessment of depleted uranium in soil of the Musandam Peninsula, Sultanate of Oman. *Turk. J. Eng. Environ. Sci.* **2012**, *36*, 236–248.
15. ISO; IEC; BIPM OIML. *Guide to the Expression of Uncertainty in Measurement*; ISO: Geneva, Switzerland, 1995.
16. International Atomic Energy Agency (IAEA). *Quantifying Uncertainty in Nuclear Analytical Measurements*; IAEA-TECDOC-1401; IAEA: Vienna, Austria, 2004.
17. Pappa, F.K.; Tsabaris, C.; Ioannidou, A.; Patiris, D.L.; Kaberi, H.; Pashalidis, I.; Eleftheriou, G.; Androulakaki, E.G.; Vlastou, R. Radioactivity and metal concentrations in marine sediments associated with mining activities in Ierissos Gulf, North Aegean Sea, Greece. *Appl. Radiat. Isot.* **2017**, *116*, 22–33. [CrossRef] [PubMed]

18. Eleftheriou, G.; Tsabaris, C.; Kapsimalis, V.; Patiris, D.L.; Androulakaki, E.G.; Pappa, F.K.; Kokkoris, M.; Vlastou, R. Radionuclides and heavy metals concentrations at the seabed of NW Piraeus, Greece. In Proceedings of the 22nd Conference of the Hellenic Nuclear Physics Society, Athens, Greece, 30 May–1 June 2013.
19. Wang, J.; Du, J.; Bi, Q. Natural radioactivity assessment of surface sediments in the Yangtze Estuary. *Mar. Pollut. Bull.* **2017**, *114*, 602–608. [CrossRef] [PubMed]
20. Hanfi, M.Y.; Masoud, M.S.; Ambrosino, F.; Mostafa, M.Y. Natural radiological characterization at the Gabal El Seila region (Egypt). *Appl. Radiat. Isot.* **2021**, *31*, 109705. [CrossRef] [PubMed]
21. Beretka, J.; Mathew, P.J. Natural radioactivity of Australian building materials, industrials wastes and by-products. *Health Phys.* **1985**, *48*, 87–95. [CrossRef] [PubMed]
22. Hamilton, E.I. The relative radioactivity of building materials. *Am. Ind. Hyg. Assoc. J.* **1971**, *32*, 398–403. [CrossRef] [PubMed]
23. NEA-OECD. *Exposure to Radiation from the Natural Radioactivity in Building Materials: Report by a Group of Exports of the OECD Nuclear Energy Agency*; NEA-OECD: Paris, France, 1979; pp. 13–19.
24. Qureshi, A.A.; Tariq, S.; Din, K.U.; Manzoor, S.; Calligaris, C.; Waheed, A. Evaluation of excessive lifetime cancer risk due to natural radioactivity in the rivers sediments of Northern Pakistan. *J. Radiat. Res. Appl. Sci.* **2014**, *7*, 438–447. [CrossRef]
25. Brown, J.E.; Alfonso, B.; Avila, R.; Beresford, N.A.; Copplestone, D.; Pröhl, G.; Ulanovsky, A. The ERICA tool. *J. Environ. Radioact.* **2008**, *99*, 1371–1383. [CrossRef] [PubMed]
26. Sirelkhatim, D.A.; Sam, A.K.; Hassona, R.K. Distribution of ^{226}Ra–^{210}Pb–^{210}Po in marine biota and surface sediments of the Red Sea, Sudan. *J. Environ. Radioact.* **2008**, *99*, 1825–1828. [CrossRef] [PubMed]
27. Sugandhi, S.; Joshi, V.M.; Ravi, P.M. Studies on natural and anthropogenic radionuclides in sediment and biota of Mumbai Harbour Bay. *J. Radioanal. Nucl. Chem.* **2014**, *300*, 67–70. [CrossRef]
28. Botwe, B.O.; Schirone, A.; Delbono, I.; Barsanti, M.; Delfanti, R.; Kelderman, P.; Nyarko, E.; Lens, P.N. Radioactivity concentrations and their radiological significance in sediments of the Tema Harbour (Greater Accra, Ghana). *J. Radiat. Res. Appl. Sci.* **2017**, *10*, 63–71. [CrossRef]

life

Article

Radon Survey in Bank Buildings of Campania Region According to the Italian Transposition of Euratom 59/2013

Vittoria D'Avino [1,2], Mariagabriella Pugliese [1,2,*], Fabrizio Ambrosino [3], Mariateresa Bifulco [4], Marco La Commara [2,5], Vincenzo Roca [6], Carlo Sabbarese [2,6] and Giuseppe La Verde [1,2]

1 Department of Physics Ettore Pancini, University of Naples Federico II, 80126 Naples, Italy; vittoria.davino@unina.it (V.D.); laverde@na.infn.it (G.L.V.)
2 National Institute for Nuclear Physics, INFN Section of Naples, 80126 Naples, Italy; marco.lacommara@unina.it (M.L.C.); carlo.sabbarese@unicampania.it (C.S.)
3 Department of Agricultural Sciences, University of Naples Federico II, 80055 Portici, Italy; fabrizio.ambrosino@unicampania.it
4 Department of Electrical Engineering and Information Technology, University of Naples Federico II, 80125 Naples, Italy; mariateresabiff@hotmail.it
5 Department of Pharmacy, University of Naples Federico II, 80131 Naples, Italy
6 Department of Mathematics and Physics of the University of Campania Luigi Vanvitelli, 81100 Caserta, Italy; vroca222@gmail.com
* Correspondence: pugliese@na.infn.it

Abstract: ^{222}Rn gas represents the major contributor to human health risk from environmental radiological exposure. In confined spaces radon can accumulate to relatively high levels so that mitigation actions are necessary. The Italian legislation on radiation protection has set a reference value for the activity concentration of radon at 300 Bq/m^3. In this study, measurements of the annual radon concentration of 62 bank buildings spread throughout the Campania region (Southern Italy) were carried out. Using devices based on CR-39 solid-state nuclear track detectors, the ^{222}Rn level was assessed in 136 confined spaces (127 at underground floors and 9 at ground floors) frequented by workers and/or the public. The survey parameters considered in the analysis of the results were: floor types, wall cladding materials, number of openings, door/window opening duration for air exchange. Radon levels were found to be between 17 and 680 Bq/m^3, with an average value of 130 Bq/m^3 and a standard deviation of 120 Bq/m^3. About 7% of the results gave a radon activity concentration above 300 Bq/m^3. The analysis showed that the floor level and air exchange have the most significant influence. This study highlighted the importance of the assessment of indoor radon levels for work environments in particular, to protect the workers and public from radon-induced health effects.

Keywords: CR-39 detector; Euratom 59/2013; Italian radiation protection legislation; radon indoor; radon survey

Citation: D'Avino, V.; Pugliese, M.; Ambrosino, F.; Bifulco, M.; La Commara, M.; Roca, V.; Sabbarese, C.; La Verde, G. Radon Survey in Bank Buildings of Campania Region According to the Italian Transposition of Euratom 59/2013. *Life* **2021**, *11*, 533. https://doi.org/10.3390/life11060533

Academic Editor: Masao Suzuki

Received: 11 May 2021
Accepted: 3 June 2021
Published: 8 June 2021

1. Introduction

Radon is the heaviest and the only radioactive noble gas present in nature everywhere, generated in rocks and soils throughout the earth's crust. Its main unstable isotopes, namely ^{222}Rn (radon), ^{219}Rn (actinon) and ^{220}Rn (thoron), are produced in the intermediate steps of the three primordial decay chains of ^{238}U, ^{235}U and ^{232}Th respectively. The radiological importance of each radon isotope depends on its relative abundance and half-life. Due to having the isotopic ratio of $^{235}U/^{238}U = 0.0072$ and a short half-life ($T_{1/2}$ = 3.98 s), ^{219}Rn is always ignored. ^{222}Rn has the greatest half-life ($T_{1/2}$ = 3.82 d) and has received the most attention from the scientific community in regard to radiation protection, followed by ^{220}Rn ($T_{1/2}$ = 56.83 s).

Radon and its progenies are amongst the major sources of the population's exposure to natural radiation; indeed, they constitute the main contributor to the annual effective radiation among all sources of ionizing radiation [1]. Radon is chemically inert and so does not react with other elements or compounds, and it can easily escape from the ground into the air where it can be inhaled. However, health hazards related to the radon issue are not caused directly by radon, but by its progenies. In fact, because the lifetime of ^{222}Rn is longer than the air change time in the human respiratory system, most of the inhaled radon is exhaled and cannot decay with the body. On the contrary, the short-lived radon progeny (^{218}Po and ^{214}Po) is solid and so reactive that it can attach to atmospheric dust and water droplets forming clusters (attached fraction). Similarly, if inhaled, the decay products of ^{222}Rn (unattached fraction) attach themselves to the epithelium of the respiratory system and, due to their short duration, decay. In this way, the alpha particles ionize the DNA structures increasing the probability that, due to the stochastic effect, they can generate carcinogenic processes [2,3].

Since 1988, based on scientific evidences, the International Agency for Research on Cancer (IARC) defined radon as a human carcinogen (group 1) [4] and some decades later the International Commission on Radiation Protection (ICRP) in 2007 [5] and the World Health Organization (WHO) in 2009 [6] identified radon as the second leading causes of lung cancer after cigarette smoking.

Due to its chemical characteristics that allow it to escape easily through rocky substrates and native soils, radon enters buildings through cracks in the foundations or walls and accumulates in indoor environments where it can be breathed by humans.

In addition to the soil, a significant contribution to the accumulation of radon in indoor environments is due to exhalation from building materials of natural origin, in particular with a porous matrix (such as tuff) [7,8].

Furthermore, its indoor concentration is affected by environmental changes such as the frequency of air exchange in a closed environment, and changes outside such as pressure, temperature, and humidity. For this reason, long-term measurements (from 3 to 12 months) that take into account daily and seasonal variations are recommended to evaluate the radon concentration inside a building [9,10]. Thus, the annual average of radon activity concentration provides a representative estimate of indoor radon levels.

Human exposure to radon occurs both in workplaces and dwellings, since people usually spend a lot of their time in these confined spaces. It has been estimated that people generally spend more than eight hours a day in their workplace; therefore, the monitoring of workers' exposure is essential [11]. In addition, it is important to assess exposure in confined spaces other than houses (such as schools, shops, offices, banks, hospitals, universities) where significant levels of radon can be observed [12].

Subsequently to the classification of radon among carcinogens, many countries and international organizations have issued norms or recommendations for managing exposure. The WHO recommends the setting of a national reference level as low as reasonably achievable in the range of 100–300 Bq/m^3 for houses, and the ICRP has also recommended a level not exceeding 300 Bq/m^3 [5,6]. In Italy, protection against the dangers arising from exposure to ionizing radiation has recently become more prominent with the Legislative Decree 101/2020 [13] which transposed the Basic Safety Standards (BSS) Directive-2013/59/Euratom Directive [14]. Compared to the previous legislation (Legislative Decree 241/2000) [15], the great novelty introduced by the Directive lies in the establishment of protection measurements against the ionizing radiation not only for workers but also for the general population in living environments (Article 19). Furthermore, the Legislative Decree 101/2020 replaced the 'national action level' of 500 Bq/m^3 with the 'reference level' of 300 Bq/m^3 for both workplaces and dwellings (Article 12 comma 1) [13]. The Italian legislation commits employers to evaluate the occupational exposure in fully underground workplaces (e.g., caves, tunnels, cellars, mines, galleries, metro stations, car parks), in thermal structures and in basements and ground floor workplaces of buildings placed in 'radon-prone areas' identified and declared by the Regions according to the National Radon Action Plan (Article 10). Remedial actions are required if the annual average activity concentration of radon exceeds the reference level. In the event that the assessment determines a level higher than the reference value, then the employer is asked to calculate the annual effective dose for workers. If the estimation results are lower than 6 mSv, (Article 12 comma 1, letter d) no further actions are required. In this context, the current work presents an extensive measurement survey of radon activity concentration in 62 buildings of a bank company throughout Campania Region, Southern Italy. Very few similar surveys can be found in the literature involving bank buildings spread over all the national territory [16,17]. This Campania region, in accordance with the 2013/59/Euratom Directive and pending for its transposition in the national regulation, approved the Regional Low No. 13/2019 [18] which establishes the reference limit level for the activity concentration of radon gas activity at 300 Bq/m^3 in all underground rooms, basements, and ground floors of closed environments open to the public, as well as in buildings intended for education and in so-called strategic buildings as declared by the Ministry of Infrastructure [19]. The measurement campaign began in October 2019 and ended in September 2020. The aim of the present study was, firstly, to estimate the annual average radon levels in the underground and ground floors of the banks and then to evaluate remedial actions for these indoor environments where necessary. In order to perform a multifactorial study, data on several factors affecting radon concentration in confined spaces were collected and analyzed. To assess the possible influences and correlations, the results of radon activity concentrations were combined with data on the building characteristics, construction standards, building materials, ventilation conditions and systems, number of doors and windows, and the habits of the occupants.

2. Materials and Methods

2.1. Study Area and Sampling Design

The banks involved in the measurements survey consisted of 62 buildings, spread across the five provinces of Campania Region: Napoli (44), Salerno (7), Caserta (6), Avellino (3), Benevento (2).

Campania is a very-interesting area, as its territory is characterized by a large variety of geological environments and a high population density. The geological features, soil characteristics and extensive use of stones of volcanic origin (yellow tuff, green tuff, etc.) in the traditional building construction systems [20] have been considered as responsible for the higher than the national average indoor radon mean activity concentration value (around 70 Bq/m^3) [21–23].

A typical bank building consists of a ground floor where the public are served, in the form of banking halls, offices, conference rooms and office spaces with a daily human occupation of at least eight hours, and often an underground floor arranged in rooms for vaults, deposits, archives and more rarely offices. The building sample object of the study consisted of a total of 136 confined environments, 9 of which were in the underground level.

A CR-39 based detector was placed at each measurement point by the person responsible for safety at work, appropriately trained by our team for correct positioning. Radon measurements were conducted following the recommendations established by the UNI ISO 11665: 2020 standard. A data collection form on building characteristics and occupants' habits (ventilation system, number of openings, floor and wall cladding materials, number of hours per day of opening doors/windows for air exchange) was requested to be completed.

2.2. Radon Activity Concentration Measurement Method

Radon concentration was measured in 136 environments of 62 bank buildings for two consecutive six-month periods. The first period was October 2019–March 2020 and the second period April–September 2020. The mean annual radon activity concentration was calculated as the time-weighted average concentration of the two periods, using the detector exposure times as weights.

Radon activity measurements were performed using Solid-State Nuclear Track Detectors (SSNTDs) of poly-allyl-diglycol-carbonate commercially known as CR-39.

The CR-39 based detector is widely used for integrated and long-term measurement of the radon levels because of its material stability, good ionization sensitivity, stability against various environmental conditions, negligible sensitivity to Thoron, and ease of use [24–27]. The detection system consists of a closed chamber (Radout®, holder for CR-39 produced by Mi.am srl) through whose walls only ^{222}Rn diffuses (not ^{220}Rn and ^{219}Rn isotope), excluding dust particles and humidity from the measurement volume. During the exposure, the α particles emitted by radon and their daughters interact with the aggregate state of the CR-39 polymer causing damage along its path. The traces of the α particles are then made visible by an optical microscope after a chemical etching of the detector. The etching process consists of immersion of the detector in 25% weight/volume sodium hydroxide (NaOH) solution at 98 °C and 1.181 g cm^{-3} density for 1 h, and then in 2% weight/volume acetic acid (CH_3COOH) solution for 30 min. Then, the detector is rinsed in distilled water for 1 h in order to stop further etching. The observed track densities were converted into radon activity concentrations using an appropriate calibration factor, which in this case was 0.00209 ± 0.00021 tracks cm^{-2} h^{-1}/Bq m^{-3}. For the exposure intervals used, we found a detection limit of 4 Bq/m^3 (obtained with an exposure time of one semester), and our maximum activity concentration resulting of about 700 Bq/m^3 is far below the saturation limit of the detector [27].

We did not perform thoron measurements in the buildings involved in the study. The device used (CR-39 mounted in a thick wall decay chamber) shows a very low sensibility to thoron (guaranteed by Mi.am srl [28]) so as to obtain radon concentrations that are not significantly affected by thoron interaction. Moreover, for the thoron measurement with CR-39, a different Radout®holder and a different etching process on the detector are required [28,29].

2.3. Statistical Analysis

Statistical analysis was carried by verifying the log-normal distribution of radon values using Kolmogorov–Smirnov test. The comparison of radon activity concentration values was performed for the categories 'ground' and 'underground' level with the non-parametric Mann–Whitney test. Descriptive statistics (median, mean, standard deviation, range, etc.) have been computed on radon annual averages estimated in the two groups. The measurement uncertainty of radon activity concentrations is expressed as expanded uncertainty with coverage factor k = 2 (95% confidence interval). This is a precautionary approach, as indicated in the ISO 11665-3:2020. The metrological relative uncertainty is equal to 14%. Thus, the rooms that showed an annual average radon activity concentration higher than the reference value were classified 'critical'. Statistical analyses were performed using the Statistical Package for the Social Sciences (IBM SPSS Statistic v.26).

3. Results

Frequency distribution of annual activity concentrations for the 136 rooms is shown in Figure 1a.

Descriptive analysis shows that data distribution is skewed (skewness = 0.45, kurtosis = 0.1), and it is well described by a log-normal model (Figure 1b), checked by the Kolmogorov–Smirnov test ($p > 0.05$, 95% confidence level). In the graph, the values of the geometric and arithmetic means are reported.

Based on the result of the Mann–Whitney test, the significant difference in the annual average radon concentrations between the ground and underground levels was observed ($p < 0.05$) at 95% confidence level. The variation of radon concentration with respect to the different floor level is reported in the box plot of Figure 2.

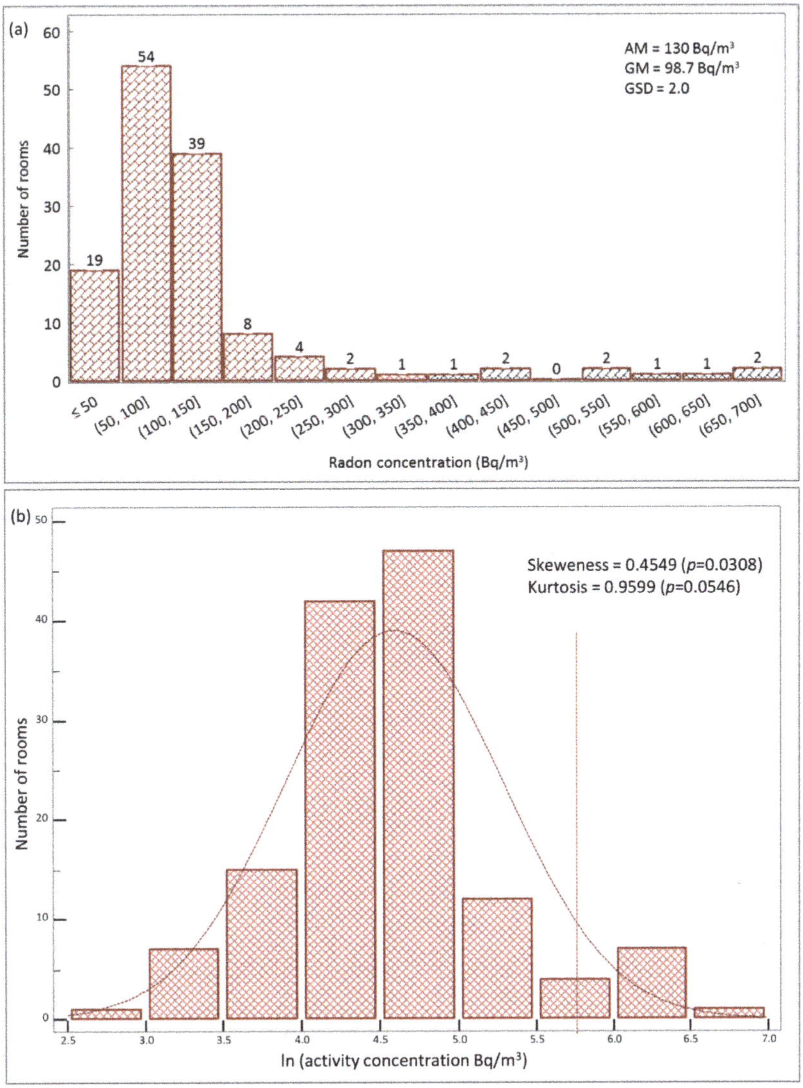

Figure 1. (**a**) Distributions of the annual average radon activity concentration for the full data set (136 bank rooms) expressed as Bq/m^3. The final bin is an overflow bin that contains all results above 300 Bq/m^3. Abbreviations: AM, arithmetic mean; GM, geometric mean; GSD, geometric standard deviation; (**b**) Normalized histogram for the natural log of radon measurements fitted with a normal distribution. Vertical dot line indicates the threshold at 300 Bq/m^3.

Frequency distributions of the separate annual specific concentrations for the ground and underground floors are reported in Figure 3.

As reported in Figure 3, a total of 10 rooms, five for each category (representing 4% and 56% of the total rooms at the ground and underground levels, respectively), belonging to 7 different buildings, showed a value of the radon concentration exceeding the reference value of 300 Bq/m^3.

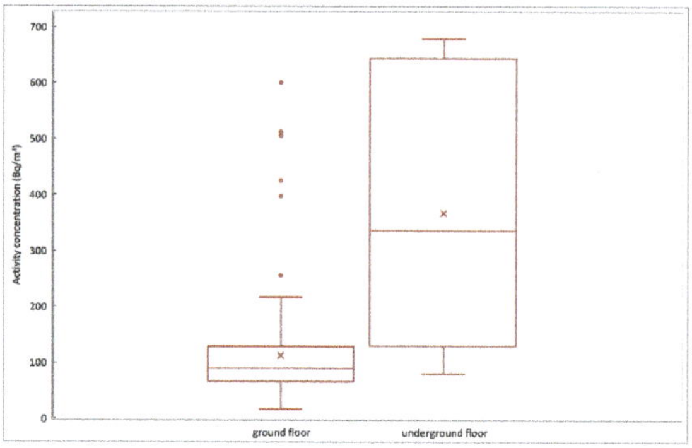

Figure 2. Comparison of annual average radon activity concentration obtained at the ground and underground floors. The graph reports the median, 25th and 75th percentile; the outside values are represented by dots. The cross marker represents the mean value.

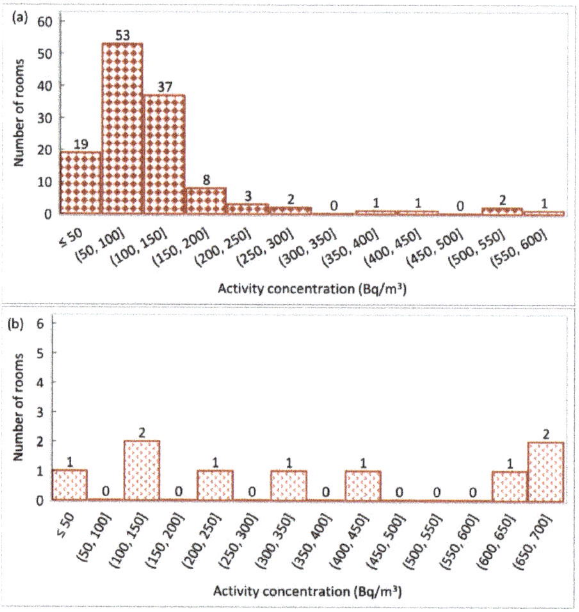

Figure 3. Distributions of the annual average specific concentrations in the (**a**) ground and underground; (**b**) floor levels expressed as Bq/m^3.

The rooms investigated at the ground and underground floors showed values of the annual average activity concentrations of 17–600 Bq/m^3 and 80–680 Bq/m^3, with an arithmetic mean of 113 $\pm$ 91 Bq/m^3 and 368 $\pm$ 242 Bq/m^3, respectively. The median values of 90 Bq/m^3 and 337 Bq/m^3 radon concentration were found for the ground and underground levels, respectively. Since the radon results distributions were skewed (Figure 3), the geometric mean was used to describe the central tendency. The results showed geometric means of 91.6 Bq/m^3 and 286.3 Bq/m^3 for the ground and underground floors, respectively. A synthesis of the statistic parameters and the number of rooms in which the radon value exceeds the reference level are shown in Table 1.

Table 1. Statistical data on annual average of indoor radon concentration (Bq/m^3) in monitored banks by floor level.

Descriptive Statistics	Ground Level	Underground Level
Range (Bq/m^3)	17–600	80–680
Median (Bq/m^3)	90	337
AM ± SD (Bq/m^3)	113 ± 91	368 ± 242
GM (Bq/m^3)	91.6	286.3
GSD	1.9	2.2
% ^{a} >300 (Bq/m^3) (No. of rooms)	4 (5)	56 (5)

Abbreviations: AM, arithmetic mean; SD, standard deviation; GM, geometric mean; GSD, geometric standard deviation; ^{a} = percentage of results exceeding 300 Bq/m^3.

The factors affecting radon concentration were investigated. Toward this aim, the rooms with a radon concentration level >300 Bq/m^3 were categorized as 'critical' for the analysis (of all the rooms investigated, 10 were considered critical and the remaining 126 were below the reference value). To verify the existence of any significant difference between the critical rooms and the other ones, the Mann–Whitney test was used. No significant difference in the distribution of the number of openings between the two groups was found, whereas significant change in the variable 'opening time' of windows/doors between 'critical' and 'noncritical' rooms are found ($p < 0.05$) at 95% confidence level. As shown in Figure 4, the range of the mean value of opening time resulted in 2.5–3 h/d in the critical rooms group and 3–5 h/d otherwise.

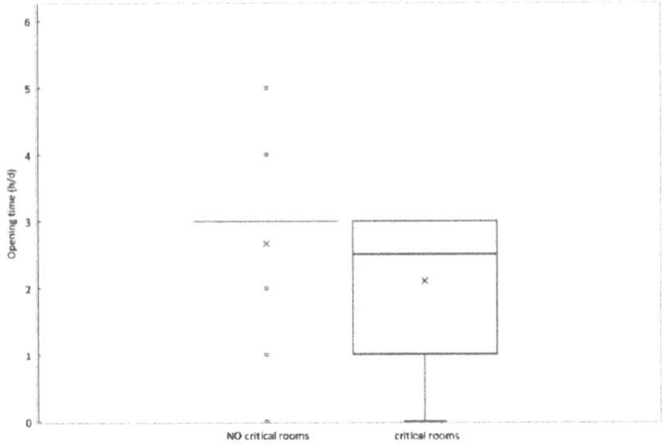

Figure 4. Variability of opening time (h/d) of windows/doors into the groups of 'critical' (radon concentration level > 300 Bq/m^3) and 'noncritical' rooms. The graph reports the median, 25th and 75th percentile; the outside values are represented by dots. The cross marker represents the mean value.

The data on the wall cladding materials showed that almost all the analyzed rooms both at the ground and underground levels are plastered (95% and 100% respectively). Similarly, no statistical significance was found with respect to the floor cladding materials for the critical and noncritical rooms of the buildings included in the analysis.

4. Discussion

In this study, we analyzed the radon activity concentration in 62 bank buildings spread on Campania region. A total of 136 measurement points (127 at ground floors and 9 at underground floors) were investigated for the annual radon monitoring. Despite that the difference in the sample sizes between the rooms at the underground and ground floors represents a limit for the statistics (it potentially induces a bias), the analyzed sample is the description of the effective distribution of the environments as the monitored buildings belong to a single bank company. The results of the overall data set, expressed in terms of annual average activity concentration, showed a skewed distribution well fitted by a log-normal curve (Figure 1), as expected [30,31]. The distribution of radon activity concentrations is comparable with the results reported in several studies available in the literature [16,21,23,31–34]. The geometric mean and the geometric standard deviation of the data have been used to describe the distribution, and this knowledge was useful for evaluation of the fraction of rooms that exceeded the reference value (300 Bq/m^3).

The legislative framework, that was the rationale for this work, plays a key role in the interpretation of the results. Campania Regional Law 13/2019 [18] requires assessment of the radon level in the underground, basement and ground floors of any building with public access, establishing the reference level of 300 Bq/m^3. According to this law, if the radon activity concentration value exceeds the reference level, the employer must implement remedial actions. Furthermore, compared to the previous Italian Legislative Decree 241/2000, the "reference level" has been introduced replacing the "action limit" and has been reduced from 500 to 300 Bq/m^3. During these measurements, the transposition of the Euratom 59/2013 directive came into force in Italy that, with respect to the regional law, incorporates all the basic safety standards for protection against the dangers arising from exposure to ionizing radiation. In particular, for radon gas exposure the annual effective dose limit has been increased from 3 to 6 mSv/y, and buildings intended for residential use are involved in the national regulation demanding that regional institutions implement an investment policy to adopt radon reduction strategies, and also for the radioprotection of people at their homes, if required.

Since the bank buildings include ground and underground confined spaces occupied both by workers and the public, according to Regional Law, a strategy for radon mitigation should be implemented in order to reduce the radon concentration. The owner of the property presents a remediation plan which will be approved by the Municipality (Article 4 comma 3) [18].

The method of choice for mitigation depends on the required reduction factor and the type of floor [35,36]. In general, the best way in which to lower radon levels is to reduce the pressure difference that draws radon into a building [35], but structural interventions are not quick to apply and their feasibility depends on several factors including construction characteristics. One practical method that is immediately applicable is passive ventilation consisting in increasing the number and frequency of opening doors and/or windows, allowing the reduction of indoor radon concentration by dilution (increased volume of fresh air dilutes radon concentration). Many studies in the literature have investigated the impact of passive ventilation through manual airing on indoor radon concentration [37–39]. In our study, the significant difference of the hours per day of opening windows and doors between 'critical' and 'noncritical' rooms supports the potential effectiveness in enhancing the ventilation of the environments. In regard to this, all buildings investigated are equipped both with air conditioners and at least one opening in each room. At the same time, it should be noted that the behaviors of occupants including window opening are influenced by building type, ventilation strategy, heating system, energy characteristics and so on [40]. However, the type of buildings investigated represents a peculiar scenario: inside banks, for security reasons, it is not possible to intervene by increasing the windows opening time, and so it is necessary to design forced ventilation systems that do not alter the degree of building security. Another aspect that plays a fundamental role in managing the risk of radon gas is the intended use of the environments. Our results showed that six rooms are bank archives (five located at the underground floor) exceeding the reference level, with staff access of 18 hours per year, so applying the criteria of Legislative Decree 101/2020 the annual effective dose is lower than the action limit of 6 mSv (Article 12 comma 1 letters c and d). Conversely, two rooms at the ground floor with high radon concentrations are occupied daily by both workers and the public, so according to the criteria of LR 13/2019 the building owner must submit a remediation plan.

From the point of view of the positioning of the rooms in respect to the floor, the statistical analysis found a significant difference between the ground and underground floors (Figure 2). The reason for high levels of radon in cellars could be the contact with soil containing uranium. Many studies in the literature have reported high radon concentration levels in underground sites nearest to the soil and that are usually poorly ventilated (mines, tunnels, underpasses, catacombs, spas, caves) [16,17,23,35,41–44]. Radon gas enters the building from the ground through cracks, crevices and other leakages or exhales from the walls of the house and, through air flows, spreads and accumulates in the internal environment. The diffusion process and the radon level in a building depends on several factors, such as concentration of radioactivity in the ground, permeability of the ground, nature of the floor and coupling of the building to the ground, ventilation conditions, and lining materials. The highest radon levels occur where each of these factors contribute to increase the radon, but small changes in one or more of them can cause appreciable differences in the radon activity concentration value, even in adjacent buildings of apparently identical construction [45]. This could be the reason for the variability of radon activity concentration found in our data set, ranging from 17 $\pm$ 7 to 680 $\pm$ 190 Bq/m^3 with a geometric mean of 98.7 Bq/m^3 and an arithmetic mean of 130 Bq/m^3 (Figure 1). In order to reduce the radon concentration to the rooms at ground floor, specific barriers between the cellar and ground floor could help to decrease the amount of radon entering the living areas [41].

Generally, the mean value of annual radon concentration found in the present investigation is higher than the mean national value (77 Bq/m^3) [30,33,46]. Furthermore, it is interesting to note that the radon values occurring in underground rooms are higher than the mean value reported in the extensive national survey on radon concentration in similar underground workplaces of bank buildings [16]. Conversely, the values of radon activity concentrations found in this study are comparable with other published results deriving from regional campaign of measurements [8,46]. We can speculate that a combination of two factors affects the radon concentration in Campania region: the complex geological and structural setting of this region [47] and the building materials of volcanic rock origin and pyroclastic sediments (i.e., lavic stones, tuffs, pozzolana) presenting high ^{226}Ra radioactivity level and used in recent and ancient constructions [8]. It is well known that the radioactivity contents of building materials contribute to radiation exposure, and radon exhalation can increase the radon level indoor, depending on the type of material [7,48–50].

In this framework, the knowledge of building materials, construction techniques, occupancy time of the space, combined with a more extensive and homogenous survey involving bank buildings spread all through region could be useful to individuate factors influencing the radon level in the geographical area involved in the survey.

While waiting to enhance the work with future measurement campaigns that could potentially target areas and dwelling types where data are currently sparse, our study provides useful results in the perspective of the imminent implementation of National Radon Action Plan as stated by the Italian Legislative Decree 101/2020 [13]. The plan defines strategies and arrangements for managing exposure to radon in workplaces and homes moving from identification of radon-prone areas (where the radon concentration in a significant number of buildings is expected to exceed the relevant national reference level) by targeted radon measurement survey. Once the radon-prone areas are identified, here the regulation will demand radon measurements in the underground, basement and ground environments both in workplaces and dwellings and, if necessary, the reduction of radon levels within the reference values established for existing and new buildings (Article 12) [13]. In this context, the work focused on the radioprotection issue of workers and the general population in underground and ground environments of buildings opened to the public and where different working activities are performed.

5. Conclusions

This study reports the results of a survey carried out to evaluate the radon concentration in bank buildings in the Campania region of southwestern Italy. The survey covered 62 bank buildings in the five provinces, including 136 closed environments in underground and ground floors. In each room, the radon device was exposed for a period of 12 months. In the underground rooms (such as archives and other rooms not occupied daily by workers) and in poorly ventilated rooms located at ground floors, the average annual radon concentrations were found to be higher than regularly ventilated rooms or those on the ground floor. The difference in radon concentration levels between the two investigated floors confirmed that soil is the main source of indoor radon, and the results also show the effectiveness of increased aeration turnover as a radon reduction strategy. About 93% of the radon activity concentration is below the national reference level of 300 Bq/m^3. Rooms that exceed the level of 300 Bq/m^3 (7%) will need remedial actions, such as forced ventilation and specially designed barriers, which could be useful to reduce the radon level. In conclusion, the results highlighted the necessity to increase the radon monitoring in workplaces with a high occupancy factor to ensure the staff and public protection against exposure. Furthermore, the work suggests that the identification of radon-prone areas will provide valuable criteria for implementing targeted radon surveillance and mitigation in workplaces and dwellings in accordance with the Italian radiation protection regulation.

Author Contributions: Conceptualization, G.L.V. and M.P.; Data Curation, V.D. and G.L.V.; Formal Analysis, V.D. and G.L.V.; Methodology, V.D., G.L.V., M.P., V.R. and C.S.; Project Administration, M.P.; Resources, M.P.; Software, V.D. and G.L.V.; Supervision, M.P.; Visualization, F.A., M.B., V.D., M.L.C., G.L.V., M.P., V.R. and C.S.; Writing—Original Draft Preparation, V.D. and G.L.V.; Writing—Review and Editing, F.A., M.B., V.D., M.L.C., G.L.V., M.P., V.R. and C.S. All authors have read and agreed to the published version of the manuscript.

Funding: This research received no external funding.

Institutional Review Board Statement: Not applicable.

Informed Consent Statement: Not applicable.

Data Availability Statement: Not applicable.

Conflicts of Interest: The authors declare no conflict of interest.

References

1. United Nations. Scientific Committee on the Effects of Atomic Radiation. In *Sources and Effects of Ionizing Radiation: United Nations Scientific Committee on the Effects of Atomic Radiation: UNSCEAR 2008 Report to the General Assembly, with Scientific Annexes*; United Nations: New York, NY, USA, 2010.
2. Durante, M.; Grossi, G.; Pugliese, M.; Manti, L.; Nappo, M.; Gialanella, G. Single charged-particle damage to living cells: A new method based on track-etch detectors. *Nucl. Instrum. Methods Phys. Res. Sect. B Beam Interact. Mater. At.* **1994**, *94*, 251–258. [CrossRef]
3. Durante, M.; Grossi, G.F.; Napolitano, M.; Pugliese, M.; Gialanella, G. Chromosome-Damage Induced by High-Let Alpha-Particles in Plateau-Phase C3h 10t1/2 Cells. *Int. J. Radiat. Biol.* **1992**, *62*, 571–580. [CrossRef]
4. International Agency for Research on Cancer (IARC). Man-made Mineral Fibres and Radon. *IARC Monogr. Eval. Carcinog. Risks Hum.* **1988**, *43*, 1–300.
5. The 2007 Recommendations of the International Commission on Radiological Protection. ICRP publication 103. *Ann. ICRP* **2007**, *37*, 1–332. [CrossRef]
6. Zeeb, H.; Shannoum, F. WHO Handbook on Indoor Radon: A Public Health Perspective. In *WHO Guidelines Approved by the Guidelines Review Committee*; World Health Organization: Geneva, Switzerland, 2009.
7. La Verde, G.; Raulo, A.; D'Avino, V.; Roca, V.; Pugliese, M. Radioactivity content in natural stones used as building materials in Puglia region analysed 1 by high resolution gamma-ray spectroscopy: Preliminary results. *Const. Build. Mater.* **2020**, *239*. [CrossRef]
8. Sabbarese, C.; Ambrosino, F.; D'Onofrio, A.; Pugliese, M.; La Verde, G.; D'Avino, V.; Roca, V. The first radon potential map of the Campania region (southern Italy). *Appl. Geochem.* **2021**, *126*, 104890. [CrossRef]
9. Baltrénas, P.; Grubliauskas, R.; Danila, V. Seasonal Variation of Indoor Radon Concentration Levels in Different Premises of a University Building. *Sustainability* **2020**, *12*, 6174. [CrossRef]
10. Miles, J.C. Temporal variation of radon levels in houses and implications for radon measurement strategies. *Radiat. Prot. Dosim.* **2001**, *93*, 369–376. [CrossRef] [PubMed]
11. ICRP. Protection against radon-222 at home and at work. A report of a task group of the International Commission on Radiological Protection. *Ann. ICRP* **1993**, *23*, 1–45.
12. Chen, J. Risk Assessment for Radon Exposure in Various Indoor Environments. *Radiat. Prot. Dosim.* **2019**, *185*, 143–150. [CrossRef]

13. Italian Governement. Decreto Legislativo n. 101 del 31 Luglio 2020, Attuazione della Direttiva 2013/59/Euratom, che Stabilisce Norme Fondamentali di Sicurezza Relative alla Protezione Contro i Pericoli Derivanti Dall'esposizione alle Radiazioni Ionizzanti, e che Abroga le Direttive 89/618/Euratom, 90/641/Euratom, 96/29/Euratom, 97/43/Euratom e 2003/122/Euratom e Riordino della Normativa di Settore in Attuazione Dell'articolo 20, comma 1, lettera a, della Legge 4 Ottobre 2019, n. 117; Gazz. Uff n. 201: Roma, Italy. 2020. Available online: https://www.governo.it/sites/new.governo.it/files/DLGS_59_2013_EURATOM-2020_01_29.pdf (accessed on 8 April 2021).
14. European Council. Council Directive 2013/59/Euratom on basic safety standards for protection against the dangers arising from exposure to ionising radiation and repealing Directives 89/618/Euratom, 90/641/Euratom, 96/29/Euratom, 97/43/Euratom and 2003/122/Euratom. *Off. J. Eur. Union* **2014**, *57*, 1–73.
15. Italian Governement. Decreto Legislativo n. 241 del 26 maggio 2000, Attuazione della direttiva 96/29/EURATOM in materia di protezione sanitaria della popolazione e dei lavoratori contro i rischi derivanti dalle radiazioni ionizzanti. *Official Gazette No. 203*, 31 August 2000.
16. Trevisi, R.; Orlando, C.; Orlando, P.; Amici, M.; Simeoni, C. Radon levels in underground workplaces e results of a nationwide survey in Italy. *Radiat. Meas.* **2012**, *47*, 178–181. [CrossRef]
17. Urso, P.; Ronchin, M.; Lietti, B.; Izzo, A.; Colloca, G.; Russignaga, D.; Carrer, P. Evaluation of radon levels in bank buildings: Results of a survey on a major Italian banking group. *Medicina Lavoro* **2008**, *99*, 216–233.
18. Campania Region. "Norme in materia di riduzione dalle esposizioni alla radioattività naturale derivante dal gas radon in ambiente confinato chiuso", Legge regionale 8 luglio 2019, n. 13. *Bollettino Ufficiale della Regione Ccampania n. 40 del 15*, 8 July 2019.
19. Ministero delle Infrastrutture. Decreto del Ministero delle infrastrutture 14 gennaio 2008. Approvazione delle nuove norme tecniche per le costruzioni. *G. U. n° 29 del 04/02/2008-Suppl. Ord. n° 30*, 14 January 2008.
20. Langella, A.; Calcaterrra, D.; Cappelletti, P.; Colella, A.; D'Albora, M.P.; Morra, V.; de Gennaro, M. Lava stones from Neapolitan volcanic districts in the architecture of Campania region, Italy. *Environ. Earth Sci.* **2009**, *59*, 145–160. [CrossRef]
21. Bochicchio, F.; Campos Venuti, G.; Nuccetelli, C.; Piermattei, S.; Risica, S.; Tommasino, L.; Torri, G. Results of the representative Italian national survey on radon indoors. *Health Phys.* **1996**, *71*, 741–748. [CrossRef] [PubMed]
22. Pugliese, M.; Roca, V.; Gialanella, G. ^{222}Rn indoor concentration in Campania. *Phys. Med.* **1994**, *10*, 118–119.
23. Quarto, M.; Pugliese, M.; Loffredo, F.; Roca, V. Indoor radon concentration measurements in some dwellings of the Penisola Sorrentina, South Italy. *Radiat. Prot. Dosim.* **2013**, *156*, 207–212. [CrossRef]
24. Bing, S. Cr-39 Radon Detector. *Nucl. Tracks Rad. Meas.* **1993**, *22*, 451–454. [CrossRef]
25. Danis, A.; Oncescu, M.; Ciubotariu, M. System for calibration of track detectors use in gaseous and solid alpha radionuclides monitoring. *Radiat. Meas.* **2001**, *34*, 155–159. [CrossRef]
26. Farid, S.M. Measurement of Concentrations of Radon and Its Daughters in Indoor Atmosphere Using Cr-39 Nuclear Track Detector. *Nucl. Tracks Rad. Meas.* **1993**, *22*, 331–334. [CrossRef]
27. Sabbarese, C.; Ambrosino, F.; Roca, V. Analysis by Scanner of Tracks Produced by Radon Alpha Particles in Cr-39 Detectors. *Radiat. Prot. Dosim.* **2020**, *191*, 154–159. [CrossRef]
28. Calamosca, M.; Penzo, S. A new CR-39 nuclear track passive thoron measuring device. *Radiat. Meas.* **2009**, *44*, 1013–1018. [CrossRef]
29. Caresana, M.; Cortesi, F.; Coria, S. Study of a discriminative technique between radon and thoron in the *Radout* detector. *Radiat. Meas.* **2020**, *138*, 106429. [CrossRef]
30. Bochicchio, F.; Campos-Venuti, G.; Piermattei, S.; Nuccetelli, C.; Risica, S.; Tommasino, L.; Torri, G.; Magnoni, M.; Agnesod, G.; Sgorbati, G.; et al. Annual average and seasonal variations of residential radon concentration for all the Italian Regions. *Radiat. Meas.* **2005**, *40*, 686–694. [CrossRef]
31. Venoso, G.; De Cicco, F.; Flores, B.; Gialanella, L.; Pugliese, M.; Roca, V.; Sabbarese, C. Radon concentrations in schools of the Neapolitan area. *Radiat. Meas.* **2009**, *44*, 127–130. [CrossRef]
32. Whyte, J.; Falcomer, R.; Chen, J. A Comparative Study of Radon Levels in Federal Buildings and Residential Homes in Canada. *Health Phys.* **2019**, *117*, 242–247. [CrossRef]
33. Bochicchio, F.; Campos-Venuti, G.; Nuccetelli, C.; Piermattei, S.; Risica, S.; Tommasi, R.; Tommasino, L.; Torri, G. The Italian Survey as the Basis of the National Radon Policy. *Radiat. Prot. Dosim.* **1994**, *56*, 1–4. [CrossRef]
34. Obed, R.I.; Ademola, A.K.; Vascotto, M.; Giannini, G. Radon measurements by nuclear track detectors in secondary schools in Oke-Ogun region, Nigeria. *J. Environ. Radioact.* **2011**, *102*, 1012–1017. [CrossRef] [PubMed]
35. Dixon, D. Understanding radon sources and mitigation in buildings. *J. Build. Apprais.* **2005**, *1*, 164–176. [CrossRef]
36. Khan, S.M.; Gomes, J.; Krewski, D.R. Radon interventions around the globe: A systematic review. *Heliyon* **2019**, *5*, e01737. [CrossRef]
37. Chauhan, R.P.; Kant, K.; Nain, M.; Chakarvarti, S.K. Indoor radon remediation: Effect of ventilation. *Environ. Geochem.* **2006**, *9*, 100–104.
38. D'Avino, V.; La Verde, G.; Pugliese, M. Effectiveness of passive ventilation on radon indoor level in Puglia Region according to European Directive 2013/59/EURATOM. *Indoor Built Environ.* **2020**, 1–7. [CrossRef]
39. Stabile, L.; Dell'Isola, M.; Frattolillo, A.; Massimo a, A.; Russi, A. Effect of natural ventilation and manual airing on indoor air quality in naturally ventilated Italian classrooms. *Build. Environ.* **2016**, *98*, 180–189. [CrossRef]

40. Roetzel, A.; Tsangrassoulis, A.; Dietrich, U.; Busching, S. A review of occupant control on natural ventilation. *Renew. Sustain. Energy Rev.* **2010**, *14*, 1001–1013. [CrossRef]
41. Crameri, R.; Furrer, D.; Burkart, W. Basement structure and barriers between the floors as main building characteristics affecting the indoor radon level of dwellings in the Swiss alpine areas. *Environ. Int.* **1991**, *17*, 337–341. [CrossRef]
42. Pugliese, M.; Quarto, M.; Roca, V. Radon concentrations in air and water in the thermal spas of ischia island. *Indoor Built Environ.* **2013**. [CrossRef]
43. Quarto, M.; Pugliese, M.; Loffredo, F.; La Verde, G.; Roca, V. Indoor radon activity concentration measurements in the great historical museums of University of Naples, Italy. *Radiat. Prot. Dosim.* **2016**, *168*, 116–123. [CrossRef]
44. Quarto, M.; Pugliese, M.; Loffredo, F.; Zambella, C.; Roca, V. Radon measurements and effective dose from radon inhalation estimation in the Neapolitan catacombs. *Radiat. Prot. Dosim.* **2014**, *158*, 442–446. [CrossRef]
45. Cliff, K.D.; Miles, J.C.H.; Brown, K. *The Incidence and Origin of Radon and Its Decay Products in Buildings*; National Radiological Protection Board: Chilton, UK, 1984; Volume 15.
46. Bochicchio, F.; Campos Venuti, G.; Piermattei, S.; Torri, G.; Nuccetelli, C.; Risica, S.; Tommasino, L. Results of the national survey on radon indoors in all the 21 italian regions. In Proceedings of the Radon in the Living Environment, Athens, Greece, 19–23 April 1999.
47. Vitale, S.; Ciarcia, S. Tectono-stratigraphic setting of the Campania region (southern Italy). *J. Maps* **2018**, *14*, 9–21. [CrossRef]
48. Nuccetelli, C.; Risica, S.; Onisei, S.; Leonardi, F.; Trevisi, R. Natural radioactivity in building materials in the Eurpean Union: A database of activity concentrations, radon emanations and radon exhalation rates. In *Rapporti ISTISAN 17/36*; Istituto Superiore di Sanità: Rome, Italy, 2012.
49. Sabbarese, C.; Ambrosino, F.; D'Onofrio, F.; Roca, V. Radiological characterization of natural building materials from the Campania region (Southern Italy). *Constr. Build. Mater.* **2021**, *268*, 121087. [CrossRef]
50. Sciocchetti, G.; Clemente, G.F.; Ingrao, G.; Scacco, F. Results of a survey on radioactivity of building materials in Italy. *Health Phys.* **1983**, *45*, 385–388. [CrossRef] [PubMed]

life

Review

Multi-Omics Approaches and Radiation on Lipid Metabolism in Toothed Whales

Jayan D. M. Senevirathna [1,2,*] and Shuichi Asakawa [1]

1 Laboratory of Aquatic Molecular Biology and Biotechnology, Department of Aquatic Bioscience, Graduate School of Agricultural and Life Sciences, The University of Tokyo, Tokyo 113-8657, Japan; asakawa@g.ecc.u-tokyo.ac.jp
2 Department of Animal Science, Faculty of Animal Science and Export Agriculture, Uva Wellassa University, Badulla 90000, Sri Lanka
* Correspondence: duminda@uwu.ac.lk

Abstract: Lipid synthesis pathways of toothed whales have evolved since their movement from the terrestrial to marine environment. The synthesis and function of these endogenous lipids and affecting factors are still little understood. In this review, we focused on different omics approaches and techniques to investigate lipid metabolism and radiation impacts on lipids in toothed whales. The selected literature was screened, and capacities, possibilities, and future approaches for identifying unusual lipid synthesis pathways by omics were evaluated. Omics approaches were categorized into the four major disciplines: lipidomics, transcriptomics, genomics, and proteomics. Genomics and transcriptomics can together identify genes related to unique lipid synthesis. As lipids interact with proteins in the animal body, lipidomics, and proteomics can correlate by creating lipid-binding proteome maps to elucidate metabolism pathways. In lipidomics studies, recent mass spectroscopic methods can address lipid profiles; however, the determination of structures of lipids are challenging. As an environmental stress, the acoustic radiation has a significant effect on the alteration of lipid profiles. Radiation studies in different omics approaches revealed the necessity of multi-omics applications. This review concluded that a combination of many of the omics areas may elucidate the metabolism of lipids and possible hazards on lipids in toothed whales by radiation.

Keywords: genomics; transcriptomics; proteomics; lipidomics; radiation; cetaceans

Citation: Senevirathna, J.D.M.; Asakawa, S. Multi-Omics Approaches and Radiation on Lipid Metabolism in Toothed Whales. *Life* **2021**, *11*, 364. https://doi.org/10.3390/life11040364

Academic Editors: Fabrizio Ambrosino and Supitcha Chanyotha

Received: 13 March 2021
Accepted: 17 April 2021
Published: 20 April 2021

1. Introduction

Marine mammals are megafauna that live in the deep sea and coastal environments after leaving the land between 50 and 25 million years ago; the modern whale fully adapted to the aquatic life around 10 million years ago [1]. According to their body features, feeding habits, and other factors, they are divided into three major orders; namely, Pinnipidia (seals, sea lions, walruses), Cetacea (whales, dolphins, and porpoises), and Sirenia (manatee and dugong) [2]. With this transition, they have developed numerous adaptations to survive in the ocean. A significant feature of these animals is the presence of a fat layer around the body, called the blubber. This blubber of marine mammals consists of subcutaneous adipose tissue and connective tissue that has three major functions: to store energy, to control buoyancy, and body heat regulation [3]. Moreover, a study has suggested that the brown adipose tissue of blubber and its protein, uncoupling protein 1, have a significant role in functioning like an insulation blanket in the cetacean body [4]. Toothed whales are a group in the order cetacea that includes some whales, and all dolphins and porpoises. They have special adaptations for the marine environment, such as echolocation for foraging and communication. Toothed whales have unique acoustic fats in the head region, namely, melon and jaw/mandibular fats, and it has been documented that those fats are involved in echolocation [5–9]. The composition and formation of these acoustic fats in toothed whales have differences with the blubber in the same animal. These specialized fats may be involved in the metabolism for various adaptations to their aquatic life. However, the origin of these specialized acoustic adipose tissue and the composition of these fats is still debated with limited scientific facts; therefore, comprehensive multi-scale approaches are needed to elucidate this kind of biological phenomena.

Lipids are a source of energy, commonly stored in specialized cells called adipocytes, as a form of triacylglycerol (TAG), in vertebrates [10]. TAG and wax esters (WE) are two common lipids usually stored in toothed whales' acoustic adipose tissues. The storing of wax esters in these animals is unusual compared to other mammals. Examples of naturally occurring wax esters are found in plant epicuticles, bees' wax, and the spermaceti oil of toothed whales. Waxes of toothed whales consist of uncommon fatty acids (FA) and fatty alcohols (FAlc), which contain special features that can be used in various industries. However, the synthesis of these lipids is currently not well understood, and the number of studies has been limited [11]. Usually, lipids are stored in adipocytes by various processes like lipoprotein hydrolysis, fatty acid uptake, *de novo* synthesis, and etherification [12]. The composition of TAG and WE in toothed whale species varies; therefore, the biological synthesis of these lipids could occur via different pathways [8]. In addition, the mammalian adipose tissue consists of many kinds of cells, collectively called adipokines, and is a complex tissue for the study of metabolism [13]. Therefore, more studies on understanding the uniformity, physical properties of lipid distribution in toothed whales, complete ontogenetic series, and metabolism of these fats have been recommended [8,14]. With this complexity of lipid species in toothed whales, a comprehensive study of the metabolism of these lipids will be highly interesting; however, it has not been tested yet.

The lipidome of the cell or cell compartment of an organism is an indicator of synthetic pathways of the organism and its environment [15]. There are inborn mechanisms in any organism to react to environmental stresses at a molecular level, such as fatty acids, where scientists can observe, measure, analyze and predict. These reactions are also important in understanding human and environmental interactions, and for predicting environmental changes, climate change, environmental toxicology, biotechnology and bioengineering, and many other aspects. Several factors, such as genes, environmental factors, and gene and environment interactions, can also affect the synthesis of the endogenous lipid profiles in mammals. Therefore, the lipid profile of toothed whales can be changed with dietary lipids, *de novo* lipogenesis, enzymes, desaturation, physical forces, and the incorporation of fatty acids into more complex lipid molecules [11,16,17]. In addition, changes in lipid metabolism in animals can occur due to anthropogenic or natural factors like radiation [18].

Radiation is the releasing of energy as waves or particles in any environment, that can cause biological impacts in any form of life. There are two main types of radiations, including ionizing and non-ionizing (extremely low frequency and radiofrequency) [19]. Sources of radiation are man-made (industries, power plants, telecommunication, and sonar) or natural (solar, space, and cosmic). A study has found that ultraviolet radiation affects cutaneous lipids and bioactive lipids for skin inflammation in humans [20]. Several types of anthropogenic radiation and radioactivities can impact marine life, including cetaceans. Moreover, different types of radiation and radioactivity underwater are environmental factors that can alter the specialized lipids of toothed whales. The ionizing radiation can also impact biological materials on cellular levels and organ levels, and have other low-level effects, such as cancers [21]. Although we could find direct evidence for the biological impacts of ionizing radiation in the literature, it is possible that ionizing radiation can also impact lipid metabolism in toothed whales. Particularly, a novel study puts forward that ionizing radiation induces reactive oxygen species and ACSL4 expression, a lipid metabolism enzyme resulting in elevated lipid peroxidation and ferroptosis in cancer cell-lines [22].

This review is mainly focused on acoustic radiation as an environmental factor that can influence lipids in toothed whales. Especially in toothed whales, this type of radiation may interrupt echolocation behavior while damaging special acoustic fats in the head region. For example, non-ionizing radiation, such as acoustic pollution, can impact various organs in whales' bodies and diminish the reproduction of these marine mammals [23]. The effects of noise pollution in various gene functions from the hypothalamic–pituitary–adrenal and hypothalamic–pituitary–gonadal (HPA) axes in animal bodies have been largely discussed; however, more studies are needed on toothed whales. Sonar and its underwater acoustic pollution is a highly concerning environmental stressor. The acoustic radiation of sonar can damage melon, and other lipid tissues in toothed whales and cause hearing loss [24]. Moreover, they have suggested that this anthropogenic radiation may also damage body tissues, effect behavioral changes, damage the dive cycle, damage breathing patterns, damage the sound production rate, and have negative effects on the energy budget. These changes are very complex to study at the molecular level and hard to understand metabolically, hence, the integration of several fields of studies like physical, chemical, and biological in the micro-environment is recommended. Experimental research on a model animal (female and male infants of rabbits) has been conducted to identify the effects of radiofrequency (RF) radiation on DNA and lipid damage [25]. Finally, they have identified that the global system for mobile telecommunication (GSM-like) RF radiation can cause attacking free radicals to DNA, and lipids, ultimately changing their biochemical structures [25]. Accordingly, we also suggest that radiation can act as a highly potential environmental stressor on lipid metabolism in other animals. Therefore, we should pay more attention to the research on the effects of radiation on lipids of toothed whales to identify its biological impacts in the polluted marine environment.

Several factors that can produce big data need to be concerned in studies of animal lipid metabolism; hence, we recommend that the integration of various kinds of molecular data is suitable and more accurate in studies on complex lipid metabolism in toothed whales and its changes due to environmental stressors. A comprehensive understanding of the lipid metabolism of these animals requires identification of molecular changes in different levels, such as genomics, transcriptomics, proteomics, and metabolomics (lipidomics), commonly called "multi-omics". In this approach, a variety of data generated from different omics tools are concerned for analysis and predictions [26]. Toothed whales have a rich diversity, including around 70 species, and they live mainly in different habitats of the marine environment; some are cosmopolitans, migratory, different sizes, and dive into various depths (sperm whale is the deepest diver). Single omics approaches can generate data comparative to many factors, according to the interest of the research design; then, integration of data in the multi-omics approach gives an overview of the experimental results as a whole for better interpretation. Moreover, these omics data are useful to understand marine mammals' health, and their environmental health [27].

In the past decades, lipidomics studies have been conducted to uncover the different aspects of lipid diversity, composition, evolution, and metabolism in toothed whales. Integrated multi-omics approaches could be tested to understand the complete set of genetics, transcriptomics, and to uncover the lipid species [28]. A recent study has identified abnormal lipid metabolisms, such as suppression of fatty acid metabolism, promotion and metabolism of glycolipids and phospholipids, using multi-omics analysis [29]. However, detailed omics projects on cetaceans on metabolism are limited due to constraints on sampling fresh tissues. Due to the limited number of combined omics data on toothed whales for functional enrichments, we had to rely on the data, methods, and technologies tested for other species. The advantage of multi-omics' is the various types of big data that can be integrated to collect a variety of biological information, unrevealing hidden mechanisms in lives. A wide variety of bioinformatics, including mathematical arrays, algorithms, statistics, computational skills, programming, artificial intelligence (AI), deep learning (DL), and machine learning (ML) techniques, are currently adapting to analyze and integrate omics data into a single platform for understanding metabolic pathways and diseases in various animals, as well as humans. Mostly these techniques were used for the identification of cetacean in the wild [30,31]. Recently, the ML technique has been applied in the detection and classification of sperm whales using bioacoustics [32]. Another study has identified the impact of polychlorinated biphenyls (PCBs) on the transcriptome of the common bottlenose dolphin [33]. Therefore, it is evident that the ML has a potential to identify lipid profile changes in toothed whales due to various factors. In addition, novel omics technologies can be applied to understand changes of molecules, from DNA to protein and their modifications, because the effects of radiation on biological systems can clearly be understood [34]. Another important aspect of multi-omics is a deeper understanding of processes, dynamic interactions, molecular frameworks, novel molecular mechanisms, novel pathways, biomarkers, and drug targets. Multi-omics can identify the activation and inactivation of enzymes in adipocytes, and ML can link the affecting factors for these biomolecular changes. Therefore, the biomolecular damages caused by radiation on toothed whales can be analyzed by applying the combined approach of multi-omics and ML in the future.

2. Methodology

In this review, we focused on identifying technologies in lipid-related genomics, transcriptomics, proteomics, lipidomics, and radiation studies that could elucidate the relationships, gaps, and opportunities for further studies on lipid metabolism in toothed whales. We searched articles in Google Scholar, and PubMed databases using various combinations of keywords, such as "-omics", "lipids", "fats", "adipose tissue", "toothed whales", "lipid metabolism", "acoustic radiation", and "radiation", and around 150 references were selected based on relatedness to lipid metabolism of toothed whales to discuss in this review. During the literature survey, we identified current approaches to study lipid metabolism, research gaps, limitations, and future aspects. Based on the information gathered, we propose a multi-omics approach by integrating omics technologies to study the unique lipid synthesis pathways and the possible impacts of radiation on the lipid metabolism of toothed whales and achieve further understanding (Figure 1). Moreover, future focuses in different omics studies were summarized to understand the applicability of multi-omics applications on lipid metabolism in the future (Supplementary Table S1).

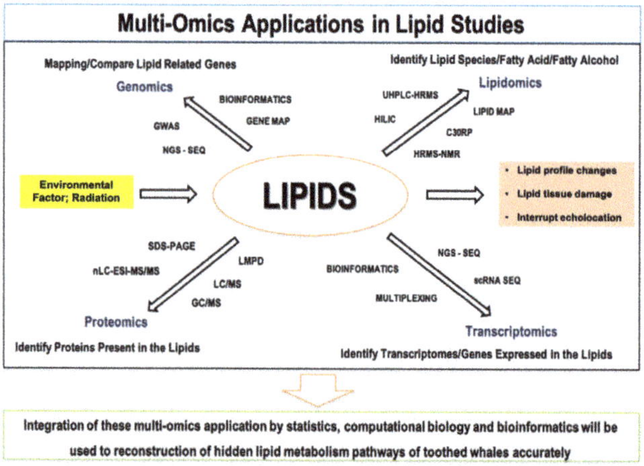

Figure 1. Integrated multi-omics approach for identification of lipid metabolic pathways. Note; GWAS (Genome-wide association study), SDS-PAGE (sodium dodecyl sulphate-polyacrylamide gel electrophoresis), GC/MS (gas chromatography/mass spectrometry), LC/MS (liquid chromatography/mass spectrometry), UHPLC-HRMS (ultra-high-performance liquid chromatography-high-resolution mass spectrometry), HILIC (hydrophilic interaction liquid chromatography), C30RP (C30 reversed-phase chromatography), HRMS-NMR (high-resolution mass spectrometry-nuclear magnetic resonance spectroscopy), LMPD (LIPID MAPS proteome database), LC/ESI-MS/MS (liquid chromatography electrospray ionization tandem mass spectrometric), scRNA-seq (single-cell RNA sequencing), NGS-SEQ (next generation sequencing).

3. Results and Discussion

Lipid patterns in toothed whales have been studied for many species during the past decades. A study found high levels of isovaleric lipids in the families Delphinidae, Phocoenidae, and Monodontidae while long-chain lipids were found in Ziphiidae, Physeteridae, and Platanistidae [7]. Later, the distribution of unusual branched-chain and wax ester lipids was identified in the mandibular fats of bottlenose dolphins [35]. Recently, Koopman (2018) has suggested that the importance of identifying specific molecular mechanisms in the synthesis of wax esters and branched-chained fatty acids in toothed whales [11]. It is also evident that some enzymes known to participate in wax ester biosynthesis have been showed pseudogenization in cetaceans [36]. As indicated in Figure 1, multi-omics is a combined approach of high-throughput technologies in genomics, transcriptomics, proteomics, and metabolomics [37]. Due to the limitations of findings from fresh samples of toothed whales, the design of this kind of study is critical. However, sample collections from strandings, fisheries, and biopsy collections are possible ways to study the toothed whales' metabolism. Therefore, a single tissue sample can be used for several omics analyses, such as isolating DNA for genome analysis, isolating RNAs for making tissue-specific cDNA libraries and de novo transcriptomics analysis, protein extractions for identifying enzymes involved in lipid metabolism, and extracting TAG, and wax esters to elucidate lipidomics. The big data produced in these omics' platforms can be integrated by bioinformatics tools and, also by applying DL, and ML techniques with the support of other AIs. Moreover, lipid profile changes, and biomolecular damages of stranding toothed whales can be subjected to comparative analyses to identify the possible impacts of acoustic radiation and further developments. Based on the present knowledge, and technology, factors affecting lipid profile changes in toothed whales can be predicted, and the unique lipid metabolism pathways of these animals will be revealed in the future.

3.1. Omics Applications on Lipid Metabolism and Radiation

3.1.1. Genomics

Genomics is one of the fundamental omics approaches. It is the science of studying genomes (DNA sequences) and nucleotide variants in their coding, and non-coding regions that provide central information of the metabolism of an organism. It has expanded to more functional levels to study gene expression, and gene interactions with proteins [38]. Toothed whales show convergent evolutionary adaptations. A study has highlighted that the parallel molecular changes in coding genes cause phenotypic changes of the animal that are favorable for aquatic life [39]. Interestingly they have identified that MYH7B, S100A9, and GPR97 genes were specific to toothed whales and not present in baleen whales. The evolutionary context of lipid metabolisms in cetaceans has been described, and 144 genes have been identified in the lipid metabolism of cetartiodactyl (cetaceans and artiodactyls), including toothed whales by genome comparison [40]. The triacylglycerol metabolism of cetaceans has been investigated using 88 related genes, and 41 were identified as being involved with triacylglycerol synthesis and lipolysis processes [41]. The importance of genes of cetacean species were described for aquatic adaptations. Several studies have been carried out to elucidate the gene loss in marine mammals in transforming to the aquatic environment, including the loss of the adenosine monophosphate deaminase gene family (AMPD3), a change that may have improved oxygen transport in sperm whales [42]. A study on bottlenose dolphins has identified significantly enriched genes for lipid transportation and localization [43]. The gene leptin (Lep) is expressed differently in seasonal and age-related variations in lipolysis in bowhead and beluga whales [3]. Additionally, the results of studies on humpback whale adaptations have been considered in human cancer therapy research [44].

Genes for lipid metabolism and de novo synthesis have been identified in many organisms. The FadR gene in *Escherichia coli* is a transcriptional regulator of fatty acid biosynthesis [45]. Expression of the WSD1 gene in *E. coli* and *Saccharomyces cerevisiae* contributes to the production of wax esters, indicating that WSD1 mainly functions as a wax synthase [46]. The expression of three genes is important in lipid metabolisms, such as lipoprotein lipase (LPL), muscle carnitine palmitoyl transferase-1 (mCPT1), and fatty acid-binding protein (FABP), as they correlate with the expression of PPAR-gamma in muscle samples of human [24]. The mesoderm-specific transcript (Mest) is identified in mice as a lipase; high expression of Mest leads to excess lipase activity, and lipid accumulation. If Mest occurs in non-lipid tissue, it can cause lipotoxicity and could be a subject for study in de novo lipid synthesis [47]. Genetic screens of *Caenorhabditis elegans* show that the fat regulation is performed by a complex gene network with some other functions, and that the fat regulatory pathways are similar to those in mammals [48].

Next-generation sequencing (NGS) provides new insight for genomics analyses to uncover genomes in de novo. In a detailed analysis of genomics, it is advisable to use more than one genome as a reference to get effective results of gene expression and function [49]. Therefore, reference genomes have been created for many species, including toothed whales with current advancements of short-read and long-read sequencing techniques. Genome-wide association studies (GWAS) are used in comparative genomics to identify potential genes for lipid synthesis. Several lipid metabolism related genes, such as LDLRAP1, APOA5, ANGPLT3/4, and PCSK9, are currently being revealed by GWAS. In the unsolved main lipid pathways, these adaptor proteins were discovered, and GWAS have shown many new loci [50]. However, monogenic approaches for identifying complex lipid traits are not suitable because of the polygenic origin of these complex traits. Therefore, the combination of large-scale data and sophisticated computer analyses using novel bioinformatics techniques are needed to identify candidate genes in lipid metabolism [50]. The identification of interaction of genomics and phenomics is an approach for understanding different lipid species. Trans-omics (reconstruction of global biochemical network with multi-omics) analyses have been used to identify PSMD9 as a previously unknown lipid regulator [51]. In contrast, this information shows the importance of genomics and combinations with other layers to future studies of genomics on lipids in toothed whales.

For many years, the genotoxic effects of radiation have been discussed. Its impacts on changing nucleic acid chemistry and biological consequences are also identified in different levels [34,52,53] and sometimes called radiogenomics [54–56]. Rare genetic diseases have been recorded due to radiation, such as ataxia-telangiectasia, Nijmegen breakage syndrome, ataxia telangiectasia-like disorder and DNA ligase IV deficiency in humans [57]. In another study, the effect of radiation in the gene expression of IL4I1, SERPINE1, TP53, RELA, and CDKN1A and their effect on the NF-kB pathway have been revealed [58–60]. A study has identified RUNX1 (a regulator of the NF-kB pathway), which plays a critical role in lipopolysaccharide-induced lung inflammation [61]. The mechanism of action of RF radiation on DNA/protein expression (heat–shock protein) has been largely discussed [62]. The radical-pair photoreceptor hypothesis is another mechanism that explains possible ways of radiation can inhibit DNA synthesis and transcription and finally, cause carcinogenesis [25]. However, we did not find any direct references related to the effects of any radiation on lipid metabolic genes in toothed whales, and hence, future investigations are needed in this aspect. Genome level mutations and alterations can happen by radiation, such as in the form of micro-waves; therefore, research on the physio-chemical changes of DNA should be conducted in the future. Genomics research on radiation should be implemented by applying multi-scale new approaches.

3.1.2. Transcriptomics

Currently, transcriptomics (the study of RNAs) has the potential to study various types of RNA expressed at the cellular level. The transcriptomics analysis of whales is limited due to difficulties in finding fresh tissue samples. No cDNA library from fats or liver is currently available for any cetacean species to facilitate the identification of up-regulated genes in the lipid metabolism. Skin transcriptomes of North Atlantic right whales have described diverse functions of the skin of the whales and metabolic pathways [63]. In addition, de novo blood transcriptomes have been identified in beluga whales and their functional roles have been described [64]. Studies of transcriptomes in bowhead whales have identified biological adaptations that prolong life in mammals [65]. Finally, DKK1, CEBPD, DDIT4, and ID1 have been identified as potential markers of acute hypothalamic–pituitary–adrenal (HPA) axis activation in marine mammal blubber, e.g., elephant seal [66].

Various lipid transcriptomics analyses have been conducted for farm animals. Regarding mammalian adipose tissue transcriptomes, fatty acid synthesis in pigs has been described compared to other tissues [67–69]. Studies of intramuscular fat (IMF) in pigs have identified potential genes and signaling pathways, such as AMP-activated protein kinase (AMPK), which plays a critical role in controlling lipid metabolism [70,71] and incRNA expression of IMF deposition [68]. In cattle, global transcriptomes in adipogenesis have given insights into the regulation of bovine adipogenic differentiation [72]. Twenty-five differentially expressed genes involved in the metabolism of lipids have been identified in chicken by transcriptomes [73]. Human adipose tissue gene expression studies have revealed the importance of environmental and individual factors in controlling the expression of human adipose tissue genes [74]. According to another transcriptome study, some adipose tissues of the Bactrian camel was shown to have functions in the immune and endocrine systems [75]. The mRNA expression patterns of dietary lipid levels controlling genes in fish give information about endoplasmic reticulum (ER) stress, unfolded protein response (UPR), and lipid metabolism [76–78]. Diglyceride acyltransferase (DGAT) is identified as an important catalyst in the metabolism of cellular diacylglycerol. In higher eukaryotes, such as mammals, it is involved with triacylglycerol metabolisms, such as intestinal fat absorption, lipoprotein assembly, adipose tissue formation, and lactation [79]. However, the functional and gene ontology pathways identified in other studies are linked with the lipid metabolism in particular animals, and there is no such study on marine mammals including toothed whales. Transcriptomics research on the special fats of toothed whales has the potential to identify significant genes for lipid synthesis and degradation.

Technologies on transcriptomics are still developing day by day by implementing various new tools, algorithms, and techniques. The extraction of RNA, and lipids together is an easy approach for limited tissues like from cetaceans. It has been tested in mice, lipids and associated gene expression patterns has been identified using a single sample [80]; this will be a useful approach in the future. Recent technologies have extended to high-throughput next-generation sequencing (NGS) technologies to apply in single-cell transcriptomics analyses, and biological features of organ development have been revealed, including the human heart [81]. Furthermore, the best practices in single-cell RNA-seq analysis and sample multiplexing have been reviewed and developed into an analysis pipeline [82,83]. Spatial transcriptomics is one of the newest technology introduced in the Visium platform for the identification of spatial topography of gene expression [84]. There are a few important challenges for RNA analysis, such as RNA-seq data analysis, the development of bioinformatics tools, and applications [85]. The identification of a single analysis pipeline is a major concern for the future development of transcriptomics [86]. RNA can modify due to various factors, and RNA-based modifications can happen in proteins [87], hence new developments also need to focus on epitranscriptomics, targeting the toothed whales' metabolism. Interestingly, a dolphin blood transcriptomics study on PCBs was conducted by applying machine learning approaches and they identified changes in gene expression under the chemical exposure in the environment [33].

The transcriptome of an animal body can change due to external environmental conditions [27]. The impacts of radiation on animals' transcriptomic levels have been studied. The potential impacts of space radiation were investigated at the transcriptomics level, concerning NF-kB pathways in different cells (mouse and human) [88]. Some cell-line studies have revealed that p53 mediated transcription modulation, induced by ionizing radiation [89–92]. Integrated global transcriptomics, and proteomic analyses have identified the induction of transforming growth factor (TGF) beta and the inactivation of peroxisome proliferator-activated receptor (PPAR) alpha signaling pathways in humans due to radiation [93]. A study on blood gene expression suggested the activation of apoptosis and the p53 signaling pathway due to ionization radiation damage by various analyses, such as differentially expressed genes analysis, weighted gene correlation network analysis, functional enrichment analysis, hypergeometric test, gene set enrichment analysis, and gene set variation analysis [94]. Further, the radiation-induced bystander effect (RIBE) has been investigated on gene expression for multiple types of RNA (mRNA, microRNA, mitochondrial RNA, long non-coding RNA, and small nucleolar RNA). Inhibition of RNA synthesis, processing, and translation can be caused by RF radiation [25]. A transcriptomics analysis on changes in lipid metabolisms of toothed whales due to radiation was not found in the searched databases. There is a high potential for radiation effects on RNA modification and changes that may cause cancers or malformations in whales. Lipid metabolism in toothed whales may involve many RNA functions; therefore, we recommend conducting this kind of lipid transcriptomics studies on toothed whales under potential environmental stressors, such as radiation.

3.1.3. Proteomics

Proteomic studies on lipid droplets in different organisms can give evolutionary perspectives for different organs [95]. Compared to genomics and transcriptomics, a significant number of lipid-related proteomics studies have been conducted on toothed whales in the past. For example, LC/MS/MS has been used to profile steroid hormones in dolphin species [66]. Brown adipose tissue (BAT) of four species of delphinoid cetacean, *Lagenorhynchus obliquidens*, *Tursiops truncates*, *Phocoenoides dalli*, and *Phocoena phocoena* has been analyzed; the brown adipocyte-specific mitochondrial protein, uncoupling protein 1 (UCP1), is only expressed in BAT for metabolic thermogenesis [96]. Proteomes in the blubber of harbor porpoises have been studied, and functional roles, such as metabolism, immune response, inflammation, and lipid metabolism, have been identified by SDS-PAGE and nLC-ESI-MS/MS technologies [97]. Recently, the expression of arachidonic acid 12-lipoxygenating ALOX15 orthologs have been recorded in marine mammals, including toothed whales, and as a function, arachidonic acid 15-lipoxygenation was identified [98]. Another study has identified that eight toll-like receptors (TLR) signaling pathway genes were under positive selection in cetaceans; interestingly, cetacean TLR4 was less responsive to lipopolysaccharides which suggest evolutionary changes between cetaceans and ungulates [99–101].

Lipid metabolism-related proteomic studies have widely been conducted on other animals, too. The recent proteomics analysis of lipid droplets from an alga revealed seven novel proteins important in lipid metabolism [102]. The proteins of broiler chickens have been investigated, and creatine pyruvate (CrPyr) has been found to have a more pronounced effect on lipid and protein metabolisms than Cr or Pyr [103]. A comparative proteomics analysis has been performed with random mutants (CR12 and CR48) with altered lipids, and the study identified significant up-regulated and down-regulated proteins in microalgal lipid biosynthesis pathways [104]. The milk fat globule membrane proteins of yak exhibit high lipid accumulation-lowering efficacy and contain potential bioactive ingredients for improving metabolism [105]. Proteins related to lipid metabolism and other functions are altered by major depressive disorder mechanisms of the chronic unpredictable mild stress mouse model [106]. The induction of lipogenic genes, and proteins has been observed in Atlantic cod by monitoring the increased plasma triglyceride levels in PCB 153-treated fish [107]. However, integrated proteomics on lipid metabolism of toothed whales are still lacking; therefore, the above findings may give insight into the future studies.

Lipidomics and proteomics analyses to find structural changes and species can easily be done with recent LC/MS or GC/MS technologies. The LIPID MAPS Proteome Database is useful for identifying lipid-associated protein sequences and annotations, and makes it possible to develop lipid interaction networks and integrate them with lipid metabolism pathways [108]. A high-resolution isotomic approach has been tested in a study to identify stable isotopes in the mammalian metabolism, including cetaceans, and they identified that extensive 13C enrichment was likely associated with fasting in the humpback whale [109]. Machine learning has been applied to proteomics to identify suitable biomarkers of diseases and predict proteins for treatment [110].

Radiation proteomics is an interesting field to discover molecules of radiation, biomarkers, acute and chronic physiological and health effects under different levels of exposure [111]. They have identified that ionizing gamma radiation could deregulate numerous proteins. Therefore, radiation can affect the function, structure, modifications, and interactions of proteins in an animal body. Under an integrated proteome and miRNA study, a decrease in the miR-21 level was observed and it was affected to alter several target proteins [112]. The fatty acid-binding protein 5 (FABP5) was investigated in human skin, and the increased level was investigated with intensifying TGF-β signaling pathway due to radiation [113]. Enhanced retinoic acid-induced signaling of PPARβ/δ and DHA-induced activation of RXR were also identified as an effect of radiation [114]. A similar study revealed that radiation-induced skin fibrosis was enhanced due to an overexpression of ZIP9 via DNA demethylation by radiation [115]. Another study of mitochondrial proteomics investigated a mechanism of apoptosis of spermatogenic cells in zebrafish caused by radiation [116]. Several technologies are currently famous for proteomics studies on radiation, such as ICPL, iTRAQ, and SILAC; however, comparative proteomics with bioinformatics is recommended for deep study on biological changes [117]. RF radiation causes denaturation of proteins and may act as a stressor to overexpress heat-shock proteins harmfully [62]. Proteome analysis of serum, and plasma has advantages for studies of radiation biology, and biomarker discovery [118]. In contrast, there is a huge gap of knowledge on proteomics of lipid metabolism, and radiation on toothed whales. Therefore, in the future, scientists should focus on addressing this gap by identifying potential radiation-related threats and their impacts. Accordingly, these findings will be useful in sustainable decision making to conserve these large animals in the ocean environment.

3.1.4. Lipidomics

Lipidomics is the comprehensive study of fats, and their components in the cells of organisms and also of the different biosynthetic pathways [119]. There are eight categories of lipids and more than 1 million lipid species [120]. In invertebrates, fats play a major role in energy storage and homeostasis; however, in vertebrates, fats are involved with many other functions such as immunity, protein synthesis, and metabolism [12]. Lipidome can affect the maximum lifespan of a species; however, identification of these lipidome determinants of a species' maximum lifespan is still at the development stage [121]. Lipids can be utilized for quantitative diagnosis, monitoring treatment response, and patient stratification as peripheral biomarkers [122].

During the past five centuries, research interest in lipids in toothed whales has gradually increased compared to genomics, transcriptomics, and proteomics. In toothed whales, several fat bodies have been identified. The development of the melon in toothed whales is still unclear; however, biochemical analysis has revealed that a portion of the isovalerate differs with the development of the body [123]. Mandibular fats of toothed whales perform acoustic functions through the expression of branched-chain fatty acids, which can alter the properties of the sound [8]. On the other hand, GC/MS, and HPLC techniques have identified high levels of fatty acid stratification in the outer blubber rather than the inner blubber in white whales, and killer whales [124]. Koopman has emphasized the necessity to conduct integrated omics studies to achieve a clear understanding of the lipid metabolism in toothed whales [11]. According to an ingested lipids analysis, right whales have evolved an unusual metabolic capability [14]. The fatty acid composition of the blubber of minke whales was investigated, and evidence of high metabolic activity was provided by the presence of long-chain polyunsaturated fatty acids (PUFAs) [125].

Lipid synthesis in many different organisms has been investigated. Triacylglycerol esters store metabolic energy as fats, and the synthesis of TAG was described 50 years ago; however, more research is needed to complete the understanding of the molecular mechanisms of TAG synthesis [126]. For example, in corals, lipogenesis in lipid bodies are controlled by coral hosts and metabolites such as triglycerols, sterol esters, and free fatty acids [127]. Further, the cell membrane is an important region for the study of lipid and protein interactions [128]. Phospholipid separation with high resolution, specificity, and signal-to-noise ratio can be done via Search Results DMS (Differential Mobility Spectrometry) with the combination of current liquid chromatography (LC) methods [129]. Microbial productions of triacylglycerols or fatty acid ethyl esters are concerned, these days, as fuel replacements or other compounds [130]. A recent study confirmed that information about wax ester metabolism was obtained by a simple gene (ADP1) insertion in bacteria [131]. Wax ester metabolism is also an important topic to discuss, especially in toothed whales.

Lipidomics is a part of metabolomics, a still-developing field with many techniques to study various types of lipid species that have unique characteristics [132]. The Folch method is the widely used lipid extraction protocol for animal tissues [133]. A newly introduced butanol and methanol (BUME) method has several advantages: simplicity, throughput, automation, solvent consumption, lower cost, health, and environmental aspects, over the Folch method for lipid extraction [134]. The use of greener solvents for lipid extractions is also considered to be in the development stage [135]. According to recent advances in lipid-related technology, two approaches have been adopted to understand the utility of sphingolipidomics, such as LC-MS- or LC-MS/MS-based, and shotgun lipidomics-based approaches [136,137]. New pulse programs have been developed for measuring lipids by nuclear magnetic resonance (NMR) [138] or HR-MAS NMR [139]. The lipid composition can vary significantly between cells, tissues, and organs; therefore, there is a need for technology with high mass accuracy, high resolution, and space atmospheric pressure, such as reliable and reproducible methods, such as MALDI MSI, for the identification of lipids [140]. Cryo-electron microscopy, and X-ray crystallography are also techniques for observing the interaction of lipids and membrane proteins [141]. Ultra-high performance liquid chromatography high-resolution mass spectrometry (UHPLC-HRMS) is a novel technology for lipidome analysis [119]. Another approach involves combining hydrophilic interactions (HILIC) with C30 reversed-phase chromatography (C30RP) coupled to high-resolution mass spectrometry (HRMS) as high-throughput platforms to analyze complex lipid mixtures [142]. Other ways are 'top-down' lipidomics, and 'bottom-up' lipidomics (Shevchenko and Simons, 2010). Flow field–flow fractionation (FlFFF)-MS is a high-speed technique for the analysis of lipoproteinic lipids as a top-down method [143]. Nanoflow liquid chromatography-ESI-MS/MS (nLC-ESI-MS/MS) analysis is a novel approach for a bottom-up approach. Lipidomics has generated a large amount of data; however, there is a need for bioinformatics and statistics for meaningful analysis. The development of chromatography and mass spectrometry are driving ongoing improvements in analytical performance to identify lipid species [144]. These researchers also noted the inadequacy of a single analytical platform for metabolomics analysis and, therefore, described the application of multi-omics approaches. Such approaches should be integrated with methods, and technologies, extraction protocols, and molecular systems-based analytical, and bioinformatics tools [120].

Understanding lipid dynamics and the role of enzymes in the metabolism pathway in toothed whales is still a challenge. Recently, scientists have looked at the structures, functions, interactions, and dynamics of single cellular lipids; therefore, the integration of lipidomics data with multi-disciplinary information is necessary for the analysis of metabolic pathways [145]. Lipids are important for membrane dynamics, energy homeostasis, and regulation of the molecular machinery, which means that lipids are a source of biological information, and biomarkers for cancer studies [146]. Lipid biomarkers are also becoming familiar in many fields and the adipocytes index in humpback whales is considered a non-destructive biomarker [147]. Therefore, integrated lipidomics has a potential to identify novel biomarkers in toothed whales in the future.

Lipids are sensitive molecules for environmental science studies since their abundance in any form of life and the lipid profiles can easily change with external stimuli [132,148], such as radiation. Metabolomic studies on the effects of radiation have been widely done; however, in this review, we especially focused on lipidomics studies for radiation effects. Exposure to radiation may cause reactive oxygen species (ROS) and damage cellular lipids [149]. Another lipidomics study found that oxidized phospholipids by radiation produce apoptotic signaling, causing lung damage in animals [150], emphasizing redox lipidomics, a growing field for the study of programmed cell death—apoptosis and ferroptosis [151]. Modifications of radiation effects can be observed by lipid peroxidation [152]. Radiation biodosimetry is another developing field and radiation-responsive lipids biomarkers identified in lipidomics studies are important in this development [153]. A recent study has investigated seven radiation-responsive lipids, including PC (18:2/18:2), PC (18:0/18:2), Lyso PC 18:1, PC (18:0/20:4), SM (D18:0/24:1), PC (16:0/18:1), and Lyso PC 18:2 which are useful in radiation biodosimeters [154]. In this study, they have observed that ionizing radiation causes changes in phospholipid metabolism by increasing the amount of phosphatidylethanolamine (PE) and phosphatidylserine (PS). The biological actions of n-3 PUFA (Omega-3 polyunsaturated fatty acids) under UV radiation has been investigated and its potential to use as a nutrient for skin health has been revealed [155]. Effects of radiation on damaging lipid profiles in cancer patients have been investigated under laboratory conditions; however, there is still more to do regarding a study on the effects of radiation on other wild animals in nature [156]. Recently, potential lipid biomarkers (distributed in linoleic acid metabolism, glycerophospholipid metabolism, glycerolipid metabolism, and glycosylphosphatidylinositol (GPI)-anchor biosynthesis) were identified as a result of microwave radiation; therefore, lipid metabolism studies under radiation impacts give insight into future research targets [18]. Based on many lipidomics and radiation studies, it is evident that various types of radiation have the potential to change lipid profiles in animals and damage the natural metabolism. Therefore, we can predict that there is a high potential for the impact of radiation on toothed whales' lipid metabolism. We recommend qualitative and quantitative measurements of lipid molecules under radiation exposure as a promising technique for lipidomics. Comparative lipidomics studies of toothed whales' populations in different healthy and unhealthy environments may also give useful information to identify the effects of environmental stressors on their lipids.

3.1.5. Integration of Multi-Omics

Many studies in the past have suggested that the integration of several omics' platforms is suitable for lipid metabolism analysis for different purposes (Supplementary Table S1). DNA or RNA isolation and sequencing have progressively become a component of multi-omics approaches, in which several layers of biology are simultaneously captured and analyzed. Multi-omics has many advantages, such as data integration and modeling pipelines, to predict metabolic pathways in animal metabolism. We believe that multi-omics is a highly potential approach to elucidate unique fatty acid synthesis and metabolism in toothed whales. For example, a recent study investigated the role of coiled-coil domain-containing protein 80 (CCDC80) gene in lipid metabolism by applying multi-omics integrative analysis and they found new roles of CCDC80 in fatty acid metabolism [157]. Moreover, multi-omics has been widely used for identifying molecular and non-molecular biomarkers [158,159]. Therefore, another benefit of large-scale multi-omics is that it may predict lipid biomarkers, and it can be useful in precision medicine. Omics data generation and integration are enhancing, in molecular biology and biotechnology, discoveries, such as new biomarkers. One of the driving forces behind the emergence of multi-omics approaches is an increasing appreciation of the contribution of multiple regulatory pathways and paradigms and of their relevance to processes that govern development, health, and disease. The combined studies of genetics, and transcriptomics have become popular in human medicine, e.g., for diagnosis and treatment of systemic lupus erythematosus [160]. In precision medicine, the integration of genomics, proteomics, and metabolomics has been identified as a potential approach for the robust characterization of biochemical signatures reflective of organismal phenotypes [161].

The combination of several omics platforms has been investigated in different animals successfully. In the arthropod metabolism, genes control the wax-secreting organ; therefore, proteins, and pathways in the lipid metabolism have been identified in ticks by transcriptomes, and proteomes [162]. Large amounts of omics data have been generated around the world; therefore, a global-integrative analytical approach needs to be established for future research [163]. Basic ML techniques, such as neural network analysis, have been widely used in studies of cetaceans for different purposes [164,165]. A combination of ML with multi-omics data may uncover relationships between molecules [166]. Therefore, we believe that the use of this combination of technologies with ML could help in finding the hidden pathways of complex lipid metabolism in toothed whales.

Applying multi-omics on the biological impacts of radiation is still an innovative field of study. A recent investigation experimentally revealed that the gut microbiota-metabolic axis acts as a shield for radiation in the animal body using the multi-omics approach, and this microbiota is important in producing the short-chain fatty acids that are used in various metabolic reactions in the host [167]. Several multi-omics techniques, such as RNA-seq, exome-seq, and H3K27ac ChIP-seq, were tested in a study to investigate the effects of UV radiation on human skin homeostasis and they found global gene dysregulation in skin cells under UV radiation [168]. Another non-animal study found that rice seed has adapted its biology to the low-level radioactive environment by multi-omics analysis [169]. The use of a combined approach of multi-omics to identify radiation effects on cetacean, including toothed whales, has not been conducted. Therefore, in this review, we also suggest focusing research on radiation by linking genomics, transcriptomics, proteomics, metabolomics, epigenetics, and environmental genomics applications to understand the impacts of radiation on lipids in toothed whales comprehensively.

3.2. *Acoustic Radiation on Lipid Metabolism in Toothed Whales*

From microorganisms to large animals/plants, every living organism consists of lipids, and individual lipid profiles can be changed by environmental stress [132]. Lipids play an important role in toothed whales' echolocation; however, the combination and profile of these specialized lipids may interact with different stressors that may cause misbehaviors. Several environmental stressors include depletion in ozone, increasing UV radiation, loss of biodiversity, and temperature [170]. Different types of radiation, such as isotopic, electromagnetic, and shortwave, have been largely discussed on biological aspects. Therefore, in this review, we selected acoustic radiation as a high impact environmental stressor to toothed whales.

A study found that low-intensity pulsed ultrasound (LIPUS) can cause visceral adipocyte differentiation in an experimental manner [171]. One study investigated that shock waves can interact with lipid membranes and, therefore, they used this technique for drug delivery by Lipid-mRNA nanoparticle [172]. Another approach is the use of acoustic radiation to investigate lipids for therapeutic purposes. Lipid particles, such as bubble liposomes, have been investigated for gene and drug delivery in the presence of ultrasound acoustic radiation [173]. The role of microwave radiation in changing lipid metabolism has also been studied recently, providing new therapeutic strategies [18]. The acoustic radiation force impulse imaging (ARFI) is a concept that is currently practiced as a liver biomarker to assess liver fibrosis, which produces shear waves in damaged tissue for monitoring [174]. Invertebrates, crustaceans, fish, and marine mammals produce sounds in a range of frequencies for foraging, mating, and complex communications naturally in water [175]. However, due to the natural and anthropogenic underwater noises, a shrinking communication space in whales has been observed [176]. Underwater noise pollution and its effects on whales have been widely discussed. For example, this noise may cause auditory masking, behavioral changes, hearing damage, and death for marine mammals, as they are very sensitive to acoustic radiation [177]. Therefore, it is evident that underwater noises, and acoustic radiation pollution should have a potential impact on sensitive tissue alterations in underwater animals, especially large animals, such as cetaceans. Currently, information is emerging on how acoustic radiation or underwater noise influence lipid metabolism in animals.

There are limited data on how metabolic alterations can happen in the animal body due to the noise pollution. Impacts of non-ionization radiation on lipids can be elucidated in laboratory condition using cetaceans' cell-lines, such as fibroblasts, adipocytes, and blood samples. In the natural environment, more comprehensive studies, including novel omics approaches, are needed in the future to investigate the effects of acoustic pollution on specialized lipid metabolism and to understand echolocation misbehaviors in toothed whales. However, understanding the whole process of these changes is more complicated to recognize on an individual level, and, therefore, integration of different omics-based studies can contribute significant results for making predictions and management decisions. ML and DL have been used for detection of acoustic sounds in cetaceans for identification purposes [178–180]. Therefore, recent ML tools may have clear potential to detect acoustic radiation impacts on toothed whale by detecting changes from their normal behaviors. ML is also useful for problem-solving in healthcare, and radiation oncology [181]. Therefore, this review also suggests implementing machine learning approaches for further understanding the lipid metabolism in toothed whales under environmental stresses, such as underwater acoustic radiation pollution. In the natural environment, establishment of this kind of study needs more innovative techniques. Molecular tagging methods has some potential to develop into a data collection tool in applying multi-omics on investigating acoustic radiation impacts in toothed whales in the wild.

4. Conclusions

Aquatic environments in the world are highly threatened by various natural and anthropogenic factors; therefore, understanding the interactions between animals and their environment, and possible stressors are highly important for sustainable ecosystem management. In this review, we evaluated lipid-related studies that have applied omics technologies and impacts of radiation to elucidate lipid metabolism in toothed whales. We identified that lipidomics, and proteomics use similar technologies, such as LC/GC/MS, while genomics, and transcriptomics also use the same technologies, such as NGS. The use of small single sample for various omics approaches has rarely been a concern in cetacean research, and, therefore, it has high potential to generate various big data for integrative analysis. There are examples for single sample preparation for lipidomics and transcriptomics analyses, which will be important for studies in toothed whale species due to the difficulties of finding fresh samples. The combination of omics approaches in toothed whales' lipid metabolisms is very limited and we identified it as a research gap in this review. Multi-omics approaches are useful to study the environmental effects on lipid metabolism in toothed whales due to their complexity. We discussed the effects of radiation as an environmental stressor that may change the lipid profiles of toothed whales. Radiation effects, including acoustic pollution, on toothed whales and their specialized lipid metabolism changes have been not studied largely. Therefore, this review suggests applying omics techniques and machine learning in this kind of research to identify mechanisms on alteration of lipid metabolism in toothed whales by external stressors. Considering all the aspects, an integrated multi-omics application will be the most reliable approach for the future studies of the complex lipid metabolism of toothed whales and to understand the impacts of radiation on their lipid profile changes in the underwater environment.

Supplementary Materials: The following are available online at https://www.mdpi.com/article/10.3390/life11040364/s1, Supplementary Table S1. Reference list used for understanding future approaches to understand lipid metabolism in toothed whales.

Author Contributions: Conceptualization, J.D.M.S. and S.A.; resources, J.D.M.S. and S.A.; data curation, J.D.M.S.; writing—original draft preparation, J.D.M.S.; writing—review and editing, J.D.M.S. and S.A.; visualization, J.D.M.S.; supervision, S.A.; funding acquisition, S.A. All authors have read and agreed to the published version of the manuscript.

Funding: This research received no external funding.

Institutional Review Board Statement: Not applicable.

Informed Consent Statement: Not applicable.

Data Availability Statement: No new data were created or analyzed in this study. Data sharing is not applicable to this article.

Conflicts of Interest: The authors declare no conflict of interest.

References

1. Mancia, A. On the Revolution of Cetacean Evolution. *Mar. Genom.* **2018**, *41*, 1–5. [CrossRef] [PubMed]
2. Berta, A.; Sumich, J.L.; Kovacs, K.M.; Folkens, P.A.; Adam, P.J. Cetacean Evolution and Systematics. *Mar. Mamm.* **2007**, 51–87. [CrossRef]
3. Ball, H.C.; Londraville, R.L.; Prokop, J.W.; George, J.C.; Suydam, R.S.; Vinyard, C.; Thewissen, J.G.M.; Duff, R.J. Beyond Thermoregulation: Metabolic Function of Cetacean Blubber in Migrating Bowhead and Beluga Whales. *J. Comp. Physiol. B Biochem. Syst. Environ. Physiol.* **2017**, *187*, 235–252. [CrossRef] [PubMed]
4. Hashimoto, O.; Ohtsuki, H.; Kakizaki, T.; Amou, K.; Sato, R.; Doi, S.; Kobayashi, S.; Matsuda, A.; Sugiyama, M.; Funaba, M.; et al. Brown Adipose Tissue in Cetacean Blubber. *PLoS ONE* **2015**, *10*, 1–14. [CrossRef]
5. Norris, K.S. Sound Transmission in the Porpoise Head. *J. Acoust. Soc. Am.* **1974**, *56*, 659–664. [CrossRef]
6. Litchfield, C.; Greenberg, A.J. Comparative Lipid Patterns in the Melon Fats of Dolphins, Porpoises and Toothed Whales. *Comp. Biochem. Physiol. Part B Biochem.* **1974**, *47*, 401–407. [CrossRef]
7. Litchfield, C.; Greenberg, A.J.; Caldwell, D.K.; Caldwell, M.C.; Sipos, J.C.; Ackman, R.G. Comparative Lipid Patterns in Acoustical and Nonacoustical Fatty Tissues of Dolphins, Porpoises and Toothed Whales. *Comp. Biochem. Physiol. Part B Biochem.* **1975**. [CrossRef]
8. Koopman, H.N.; Budge, S.M.; Ketten, D.R.; Iverson, S.J. Topographical Distribution of Lipids Inside the Mandibular Fat Bodies of Odontocetes: Remarkable Complexity and Consistency. *IEEE J. Ocean. Eng.* **2006**, *31*, 95–106. [CrossRef]
9. Yamato, M.; Koopman, H.; Niemeyer, M.; Ketten, D. Characterization of Lipids in Adipose Depots Associated with Minke and Fin Whale Ears: Comparison with "Acoustic Fats" of Toothed Whales. *Mar. Mammal Sci.* **2014**, *30*, 1549–1563. [CrossRef]
10. Birsoy, K.; Festuccia, W.T.; Laplante, M. A Comparative Perspective on Lipid Storage in Animals. *J. Cell Sci.* **2013**, *126*, 1541–1552. [CrossRef]
11. Koopman, H.N. Function and Evolution of Specialized Endogenous Lipids in Toothed Whales. *J. Exp. Biol.* **2018**, *221*, jeb161471. [CrossRef] [PubMed]
12. Azeez, O.I.; Meintjes, R.; Chamunorwa, J.P. Fat Body, Fat Pad and Adipose Tissues in Invertebrates and Vertebrates: The Nexus. *Lipids Health Dis.* **2014**, *13*, 1–13. [CrossRef] [PubMed]
13. Zheng, Z.; Dai, Z.; Cao, Y. Isolation, Purification of DPAn-3 from the Seal Oil Ethyl Ester. *Eur. J. Lipid Sci. Technol.* **2018**, *120*, 1–17. [CrossRef]
14. Swaim, Z.T.; Westgate, A.J.; Koopman, H.N.; Rolland, R.M.; Kraus, S.D. Metabolism of Ingested Lipids by North Atlantic Right Whales. *Endanger. Species Res.* **2009**, *6*, 259–271. [CrossRef]
15. Marella, C.; Torda, A.E.; Schwudke, D. The LUX Score: A Metric for Lipidome Homology. *PLoS Comput. Biol.* **2015**, *11*, 1–20. [CrossRef]
16. Parks, E.J. Changes in Fat Synthesis Influenced by Dietary Macronutrient Content. *Proc. Nutr. Soc.* **2002**, 281–286. [CrossRef]
17. Groß, J.; Virtue, P.; Nichols, P.D.; Eisenmann, P.; Waugh, C.A.; Bengtson Nash, S. Interannual Variability in the Lipid and Fatty Acid Profiles of East Australia-migrating Humpback Whales (*Megaptera novaeangliae*) Across a 10-year Timeline. *Sci. Rep.* **2020**, *10*, 1–14. [CrossRef] [PubMed]
18. Tian, Y.; Xia, Z.; Li, M.; Zhang, G.; Cui, H.; Li, B.; Zhou, H.; Dong, J. The Relationship between Microwave Radiation Injury and Abnormal Lipid Metabolism. *Chem. Phys. Lipids* **2019**, *225*, 104802. [CrossRef]
19. Kesari, K.K.; Agarwal, A.; Henkel, R. Radiations and Male Fertility. *Reprod. Biol. Endocrinol.* **2018**, *16*, 1–16. [CrossRef]
20. Nicolaou, A.; Pilkington, S.M.; Rhodes, L.E. Ultraviolet-radiation Induced Skin Inflammation: Dissecting the Role of Bioactive Lipids. *Chem. Phys. Lipids* **2011**, *164*, 535–543. [CrossRef]
21. Keith, S.; Faroon, O.; Roney, N.; Scinicariello, F.; Wilbur, S.; Ingerman, L.; Llados, F.; Plewak, D.; Wohlers, D.; Diamond, G. Overview of Basic Radiation Physics, Chemistry, and Biology. In *Toxicological Profile for Uranium*; Agency for Toxic Substances and Disease Registry: Atlanta, GA, USA, 2013.
22. Lei, G.; Zhang, Y.; Koppula, P.; Liu, X.; Zhang, J.; Lin, S.H.; Ajani, J.A.; Xiao, Q.; Liao, Z.; Wang, H.; et al. The Role of Ferroptosis in Ionizing Radiation-induced Cell Death and Tumor Suppression. *Cell Res.* **2020**, *30*, 146–162. [CrossRef] [PubMed]
23. Nabi, G.; McLaughlin, R.W.; Hao, Y.; Wang, K.; Zeng, X.; Khan, S.; Wang, D. The Possible Effects of Anthropogenic Acoustic Pollution on Marine Mammals' Reproduction: An Emerging Threat to Animal Extinction. *Environ. Sci. Pollut. Res.* **2018**, *25*, 19338–19345. [CrossRef] [PubMed]
24. Lurton, X.; Deruiter, S. Sound Radiation of Seafloor-Mapping Echosounders in the Water Column, in Relation to the Risks Posed to Marine Mammals. *Int. Hydrogr. Rev.* **2011**, 7–17.

25. Güler, G.; Tomruk, A.; Ozgur, E.; Sahin, D.; Sepici, A.; Altan, N.; Seyhan, N. The Effect of Radiofrequency Radiation on DNA and Lipid Damage in Female and Male Infant Rabbits. *Int. J. Radiat. Biol.* **2012**, *88*, 367–373. [CrossRef] [PubMed]
26. Subramanian, I.; Verma, S.; Kumar, S.; Jere, A.; Anamika, K. Multi-omics Data Integration, Interpretation, and Its Application. *Bioinform. Biol. Insights* **2020**, *14*, 7–9. [CrossRef] [PubMed]
27. Mancia, A. New Technologies for Monitoring Marine Mammal Health. In *Marine Mammal Ecotoxicology: Impacts of Multiple Stressors on Population Health*; Elsevier Inc.: Amsterdam, The Netherlands, 2018; pp. 291–320. ISBN 9780128122501.
28. Jha, P.; McDevitt, M.T.; Halilbasic, E.; Williams, E.G.; Quiros, P.M.; Gariani, K.; Sleiman, M.B.; Gupta, R.; Ulbrich, A.; Jochem, A.; et al. Genetic Regulation of Plasma Lipid Species and Their Association with Metabolic Phenotypes. *Cell Syst.* **2018**, *6*, 709–721.e6. [CrossRef]
29. Zhao, J.; Wei, W.; Yan, H.; Zhou, Y.; Li, Z.; Chen, Y.; Zhang, C.; Zeng, J.; Chen, T.; Zhou, L. Assessing Capreomycin Resistance on tlyA Deficient and Point Mutation (G695A) *Mycobacterium tuberculosis* Strains Using Multi-omics Analysis. *Int. J. Med. Microbiol.* **2019**, *309*, 151323. [CrossRef] [PubMed]
30. Guirado, E.; Tabik, S.; Rivas, M.L.; Alcaraz-Segura, D.; Herrera, F. Whale Counting in Satellite and Aerial Images with Deep Learning. *Sci. Rep.* **2019**, *9*, 1–12. [CrossRef]
31. Bogucki, R.; Cygan, M.; Khan, C.B.; Klimek, M.; Milczek, J.K.; Mucha, M. Applying Deep Learning to Right Whale Photo Identification. *Conserv. Biol.* **2019**, *33*, 676–684. [CrossRef]
32. Bermant, P.C.; Bronstein, M.M.; Wood, R.J.; Gero, S.; Gruber, D.F. Deep Machine Learning Techniques for the Detection and Classification of Sperm Whale Bioacoustics. *Sci. Rep.* **2019**, *9*, 1–10. [CrossRef]
33. Mancia, A.; Ryan, J.C.; Van Dolah, F.M.; Kucklick, J.R.; Rowles, T.K.; Wells, R.S.; Rosel, P.E.; Hohn, A.A.; Schwacke, L.H. Machine Learning Approaches to Investigate the Impact of PCBs on the Transcriptome of the Common Bottlenose Dolphin (*Tursiops truncatus*). *Mar. Environ. Res.* **2014**, *100*, 57–67. [CrossRef] [PubMed]
34. Reisz, J.A.; Bansal, N.; Qian, J.; Zhao, W.; Furdui, C.M. Effects of Ionizing Radiation on Biological Molecules–Mechanisms of Damage and Emerging Methods of Detection. *Antioxid. Redox Signal.* **2014**, *21*, 260–292. [CrossRef] [PubMed]
35. Duggan, Z.P.Z.; Koopman, H.N.; Budge, S.M. Distribution and Development of the Highly Specialized Lipids in the Sound Reception Systems of Dolphins. *J. Comp. Physiol. B Biochem. Syst. Environ. Physiol.* **2009**, *179*, 783–798. [CrossRef]
36. Lopes-Marques, M.; Machado, A.M.; Alves, L.Q.; Fonseca, M.M.; Barbosa, S.; Sinding, M.-H.S.; Rasmussen, M.H.; Iversen, M.R.; Frost Bertelsen, M.; Campos, P.F.; et al. Complete Inactivation of Sebum-Producing Genes Parallels the Loss of Sebaceous Glands in Cetacea. *Mol. Biol. Evol.* **2019**, *36*, 1270–1280. [CrossRef] [PubMed]
37. Aizat, W.M.; Ismail, I.; Noor, N.M. Recent development in omics studies. In *Advances in Experimental Medicine and Biology*; Springer New York LLC: New York, NY, USA, 2018; Volume 1102, pp. 1–9.
38. Del Giacco, L.; Cattaneo, C. Introduction to Genomics. *Methods Mol. Biol.* **2012**, *823*, 79–88. [CrossRef]
39. Foote, A.D.; Liu, Y.; Thomas, G.W.C.; Vinař, T.; Alföldi, J.; Deng, J.; Dugan, S.; Van Elk, C.E.; Hunter, M.E.; Joshi, V.; et al. Convergent Evolution of the Genomes of Marine Mammals. *Nat. Genet.* **2015**, *47*. [CrossRef] [PubMed]
40. Endo, Y.; Kamei, K.-i.; Inoue-Murayama, M. Genetic signatures of lipid metabolism evolution in Cetacea since the divergence from terrestrial ancestor. *J. Evol. Biol.* **2018**, *31*, 1655–1665. [CrossRef] [PubMed]
41. Wang, Z.; Chen, Z.; Xu, S.; Ren, W.; Zhou, K.; Yang, G. "Obesity" is Healthy for Cetaceans? Evidence from Pervasive Positive Selection in Genes Related to Triacylglycerol Metabolism. *Sci. Rep.* **2015**, *5*, 1–12. [CrossRef]
42. Sharma, V.; Hecker, N.; Roscito, J.G.; Foerster, L.; Langer, B.E.; Hiller, M. A Genomics Approach Reveals Insights into the Importance of Gene Losses for Mammalian Adaptations. *Nat. Commun.* **2018**, *9*. [CrossRef]
43. Sun, Y.B.; Zhou, W.P.; Liu, H.Q.; Irwin, D.M.; Shen, Y.Y.; Zhang, Y.P. Genome-wide Scans for Candidate Genes Involved in the Aquatic Adaptation of Dolphins. *Genome Biol. Evol.* **2013**, *5*, 130–139. [CrossRef]
44. Tollis, M.; Robbins, J.; Webb, A.E.; Kuderna, L.F.K.; Caulin, A.F.; Garcia, J.D.; Bèrubè, M.; Pourmand, N.; Marques-Bonet, T.; O'Connell, M.J.; et al. Return to the Sea, Get Huge, Beat Cancer: An Analysis of Cetacean Genomes Including an Assembly for the Humpback Whale (Megaptera novaeangliae). *Mol. Biol. Evol.* **2019**. [CrossRef]
45. Dibrova, D.V.; Galperin, M.Y.; Mulkidjanian, A.Y. Phylogenomic Reconstruction of Archaeal Fatty Acid Metabolism. *Environ. Microbiol.* **2014**, *16*, 907–918. [CrossRef]
46. Li, F.; Wu, X.; Lam, P.; Bird, D.; Zheng, H.; Samuels, L.; Jetter, R.; Kunst, L. Identification of the Wax Ester Synthase/Acyl-coenzyme A: Diacylglycerol Acyltransferase WSD1 Required for Stem Wax Ester Biosynthesis in Arabidopsis. *Plant Physiol.* **2008**, *148*, 97–107. [CrossRef]
47. Himes, K.P.; Young, A.; Koppes, E.; Stolz, D.; Barak, Y.; Sadovsky, Y.; Chaillet, J.R. Loss of Inherited Genomic Imprints in Mice Leads to Severe Disruption in Placental Lipid Metabolism. *Placenta* **2015**, *36*, 389–396. [CrossRef]
48. Ashrafi, K. Obesity and the Regulation of Fat Metabolism. *WormBook* **2007**, 1–20. [CrossRef]
49. Yang, X.; Lee, W.-P.; Ye, K.; Lee, C. One Reference Genome is not Enough. *Genome Biol.* **2019**, *20*, 104. [CrossRef]
50. Kuivenhoven, J.A.; Hegele, R.A. Mining the Genome for Lipid Genes. *Biochim. Biophys. Acta Mol. Basis Dis.* **2014**, *1842*, 1993–2009. [CrossRef] [PubMed]
51. Parker, B.L.; Calkin, A.C.; Seldin, M.M.; Keating, M.F.; Tarling, E.J.; Yang, P.; Moody, S.C.; Liu, Y.; Zerenturk, E.J.; Needham, E.J.; et al. An Integrative Systems Genetic Analysis of Mammalian Lipid Metabolism. *Nature* **2019**, *567*, 187–193. [CrossRef] [PubMed]
52. Roy, S. Impact of UV radiation on genome stability and human health. In *Advances in Experimental Medicine and Biology*; Springer New York LLC: New York, NY, USA, 2017; Volume 996, pp. 207–219.

53. Council, N.R. *Health Effects of Exposure to Low Levels of Ionizing Radiation: BEIR V*; The National Academies Press: Washington, DC, USA, 1990; ISBN 978-0-309-03995-6.
54. Kerns, S.L.; West, C.M.L.; Andreassen, C.N.; Barnett, G.C.; Bentzen, S.M.; Burnet, N.G.; Dekker, A.; De Ruysscher, D.; Dunning, A.; Parliament, M.; et al. Radiogenomics: The Search for Genetic Predictors of Radiotherapy Response. *Futur. Oncol.* **2014**, *10*, 2391–2406. [CrossRef] [PubMed]
55. Andreassen, C.N.; Schack, L.M.H.; Laursen, L.V.; Alsner, J. Radiogenomics–Current Status, Challenges and Future Directions. *Cancer Lett.* **2016**, *382*, 127–136. [CrossRef] [PubMed]
56. Roberson, J.D.; Burnett, O.L.; Robin, N. Radiogenomics: Towards a Personalized Radiation Oncology. *Curr. Opin. Pediatr.* **2016**, *28*, 713–717. [CrossRef] [PubMed]
57. Kerns, S.L.; Chuang, K.H.; Hall, W.; Werner, Z.; Chen, Y.; Ostrer, H.; West, C.; Rosenstein, B. Radiation biology and oncology in the genomic era. *Br. J. Radiol.* **2018**, *91*. [CrossRef] [PubMed]
58. Szołtysek, K.; Janus, P.; Zajac, G.; Stokowy, T.; Walaszczyk, A.; Widłak, W.; Wojtaś, B.; Gielniewski, B.; Cockell, S.; Perkins, N.D.; et al. RRAD, IL4I1, CDKN1A, and SERPINE1 Genes are Potentially Co-regulated by NF-κB and p53 Transcription Factors in Cells Exposed to High Doses of Ionizing Radiation. *BMC Genom.* **2018**, *19*. [CrossRef] [PubMed]
59. Janus, P.; Szołtysek, K.; Zając, G.; Stokowy, T.; Walaszczyk, A.; Widłak, W.; Wojtaś, B.; Gielniewski, B.; Iwanaszko, M.; Braun, R.; et al. Pro-inflammatory Cytokine and High Doses of Ionizing Radiation Have Similar Effects on the Expression of NF-kappaB-dependent Genes. *Cell. Signal.* **2018**, *46*, 23–31. [CrossRef] [PubMed]
60. Zyla, J.; Kabacik, S.; O'Brien, G.; Wakil, S.; Al-Harbi, N.; Kaprio, J.; Badie, C.; Polanska, J.; Alsbeih, G. Combining CDKN1A Gene Expression and Genome-wide SNPs in a Twin Cohort to Gain Insight into the Heritability of Individual Radiosensitivity. *Funct. Integr. Genom.* **2019**, *19*, 575–585. [CrossRef]
61. Tang, X.; Sun, L.; Wang, G.; Chen, B.; Luo, F. RUNX1: A Regulator of NF-κB Signaling in Pulmonary Diseases. *Curr. Protein Pept. Sci.* **2017**, *19*, 172–178. [CrossRef]
62. McNamee, J.P.; Chauhan, V. Radiofrequency Radiation and Gene/Protein Expression: A Review. *Radiat. Res.* **2009**, *172*, 265–287. [CrossRef]
63. Ierardi, J.L.; Mancia, A.; McMillan, J.; Lundqvist, M.L.; Romano, T.A.; Wise, J.P.; Warr, G.W.; Chapman, R.W. Sampling the Skin Transcriptome of the North Atlantic Right Whale. *Comp. Biochem. Physiol. Part D Genom. Proteom.* **2009**, *4*, 154–158. [CrossRef]
64. Morey, J.S.; Burek Huntington, K.A.; Campbell, M.; Clauss, T.M.; Goertz, C.E.; Hobbs, R.C.; Lunardi, D.; Moors, A.J.; Neely, M.G.; Schwacke, L.H.; et al. De novo Transcriptome Assembly and RNA-Seq Expression Analysis in Blood from Beluga Whales of Bristol Bay, AK. *Mar. Genom.* **2017**, *35*, 77–92. [CrossRef]
65. Ma, S.; Zhou, X.; Gerashchenko, M.V.; Lee, S.G.; George, J.C.; Bickham, J.W.; Gladyshev, V.N.; Suydam, R.; Seim, I. The Transcriptome of the Bowhead Whale *Balaena mysticetus* Reveals Adaptations of the Longest-lived Mammal. *Aging* **2014**, *6*, 879–899. [CrossRef]
66. Khudyakov, J.I.; Champagne, C.D.; Meneghetti, L.M.; Crocker, D.E. Blubber Transcriptome Response to Acute Stress Axis Activation Involves Transient Changes in Adipogenesis and Lipolysis in a Fasting-adapted Marine Mammal. *Sci. Rep.* **2017**, *7*, 1–12. [CrossRef] [PubMed]
67. Corominas, J.; Ramayo-Caldas, Y.; Puig-Oliveras, A.; Estellé, J.; Castelló, A.; Alves, E.; Pena, R.N.; Ballester, M.; Folch, J.M. Analysis of Porcine Adipose Tissue Transcriptome Reveals Differences in de novo Fatty Acid Synthesis in Pigs with Divergent Muscle Fatty Acid Composition. *BMC Genom.* **2013**, *14*, 843. [CrossRef]
68. Cui, J.X.; Zeng, Q.F.; Chen, W.; Zhang, H.; Zeng, Y.Q. Analysis and Preliminary Validation of the Molecular Mechanism of Fat Deposition in Fatty and Lean Pigs by High-throughput Sequencing. *Mamm. Genome* **2019**, *30*, 71–80. [CrossRef] [PubMed]
69. Vigors, S.; O'Doherty, J.V.; Bryan, K.; Sweeney, T. A Comparative Analysis of the Transcriptome Profiles of Liver and Muscle Tissue in Pigs Divergent for Feed Efficiency. *BMC Genom.* **2019**, *20*, 461. [CrossRef] [PubMed]
70. Yao, C.; Pang, D.; Lu, C.; Xu, A.; Huang, P.; Ouyang, H.; Yu, H. Data Mining and Validation of AMPK Pathway as a Novel Candidate Role Affecting Intramuscular Fat Content in Pigs. *Animals* **2019**, *9*, 137. [CrossRef] [PubMed]
71. Dudhate, A.; Shinde, H.; Tsugama, D.; Liu, S.; Takano, T. Transcriptomic Analysis Reveals the Differentially Expressed Genes and Pathways Involved in Drought Tolerance in Pearl Millet (Pennisetum glaucum (l.) r. Br). *PLoS ONE* **2018**, *13*, 1–14. [CrossRef] [PubMed]
72. Cai, H.; Li, M.; Sun, X.; Plath, M.; Li, C.; Lan, X.; Lei, C.; Huang, Y.; Bai, Y.; Qi, X.; et al. Global Transcriptome Analysis during Adipogenic Differentiation and Involvement of Transthyretin Gene in Adipogenesis in Cattle. *Front. Genet.* **2018**, *9*, 1–13. [CrossRef]
73. Resnyk, C.W.; Chen, C.; Huang, H.; Wu, C.H.; Simon, J.; Le Bihan-Duval, E.; Duclos, M.J.; Cogburn, L.A. RNA-seq Analysis of Abdominal Fat in Genetically Fat and Lean Chickens Highlights a Divergence in Expression of Genes Controlling Adiposity, Hemostasis, and Lipid Metabolism. *PLoS ONE* **2015**, *10*, 1–41. [CrossRef]
74. Viguerie, N.; Montastier, E.; Maoret, J.J.; Roussel, B.; Combes, M.; Valle, C.; Villa-Vialaneix, N.; Iacovoni, J.S.; Martinez, J.A.; Holst, C.; et al. Determinants of Human Adipose Tissue Gene Expression: Impact of Diet, Sex, Metabolic Status, and Cis Genetic Regulation. *PLoS Genet.* **2012**, *8*. [CrossRef] [PubMed]
75. Guo, F.; Si, R.; He, J.; Yuan, L.; Hai, L.; Ming, L.; Yi, L.; Ji, R. Comprehensive Transcriptome Analysis of Adipose Tissue in the Bactrian Camel Reveals Fore Hump Has More Specific Physiological Functions in Immune and Endocrine Systems. *Livest. Sci.* **2019**, *228*, 195–200. [CrossRef]

76. Zhang, D.G.; Cheng, J.; Tai, Z.P.; Luo, Z. Identification of Five Genes in Endoplasmic Reticulum (ER) Stress–apoptosis Pathways in Yellow Catfish Pelteobagrus fulvidraco and Their Transcriptional Responses to Dietary Lipid Levels. *Fish Physiol. Biochem.* **2019**. [CrossRef]
77. Kondo, H.; Suda, S.; Kawana, Y.; Hirono, I.; Nagasaka, R.; Kaneko, G.; Ushio, H.; Watabe, S. Effects of Feed Restriction on the Expression Profiles of the Glucose and Fatty Acid Metabolism-related Genes in Rainbow Trout Oncorhynchus mykiss Muscle. *Fish. Sci.* **2012**, *78*, 1205–1211. [CrossRef]
78. Kaneko, G.; Shirakami, H.; Yamada, T.; Ide, S.I.; Haga, Y.; Satoh, S.; Ushio, H. Short-term Fasting Increases Skeletal Muscle Lipid Content in Association with Enhanced mRNA Levels of Lipoprotein Lipase 1 in Lean Juvenile Red Seabream (Pagrus major). *Aquaculture* **2016**, *452*, 160–168. [CrossRef]
79. Cases, S.; Smith, S.J.; Zheng, Y.-W.; Myers, H.M.; Lear, S.R.; Sande, E.; Novak, S.; Collins, C.; Welch, C.B.; Lusis, A.J.; et al. Identification of a Gene Encoding an Acyl CoA:diacylglycerol Acyltransferase, a Key Enzyme in Triacylglycerol Synthesis. *Proc. Natl. Acad. Sci. USA* **1998**, *95*, 13018–13023. [CrossRef]
80. Podechard, N.; Ducheix, S.; Polizzi, A.; Lasserre, F.; Montagner, A.; Legagneux, V.; Fouché, E.; Saez, F.; Lobaccaro, J.M.; Lakhal, L.; et al. Dual Extraction of mRNA and Lipids from a Single Biological Sample. *Sci. Rep.* **2018**, *8*, 1–11. [CrossRef]
81. Cui, Y.; Zheng, Y.; Liu, X.; Yan, L.; Fan, X.; Yong, J.; Hu, Y.; Dong, J.; Li, Q.; Wu, X.; et al. Single-Cell Transcriptome Analysis Maps the Developmental Track of the Human Heart. *Cell Rep.* **2019**, *26*, 1934–1950.e5. [CrossRef]
82. Luecken, M.D.; Theis, F.J. Current Best Practices in Single-cell RNA-seq Analysis: A Tutorial. *Mol. Syst. Biol.* **2019**, *15*, e8746. [CrossRef]
83. McGinnis, C.S.; Patterson, D.M.; Winkler, J.; Conrad, D.N.; Hein, M.Y.; Srivastava, V.; Hu, J.L.; Murrow, L.M.; Weissman, J.S.; Werb, Z.; et al. MULTI-seq: Sample Multiplexing for Single-cell RNA Sequencing Using Lipid-tagged Indices. *Nat. Methods* **2019**, *16*, 619–626. [CrossRef]
84. Maynard, K.R.; Collado-Torres, L.; Weber, L.M.; Uytingco, C.; Barry, B.K.; Williams, S.R.; Catallini, J.L.; Tran, M.N.; Besich, Z.; Tippani, M.; et al. Transcriptome-scale Spatial Gene Expression in the Human Dorsolateral Prefrontal Cortex. *bioRxiv* **2020**. [CrossRef]
85. Han, Y.; Gao, S.; Muegge, K.; Zhang, W.; Zhou, B. Advanced Applications of RNA Sequencing and Challenges. *Bioinform. Biol. Insights* **2015**, *9*, 29–46. [CrossRef]
86. Conesa, A.; Madrigal, P.; Tarazona, S.; Gomez-Cabrero, D.; Cervera, A.; McPherson, A.; Szcześniak, M.W.; Gaffney, D.J.; Elo, L.L.; Zhang, X.; et al. A Survey of Best Practices for RNA-seq Data Analysis. *Genome Biol.* **2016**, *17*, 1–19. [CrossRef]
87. Frye, M.; Jaffrey, S.R.; Pan, T.; Rechavi, G.; Suzuki, T. RNA Modifications: What Have We Learned and Where Are We Headed? *Nat. Rev. Genet.* **2016**, *17*, 365–372. [CrossRef]
88. Zhang, Y.; Maria-Villanueva, M.; Krieger, S.; Ramesh, G.T.; Neelam, S.; Wu, H. Transcriptomics, NF-KB pathway, and Their Potential Spaceflight-related Health Consequences. *Int. J. Mol. Sci.* **2017**, *18*, 1166. [CrossRef]
89. Rashi-Elkeles, S.; Elkon, R.; Shavit, S.; Lerenthal, Y.; Linhart, C.; Kupershtein, A.; Amariglio, N.; Rechavi, G.; Shamir, R.; Shiloh, Y. Transcriptional Modulation Induced by Ionizing Radiation: P53 Remains a Central Player. *Mol. Oncol.* **2011**, *5*, 336–348. [CrossRef]
90. Narayanan, I.V.; Paulsen, M.T.; Bedi, K.; Berg, N.; Ljungman, E.A.; Francia, S.; Veloso, A.; Magnuson, B.; Di Fagagna, F.D.A.; Wilson, T.E.; et al. Transcriptional and Post-transcriptional Regulation of the Ionizing Radiation Response by ATM and p53. *Sci. Rep.* **2017**, *7*. [CrossRef]
91. Lieberman, H.B.; Panigrahi, S.K.; Hopkins, K.M.; Wang, L.; Broustas, C.G. P53 and RAD9, the DNA Damage Response, and Regulation of Transcription Networks. *Radiat. Res.* **2017**, *187*, 424–432. [CrossRef]
92. Tonelli, C.; Morelli, M.J.; Bianchi, S.; Rotta, L.; Capra, T.; Sabò, A.; Campaner, S.; Amati, B. Genome-wide Analysis of p53 Transcriptional Programs in B Cells upon Exposure to Genotoxic Stress in vivo. *Oncotarget* **2015**, *6*, 24611–24626. [CrossRef]
93. Subramanian, V.; Seemann, I.; Merl-Pham, J.; Hauck, S.M.; Stewart, F.A.; Atkinson, M.J.; Tapio, S.; Azimzadeh, O. Role of TGF Beta and PPAR Alpha Signaling Pathways in Radiation Response of Locally Exposed Heart: Integrated Global Transcriptomics and Proteomics Analysis. *J. Proteome Res.* **2017**, *16*, 307–318. [CrossRef]
94. He, G.; Tang, A.; Xie, M.; Xia, W.; Zhao, P.; Wei, J.; Lai, Y.; Tang, X.; Zou, Y.M.; Liu, H. Blood Gene Expression Profile Study Revealed the Activation of Apoptosis and p53 Signaling Pathway May Be the Potential Molecular Mechanisms of Ionizing Radiation Damage and Radiation-Induced Bystander Effects. *Dose Response* **2020**, *18*, 1559325820914184. [CrossRef]
95. Yang, L.; Ding, Y.; Chen, Y.; Zhang, S.; Huo, C.; Wang, Y.; Yu, J.; Zhang, P.; Na, H.; Zhang, H.; et al. The Proteomics of Lipid Droplets: Structure, Dynamics, and Functions of the Organelle Conserved from Bacteria to Humans. *J. Lipid Res.* **2012**, *53*, 1245–1253. [CrossRef]
96. Maeda, H.; Hosokawa, M.; Sashima, T.; Funayama, K.; Miyashita, K. Fucoxanthin from Edible Seaweed, Undaria pinnatifida, Shows Antiobesity Effect through UCP1 Expression in White Adipose Tissues. *Biochem. Biophys. Res. Commun.* **2005**, *332*, 392–397. [CrossRef] [PubMed]
97. Kershaw, J.L.; Botting, C.H.; Brownlow, A.; Hall, A.J. Not just Fat: Investigating the Proteome of Cetacean Blubber Tissue. *Conserv. Physiol.* **2018**, *6*, 1–15. [CrossRef] [PubMed]
98. Reisch, F.; Kakularam, K.R.; Stehling, S.; Heydeck, D.; Kuhn, H. Eicosanoid Biosynthesis in Marine Mammals. *FEBS J.* **2020**. [CrossRef]

99. Tian, R.; Seim, I.; Zhang, Z.; Yang, Y.; Ren, W.; Xu, S.; Yang, G. Distinct Evolution of Toll-like Receptor Signaling Pathway Genes in Cetaceans. *Genes Genom.* **2019**, *41*, 1417–1430. [CrossRef] [PubMed]
100. Xu, S.; Tian, R.; Lin, Y.; Yu, Z.; Zhang, Z.; Niu, X.; Wang, X.; Yang, G. Widespread Positive Selection on Cetacean TLR Extracellular Domain. *Mol. Immunol.* **2019**, *106*, 135–142. [CrossRef]
101. Shen, T.; Xu, S.; Wang, X.; Yu, W.; Zhou, K.; Yang, G. Adaptive evolution and functional constraint at TLR4 during the secondary aquatic adaptation and diversification of cetaceans. *BMC Evol. Biol.* **2012**, *12*, 39. [CrossRef] [PubMed]
102. Wang, X.; Wei, H.; Mao, X.; Liu, J. Proteomics Analysis of Lipid Droplets from the Oleaginous Alga Chromochloris zofingiensis Reveals Novel Proteins for Lipid Metabolism. *Genom. Proteom. Bioinform.* **2019**. [CrossRef]
103. Chen, J.; Huang, J.; Deng, J.; Ma, H.; Zou, S. Use of Comparative Proteomics to Identify the Effects of Creatine Pyruvate on Lipid and Protein Metabolism in Broiler Chickens. *Vet. J.* **2012**. [CrossRef]
104. Choi, Y.E.; Hwang, H.; Kim, H.S.; Ahn, J.W.; Jeong, W.J.; Yang, J.W. Comparative Proteomics Using Lipid Over-producing or Less-producing Mutants Unravels Lipid Metabolisms in Chlamydomonas reinhardtii. *Bioresour. Technol.* **2013**. [CrossRef]
105. Zhao, L.; Du, M.; Gao, J.; Zhan, B.; Mao, X. Label-free Quantitative Proteomic Analysis of Milk Fat Globule Membrane Proteins of Yak and Cow and Identification of Proteins Associated with Glucose and Lipid Metabolism. *Food Chem.* **2019**, *275*, 59–68. [CrossRef]
106. Wu, Y.; Tang, J.; Zhou, C.; Zhao, L.; Chen, J.; Zeng, L.; Rao, C.; Shi, H.; Liao, L.; Liang, Z.; et al. Quantitative Proteomics Analysis of the Liver Reveals Immune Regulation and Lipid Metabolism Dysregulation in a Mouse Model of Depression. *Behav. Brain Res.* **2016**, *311*, 330–339. [CrossRef]
107. Yadetie, F.; Oveland, E.; Døskeland, A.; Berven, F.; Goksøyr, A.; Karlsen, O.A. Quantitative Proteomics Analysis Reveals Perturbation of Lipid Metabolic Pathways in the Liver of Atlantic Cod (Gadus morhua) Treated with PCB 153. *Aquat. Toxicol.* **2017**. [CrossRef] [PubMed]
108. Cotter, D. LMPD: Lipid Maps Proteome Database. *Nucleic Acids Res.* **2005**, *34*, D507–D510. [CrossRef] [PubMed]
109. Fry, B.; Carter, J.F. Stable Carbon Isotope Diagnostics of Mammalian Metabolism, a High-resolution Isotomics Approach Using Amino Acid Carboxyl Groups. *PLoS ONE* **2019**, *14*, e0224297. [CrossRef]
110. Swan, A.L.; Mobasheri, A.; Allaway, D.; Liddell, S.; Bacardit, J. Application of Machine Learning to Proteomics Data: Classification and Biomarker Identification in Postgenomics Biology. *Omi. A J. Integr. Biol.* **2013**, *17*, 595–610. [CrossRef] [PubMed]
111. Leszczynski, D. Radiation Proteomics: A Brief Overview. *Proteomics* **2014**, *14*, 481–488. [CrossRef]
112. Barjaktarovic, Z.; Anastasov, N.; Azimzadeh, O.; Sriharshan, A.; Sarioglu, H.; Ueffing, M.; Tammio, H.; Hakanen, A.; Leszczynski, D.; Atkinson, M.J.; et al. Integrative Proteomic and microRNA Analysis of Primary Human Coronary Artery Endothelial Cells Exposed to Low-dose Gamma Radiation. *Radiat. Environ. Biophys.* **2013**, *52*, 87–98. [CrossRef]
113. Song, J.; Zhang, H.; Wang, Z.; Xu, W.; Zhong, L.; Cao, J.; Yang, J.; Tian, Y.; Yu, D.; Ji, J.; et al. The Role of FABP5 in Radiation-Induced Human Skin Fibrosis. *Radiat. Res.* **2018**, *189*, 177–186. [CrossRef] [PubMed]
114. Volakakis, N.; Joodmardi, E.; Perlmann, T. NR4A Orphan Nuclear Receptors Influence Retinoic Acid and Docosahexaenoic Acid Signaling via Up-regulation of Fatty Acid Binding Protein 5. *Biochem. Biophys. Res. Commun.* **2009**, *390*, 1186–1191. [CrossRef]
115. Qiu, Y.; Gao, Y.; Yu, D.; Zhong, L.; Cai, W.; Ji, J.; Geng, F.; Tang, G.; Zhang, H.; Cao, J.; et al. Genome-Wide Analysis Reveals Zinc Transporter ZIP9 Regulated by DNA Methylation Promotes Radiation-Induced Skin Fibrosis via the TGF-β Signaling Pathway. *J. Invest. Dermatol.* **2020**, *140*, 94–102.e7. [CrossRef]
116. Li, H.; Zhang, W.; Zhang, H.; Xie, Y.; Sun, C.; Di, C.; Si, J.; Gan, L.; Yan, J. Mitochondrial Proteomics Reveals the Mechanism of Spermatogenic Cells Apoptosis Induced by Carbon Ion Radiation in Zebrafish. *J. Cell. Physiol.* **2019**, *234*, 22439–22449. [CrossRef] [PubMed]
117. Azimzadeh, O.; Atkinson, M.J.; Tapio, S. Proteomics in Radiation Research: Present Status and Future Perspectives. *Radiat. Environ. Biophys.* **2014**, *53*, 31–38. [CrossRef] [PubMed]
118. Guipaud, O. Serum and Plasma Proteomics and Its Possible Use as Detector and Predictor of Radiation Diseases. In *Radiation Proteomics*; Springer: Dordrecht, The Netherlands, 2013; pp. 61–86.
119. Ulmer, C.Z.; Jones, C.M.; Yost, R.A.; Garrett, T.J.; Bowden, J.A. Optimization of Folch, Bligh-Dyer, and Matyash Sample-to-extraction Solvent Ratios for Human Plasma-based Lipidomics Studies. *Anal. Chim. Acta* **2018**, *1037*, 351–357. [CrossRef] [PubMed]
120. Li, L.; Han, J.; Wang, Z.; Liu, J.; Wei, J.; Xiong, S.; Zhao, Z. Mass Spectrometry Methodology in Lipid Analysis. *Int. J. Mol. Sci.* **2014**, *15*, 10492–10507. [CrossRef] [PubMed]
121. Bozek, K.; Khrameeva, E.E.; Reznick, J.; Omerbašić, D.; Bennett, N.C.; Lewin, G.R.; Azpurua, J.; Gorbunova, V.; Seluanov, A.; Regnard, P.; et al. Lipidome Determinants of Maximal Lifespan in Mammals. *Sci. Rep.* **2017**, *7*, 1–10. [CrossRef]
122. Parekh, A.; Smeeth, D.; Milner, Y.; Thuret, S. The Role of Lipid Biomarkers in Major Depression. *Healthcare* **2017**, *5*, 5. [CrossRef]
123. Gardner, S.C.; Varanasi, U. Isovaleric Acid Accumulation in Odontocete Melon during Development. *Naturwissenschaften* **2003**, *90*, 528–531. [CrossRef]
124. Krahn, M.M.; Herman, D.P.; Ylitalo, G.M.; Sloan, C.A.; Burrows, D.G.; Hobbs, R.C.; Mahoney, B.A.; Yanagida, G.K.; Calambokidis, J.; Moore, S.E. Stratification of Lipids, Fatty Acids and Organochlorine Contaminants in Blubber of White Whales and Killer Whales. *J. Cetacean Res. Manag.* **2004**, *6*, 175–189.

125. Meier, S.; Falk-Petersen, S.; Aage Gade-Sørensen, L.; Greenacre, M.; Haug, T.; Lindstrøm, U. Fatty Acids in Common Minke Whale (Balaenoptera acutorostrata) Blubber Reflect the Feeding Area and Food Selection, but also High Endogenous Metabolism. *Mar. Biol. Res.* **2016**, *12*, 221–238. [CrossRef]
126. Nelson, D.W.; Eric Yen, C.-L. Triacylglycerol Synthesis and Energy Metabolism: A Gut Reaction? *Clin. Lipidol.* **2009**, *4*, 683–686. [CrossRef]
127. Chen, H.K.; Wang, L.H.; Chen, W.N.U.; Mayfield, A.B.; Levy, O.; Lin, C.S.; Chen, C.S. Coral Lipid Bodies as the Relay Center Interconnecting Diel-dependent Lipidomic Changes in Different Cellular Compartments. *Sci. Rep.* **2017**, *7*, 1–13. [CrossRef]
128. Shevchenko, A.; Simons, K. Lipidomics: Coming to grips with lipid diversity. *Nat. Rev. Mol. Cell Biol.* **2010**, *11*, 593–598. [CrossRef]
129. Baker, P.R.S.; Armando, A.M.; Campbell, J.L.; Quehenberger, O.; Dennis, E.A. Three-dimensional Enhanced Lipidomics Analysis Combining UPLC, Differential Ion Mobility Spectrometry, and Mass Spectrometric Separation Strategies. *J. Lipid Res.* **2014**, *55*, 2432–2442. [CrossRef]
130. Röttig, A.; Zurek, P.J.; Steinbüchel, A. Assessment of Bacterial Acyltransferases for an Efficient Lipid Production in Metabolically Engineered Strains of E. coli. *Metab. Eng.* **2015**, *32*, 195–206. [CrossRef]
131. Santala, S.; Efimova, E.; Karp, M.; Santala, V. Real-Time Monitoring of Intracellular Wax Ester Metabolism. *Microb. Cell Fact.* **2011**, *10*, 75. [CrossRef]
132. Koelmel, J.P.; Napolitano, M.P.; Ulmer, C.Z.; Vasiliou, V.; Garrett, T.J.; Yost, R.A.; Prasad, M.N.V.; Godri Pollitt, K.J.; Bowden, J.A. Environmental Lipidomics: Understanding the Response of Organisms and Ecosystems to a Changing World. *Metabolomics* **2020**, *16*, 56. [CrossRef]
133. Waksman, B.H.; Porter, H.; Lees, M.D.; Adams, R.D.; Folch, J. A Study of the Chemical Nature of Components of Bovine White Matter Effective in Producing Allergic Encephalomyelitis in the Rabbit. *J. Exp. Med.* **1954**, *100*, 451–471. [CrossRef] [PubMed]
134. Löfgren, L.; Forsberg, G.B.; Ståhlman, M. The BUME Method: A New Rapid and Simple Chloroform-free Method for Total Lipid Extraction of Animal Tissue. *Sci. Rep.* **2016**, *6*, 1–11. [CrossRef]
135. Breil, C.; Abert Vian, M.; Zemb, T.; Kunz, W.; Chemat, F. "Bligh and Dyer" and Folch Methods for Solid–liquid–liquid Extraction of Lipids from Microorganisms. Comprehension of Solvatation Mechanisms and Towards Substitution with Alternative Solvents. *Int. J. Mol. Sci.* **2017**, *18*, 708. [CrossRef] [PubMed]
136. Han, X.; Jiang, X. A Review of Lipidomic Technologies Applicable to Sphingolipidomics and Their Relevant Applications. *Eur. J. Lipid Sci. Technol.* **2009**, *111*, 39–52. [CrossRef] [PubMed]
137. Sadowski, T.; Klose, C.; Gerl, M.J.; Wójcik-Maciejewicz, A.; Herzog, R.; Simons, K.; Reich, A.; Surma, M.A. Large-scale Human Skin Lipidomics by Quantitative, High-throughput Shotgun Mass Spectrometry. *Sci. Rep.* **2017**, *7*, 1–11. [CrossRef] [PubMed]
138. Barnes, S.; Benton, H.P.; Casazza, K.; Cooper, S.J.; Cui, X.; Du, X.; Engler, J.; Kabarowski, J.H.; Li, S.; Pathmasiri, W.; et al. Training in Metabolomics Research. I. Designing the Experiment, Collecting and Extracting Samples and Generating Metabolomics Data. *J. Mass Spectrom.* **2016**, 461–475. [CrossRef]
139. Jung, J.L.; Simon, G.; Alfonsi, E.; Thoraval, D.; Kervarec, N.; Ben Salem, D.; Hassani, S.; Domergue, F. Qualitative and Quantitative Study of the Highly Specialized Lipid Tissues of Cetaceans Using HR-MAS NMR and Classical GC. *PLoS ONE* **2017**, *12*, 1–23. [CrossRef] [PubMed]
140. Garikapati, V.; Karnati, S.; Bhandari, D.R.; Baumgart-Vogt, E.; Spengler, B. High-resolution Atmospheric-pressure MALDI Mass Spectrometry Imaging Workflow for Lipidomic Analysis of Late Fetal Mouse Lungs. *Sci. Rep.* **2019**, *9*, 1–14. [CrossRef]
141. Gupta, K.; Li, J.; Liko, I.; Gault, J.; Bechara, C.; Wu, D.; Hopper, J.T.S.; Giles, K.; Benesch, J.L.P.; Robinson, C.V. Identifying Key Membrane Protein Lipid Interactions Using Mass Spectrometry. *Nat. Protoc.* **2018**, *13*, 1106–1120. [CrossRef]
142. Pham, T.H.; Zaeem, M.; Fillier, T.A.; Nadeem, M.; Vidal, N.P.; Manful, C.; Cheema, S.; Cheema, M.; Thomas, R.H. Targeting Modified Lipids during Routine Lipidomics Analysis Using HILIC and C30 Reverse Phase Liquid Chromatography Coupled to Mass Spectrometry. *Sci. Rep.* **2019**, *9*, 1–15. [CrossRef]
143. Byeon, S.K.; Kim, J.Y.; Lee, J.Y.; Chung, B.C.; Seo, H.S.; Moon, M.H. Top-down and Bottom-up Lipidomic Analysis of Rabbit Lipoproteins under Different Metabolic Conditions Using Flow Field-flow Fractionation, Nanoflow Liquid Chromatography and Mass Spectrometry. *J. Chromatogr. A* **2015**, *1405*, 140–148. [CrossRef]
144. Tumanov, S.; Kamphorst, J.J. Recent Advances in Expanding the Coverage of the Lipidome. *Curr. Opin. Biotechnol.* **2017**, *43*, 127–133. [CrossRef]
145. Tyagi, M.G.; Varshney, P.; Sagar, K.D.; Nirmala, J. Recent Analytical Trends in Lipidomics; Techniques and Applications in Clinical Medicine. *J. Pharm. Biol. Sci.* **2016**, *11*, 88–92. [CrossRef]
146. Fernandis, A.Z.; Wenk, M.R. Lipid-based Biomarkers for Cancer. *J. Chromatogr. B Anal. Technol. Biomed. Life Sci.* **2009**, *877*, 2830–2835. [CrossRef]
147. Castrillon, J.; Huston, W.; Bengtson Nash, S. The Blubber Adipocyte Index: A Nondestructive Biomarker of Adiposity in Humpback Whales (*Megaptera novaeangliae*). *Ecol. Evol.* **2017**, *7*, 5131–5139. [CrossRef] [PubMed]
148. Albergamo, A.; Rigano, F.; Purcaro, G.; Mauceri, A.; Fasulo, S.; Mondello, L. Free Fatty Acid Profiling of Marine Sentinels by nanoLC-EI-MS for the Assessment of Environmental Pollution Effects. *Sci. Total Environ.* **2016**, *571*, 955–962. [CrossRef]
149. Pannkuk, E.L.; Laiakis, E.C.; Singh, V.K.; Fornace, A.J. Lipidomic Signatures of Nonhuman Primates with Radiation-Induced Hematopoietic Syndrome. *Sci. Rep.* **2017**, *7*. [CrossRef]

150. Tyurina, Y.Y.; Tyurin, V.A.; Kapralova, V.I.; Wasserloos, K.; Mosher, M.; Epperly, M.W.; Greenberger, J.S.; Pitt, B.R.; Kagan, V.E. Oxidative Lipidomics of γ-radiation-induced Lung Injury: Mass Spectrometric Characterization of Cardiolipin and Phosphatidylserine Peroxidation. *Radiat. Res.* **2011**, *175*, 610–621. [CrossRef] [PubMed]
151. Tyurina, Y.Y.; Tyurin, V.A.; Anthonymuthu, T.; Amoscato, A.A.; Sparvero, L.J.; Nesterova, A.M.; Baynard, M.L.; Sun, W.; He, R.R.; Khaitovich, P.; et al. Redox Lipidomics Technology: Looking for a Needle in a Haystack. *Chem. Phys. Lipids* **2019**, *221*, 93–107. [CrossRef]
152. Shadyro, O.I.; Yurkova, I.L.; Kisel, M.A. Radiation-induced Peroxidation and Fragmentation of Lipids in a Model Membrane. *Int. J. Radiat. Biol.* **2002**, *78*, 211–217. [CrossRef] [PubMed]
153. Xi, C.; Zhao, H.; Lu, X.; Cai, T.J.; Li, S.; Liu, K.H.; Tian, M.; Liu, Q.J. Screening of Lipids for Early Triage and Dose Estimation after Acute Radiation Exposure in Rat Plasma Based on Targeted Lipidomics Analysis. *J. Proteome Res.* **2020**. [CrossRef] [PubMed]
154. Huang, J.; Wang, Q.; Qi, Z.; Zhou, S.; Zhou, M.; Wang, Z. Lipidomic Profiling for Serum Biomarkers in Mice Exposed to Ionizing Radiation. *Dose Response* **2020**, *18*, 1559325820914209. [CrossRef]
155. Pilkington, S.M.; Watson, R.E.B.; Nicolaou, A.; Rhodes, L.E. Omega-3 polyunsaturated Fatty Acids: Photoprotective Macronutrients. *Exp. Dermatol.* **2011**, *20*, 537–543. [CrossRef]
156. Aristizabal-Henao, J.J.; Ahmadireskety, A.; Griffin, E.K.; Ferreira Da Silva, B.; Bowden, J.A. Lipidomics and Environmental Toxicology: Recent Trends. *Curr. Opin. Environ. Sci. Heal.* **2020**, *15*, 26–31. [CrossRef]
157. Li, W.; Kuang, Z.; Zheng, M.; He, G.; Liu, Y. Multi-omics Integrative Analysis to Access Role of Coiled-coil Domain-containing 80 in Lipid Metabolism. *Biochem. Biophys. Res. Commun.* **2020**, *526*, 813–819. [CrossRef]
158. Kudryashova, K.S.; Burka, K.; Kulaga, A.Y.; Vorobyeva, N.S.; Kennedy, B.K. Aging Biomarkers: From Functional Tests to Multi-Omics Approaches. *Proteomics* **2020**, *20*. [CrossRef]
159. Montaner, J.; Ramiro, L.; Simats, A.; Tiedt, S.; Makris, K.; Jickling, G.C.; Debette, S.; Sanchez, J.C.; Bustamante, A. Multilevel Omics for the Discovery of Biomarkers and Therapeutic Targets for Stroke. *Nat. Rev. Neurol.* **2020**, *16*, 247–264. [CrossRef]
160. Panousis, N.I.; Bertsias, G.K.; Ongen, H.; Gergianaki, I.; Tektonidou, M.G.; Trachana, M.; Romano-Palumbo, L.; Bielser, D.; Howald, C.; Pamfil, C.; et al. Combined Genetic and Transcriptome Analysis of Patients with SLE: Distinct, Targetable Signatures for Susceptibility and Severity. *Ann. Rheum. Dis.* **2019**, *78*, 1079–1089. [CrossRef] [PubMed]
161. Grapov, D.; Fahrmann, J.; Wanichthanarak, K.; Khoomrung, S. Rise of Deep Learning for Genomic, Proteomic, and Metabolomic Data Integration in Precision Medicine. *Omi. A J. Integr. Biol.* **2018**, *22*, 630–636. [CrossRef] [PubMed]
162. Xavier, M.A.; Tirloni, L.; Pinto, A.F.M.; Diedrich, J.K.; Yates, J.R.; Gonzales, S.; Farber, M.; da Silva Vaz, I.; Termignoni, C. Tick Gené's Organ Engagement in Lipid Metabolism Revealed by a Combined Transcriptomic and Proteomic Approach. *Ticks Tick. Borne. Dis.* **2019**, *10*, 787–797. [CrossRef]
163. Manzoni, C.; Kia, D.A.; Vandrovcova, J.; Hardy, J.; Wood, N.W.; Lewis, P.A.; Ferrari, R. Genome, Transcriptome and Proteome: The Rise of Omics Data and Their Integration in Biomedical Sciences. *Brief. Bioinform.* **2018**, *19*, 286–302. [CrossRef] [PubMed]
164. Gaetz, W.; Jantzen, K.; Weinberg, H.; Spong, P.; Symonds, H. Neural network method for recognition of individual Orcinus orca based on their acoustic behaviour: Phase 1. In Proceedings of the OCEANS'93, Victoria, Canada, 18–21 October 1993; pp. 455–457. [CrossRef]
165. Mikolov, T.; Deoras, A.; Povey, D.; Burget, L.; Černocký, J. Strategies for training large scale neural network language models. In Proceedings of the 2011 IEEE Workshop on Automatic Speech Recognition & Understanding, Big Island, HI, USA, 11–15 December 2011; pp. 196–201. [CrossRef]
166. Ma, T.; Zhang, A. Integrate Multi-omics Data with Biological Interaction Networks Using Multi-view Factorization AutoEncoder (MAE). *BMC Genom.* **2019**, *20*, 1–11. [CrossRef]
167. Guo, H.; Chou, W.C.; Lai, Y.; Liang, K.; Tam, J.W.; Brickey, W.J.; Chen, L.; Montgomery, N.D.; Li, X.; Bohannon, L.M.; et al. Multi-omics Analyses of Radiation Survivors Identify Radioprotective Microbes and Metabolites. *Science* **2020**, *370*. [CrossRef]
168. Shen, Y.; Stanislauskas, M.; Li, G.; Zheng, D.; Liu, L. Epigenetic and Genetic Dissections of UV-induced Global Gene Dysregulation in Skin Cells through Multi-omics Analyses. *Sci. Rep.* **2017**, *7*, 1–12. [CrossRef]
169. Rakwal, R.; Hayashi, G.; Shibato, J.; Deepak, S.A.; Gundimeda, S.; Simha, U.; Padmanaban, A.; Gupta, R.; Han, S.I.; Kim, S.T.; et al. Progress toward Rice Seed OMICS in Low-level Gamma Radiation Environment in Iitate Village, Fukushima. *J. Hered.* **2018**, *109*, 206–211. [CrossRef]
170. Ali, A.; Rashid, M.A.; Huang, Q.Y.; Lei, C.L. Effect of UV-A Radiation as an Environmental Stress on the Development, Longevity, and Reproduction of the Oriental Armyworm, Mythimna separata (Lepidoptera: Noctuidae). *Environ. Sci. Pollut. Res.* **2016**, *23*, 17002–17007. [CrossRef]
171. Xu, T.; Zhao, K.; Guo, X.; Tu, J.; Zhang, D.; Sun, W.; Kong, X. Low-intensity Pulsed Ultrasound Inhibits Adipogenic Differentiation via HDAC1 Signalling in Rat Visceral Preadipocytes. *Adipocyte* **2019**, *8*, 292–303. [CrossRef]
172. Zhang, J.; Shrivastava, S.; Cleveland, R.O.; Rabbitts, T.H. Lipid-mRNA Nanoparticle Designed to Enhance Intracellular Delivery Mediated by Shock Waves. *ACS Appl. Mater. Interfaces* **2019**, *11*, 10481–10491. [CrossRef] [PubMed]
173. Koda, R.; Koido, J.; Hosaka, N.; Onogi, S.; Mochizuki, T.; Masuda, K.; Suzuki, R.; Maruyama, K. Evaluation of Active Control of Bubble Liposomes in a Bifurcated Flow under Various Ultrasound Conditions. *Adv. Biomed. Eng.* **2014**, *3*, 21–28. [CrossRef]
174. Suk, K.T.; Kim, D.Y.; Sohn, K.M.; Kim, D.J. *Biomarkers of Liver Fibrosis*; Academic Press Inc.: Cambridge, MA, USA, 2013; Volume 62, pp. 33–122.
175. Barnett, K. *Underwater Noise-The Neglected Threat to Marine Life*; Coalition Clean Baltic: Uppsala, Sweden, 2020.

176. Blackwell, S.B.; Thode, A.M. Effects of Noise. In *The Bowhead Whale*; Elsevier: Amsterdam, The Netherlands, 2021; pp. 565–576.
177. Halliday, W.D.; Pine, M.K.; Insley, S.J. Underwater Noise and Arctic Marine Mammals: Review and Policy Recommendations. *Environ. Rev.* **2020**, *28*, 438–448. [CrossRef]
178. Halkias, X.C.; Paris, S.; Glotin, H. Classification of Mysticete Sounds Using Machine Learning Techniques. *J. Acoust. Soc. Am.* **2013**, *134*, 3496–3505. [CrossRef]
179. Bianco, M.J.; Gerstoft, P.; Traer, J.; Ozanich, E.; Roch, M.A.; Gannot, S.; Deledalle, C.A.; Li, W. Machine learning in acoustics: Theory and applications. *J. Acoust. Soc. Am.* **2019**, *146*, 3590–3628. [CrossRef] [PubMed]
180. Shamir, L.; Yerby, C.; Simpson, R.; von Benda-Beckmann, A.M.; Tyack, P.; Samarra, F.; Miller, P.; Wallin, J. Classification of Large Acoustic Datasets Using Machine Learning and Crowdsourcing: Application to Whale Calls. *J. Acoust. Soc. Am.* **2014**, *135*, 953–962. [CrossRef]
181. Jarrett, D.; Stride, E.; Vallis, K.; Gooding, M.J. Applications and Limitations of Machine Learning in Radiation Oncology. *Br. J. Radiol.* **2019**, *92*, 1–12. [CrossRef]

life

MDPI

Article

Radionuclides Transfer from Soil to Tea Leaves and Estimation of Committed Effective Dose to the Bangladesh Populace

Nurul Absar [1], Jainal Abedin [1], Md. Mashiur Rahman [2], Moazzem Hossain Miah [3], Naziba Siddique [3], Masud Kamal [4], Mantazul Islam Chowdhury [4], Abdelmoneim Adam Mohamed Sulieman [5], Mohammad Rashed Iqbal Faruque [6], Mayeen Uddin Khandaker [7,*], David Andrew Bradley [7,8] and Abdullah Alsubaie [9]

1 Department of Computer Science and Engineering, BGC Trust University Bangladesh, Chittagong 4381, Bangladesh; nabsar@bgctub.ac.bd (N.A.); abedinj7110@bgctub.ac.bd (J.A.)
2 Hughes Network Systems, 11717 Exploration Lane, Germantown, MD 20876, USA; mashiur.rahman@hughes.com
3 Department of Physics, University of Chittagong, Chittagong 4331, Bangladesh; mhmiah@cu.ac.bd (M.H.M.); snaziba@yahoo.com (N.S.)
4 Atomic Energy Centre-Chittagong, Radioactivity Testing and Monitoring Laboratory, Bangladesh Atomic Energy Commission, Chittagong 4209, Bangladesh; masud.kamal@gmail.com (M.K.); mantaz54@yahoo.com (M.I.C.)
5 Department of Radiology and Medical Imaging, College of Applied Medical Sciences, Prince Sattam Bin Abdulaziz University, P.O. Box 422, Alkharj 11942, Saudi Arabia; a.sulieman@psau.edu.sa
6 Space Science Centre (ANGKASA), Universiti Kebangsaan Malaysia, Bangi 43600, Malaysia; rashed@ukm.edu.my
7 Centre for Applied Physics and Radiation Technologies, School of Engineering and Technology, Sunway University, Bandar Sunway 47500, Malaysia; d.a.bradley@surrey.ac.uk
8 Centre for Nuclear and Radiation Physics, Department of Physics, University of Surrey, Guildford, Surrey GU2 7XH, UK
9 Department of Physics, College of Khurma, Taif University, P.O. Box 11099, Taif 21944, Saudi Arabia; a.alsubaie@tu.edu.sa
* Correspondence: mayeenk@sunway.edu.my

Citation: Absar, N.; Abedin, J.; Rahman, MM.; Miah, M.M.H.; Siddique, N.; Kamal, M.; Chowdhury, M.I.; Sulieman, A.A.M.; Faruque, M.R.I.; Khandaker, M.U.; et al. Radionuclides Transfer from Soil to Tea Leaves and Estimation of Committed Effective Dose to the Bangladesh Populace. *Life* **2021**, *11*, 282. https://doi.org/10.3390/life11040282

Academic Editors: Fabrizio Ambrosino and Supitcha Chanyotha

Received: 2 March 2021
Accepted: 24 March 2021
Published: 27 March 2021

Abstract: Considering the probable health risks due to radioactivity input via drinking tea, the concentrations of ^{226}Ra, ^{232}Th, ^{40}K and ^{137}Cs radionuclides in the soil and the corresponding tea leaves of a large tea plantation were measured using high purity germanium (HPGe) γ-ray spectrometry. Different layers of soil and fresh tea leaf samples were collected from the Udalia Tea Estate (UTE) in the Fatickchari area of Chittagong, Bangladesh. The mean concentrations (in Bq/kg) of radionuclides in the studied soil samples were found to be 34 $\pm$ 9 to 45 $\pm$ 3 for ^{226}Ra, 50 $\pm$ 13 to 63 $\pm$ 5 for ^{232}Th, 245 $\pm$ 30 to 635 $\pm$ 35 for ^{40}K and 3 $\pm$ 1 to 10 $\pm$ 1 for ^{137}Cs, while the respective values in the corresponding tea leaf samples were 3.6 $\pm$ 0.7 to 5.7 $\pm$ 1.0, 2.4 $\pm$ 0.5 to 5.8 $\pm$ 0.9, 132 $\pm$ 25 to 258 $\pm$ 29 and <0.4. The mean transfer factors for ^{226}Ra, ^{232}Th and ^{40}K from soil to tea leaves were calculated to be 0.12, 0.08 and 0.46, respectively, the complete range being 1.1×10^{-2} to 1.0, in accordance with IAEA values. Additionally, the most popularly consumed tea brands available in the Bangladeshi market were also analyzed and, with the exception of ^{40}K, were found to have similar concentrations to the fresh tea leaves collected from the UTE. The committed effective dose via the consumption of tea was estimated to be low in comparison with the United Nations Scientific Committee on the Effects of Atomic Radiation (UNSCEAR) reference ingestion dose limit of 290 μSv/y. Current indicative tea consumption of 4 g/day/person shows an insignificant radiological risk to public health, while cumulative dietary exposures may not be entirely negligible, because the UNSCEAR reference dose limit is derived from total dietary exposures. This study suggests a periodic monitoring of radiation levels in tea leaves in seeking to ensure the safety of human health.

Keywords: soil; tea leaves; HPGe γ-ray spectrometry; terrestrial and anthropogenic radionuclides; threshold consumption rate; committed effective dose

1. Introduction

The most common forms of ionizing radiation on earth, resulting from terrestrial, extra-terrestrial and anthropogenic sources, are α- and β-particles and γ-rays [1,2]. According to the National Council on Radiation Protection and Measurements (NCRP), environmental radiation is the most significant source of radiation exposure to humans [3]. Interestingly, although the International Atomic Energy Agency (IAEA) reported that the public exposure from natural radiation is of little health concern [4], the World Nuclear Association (WNA) states that any dose of radiation involves a possible risk to human health [5]. Since ^{238}U and ^{232}Th decay series and singly occurring ^{40}K are the most abundant radionuclides found naturally in soil, air, water, rocks, plants and foodstuffs [5], to protect the public against unwanted exposures to natural radiation, the radioactivity in environmental samples, including foodstuffs, needs to be monitored periodically [4].

The surface soil, especially the top layer in the earth, is a mixture of various components in the natural environment [6,7]. Although the main source of U, Th and K is the earth crust, their contents appear at varying levels in the soils of different regions in the world following the variation of the local geology [8]. In addition to the prevailing concentration of terrestrial radionuclides in acidic soils, at high altitudes the contribution from extra-terrestrial radioactivity may be enhanced, moreover with fallout of artificial radionuclides that may be subject to greater deposition. The latter concerns atmospheric nuclear device testing or unplanned phenomena such as the Three Mile Island power plant accident, the Chernobyl accident and the Fukushima Dai-ichi nuclear power plant accident [9–12]. Regardless of origin (i.e., natural or artificial), the radionuclides may appear in plants along with the uptake of minerals and nutrients required for their vitality, majorly by the root system [13]. Their availability in plant life enables them to be transported to humans via the daily intake of foodstuffs [14–16]. Thus, it is necessary to know the natural radioactivity in a particular area to obtain their distribution, modelling and transport processes leading to the estimation of radiation dose and hazards to the general population [6].

Due to its pleasant taste, aroma and positive physiological functions, tea is one of the most popular stimulating drinks in the world after water [17–19]. It is generally obtained by processing the leaves and leaf buds of the Camellia sinensis plant. Primarily there are four types of processed teas: green, black, oolong and brick tea [20]. Green tea is obtained by drying and roasting the tea leaves without any fermentation, whereas an additional fermentation produces black tea leaves. A partial fermentation of tea leaves results in oolong tea. Brick tea is made from the blocks of whole or finely ground black tea, compressed in a form under extremely high pressure. Asian populations generally drink the semi-fermented green tea and fermented black teas [21] as a hot drink. While, at present, tea is cultivated in more than 40 countries in the world, the major portion (90%) is produced by Asian countries [22]. Due to the presence of biologically active compounds (antitoxin, antioxidant, anti-inflammatory, antibacterial, antiviral, anti-carcinogenic, etc.) like polyphenols, amino acids and vitamins in tea [23,24], tea drinking has been promoted for centuries [25–27]. Controversies about the benefits and risks due to the consumption of tea are not completely absent, but the few reported toxic effects are outclassed by its countless health-promoting benefits [28]. Harmful effects like stomach ache, intestinal gas, heartburn and abdominal pain from over consumption of tea are identified due to the presence of caffeine, aluminum and the influence of tea polyphenols on iron bioavailability [29].

Following the suitable geographical location and weather conditions for tea plantation, tea plantations were established in the hilly areas of Sylhet, Moulavibazar, Comilla and Chittagong regions in Bangladesh, centuries ago [30]. Moreover, due to recent developments in socio-economic conditions, tea consumption is increasing at a significant level among Bangladeshi population [31]. In producing more than 95 million kg of tea in 2019, harvested from about 115,757 hectares of land, Bangladesh has become the world's tenth largest tea producer and the world's ninth tea exporting country [32]. As a member of Bangladesh Tea Board and Bangladesh Tea Association, the Udalia Tea Estate (UTE), since 1962, has played a very important role in quality tea production. In recent times, the UTE has come to be ranked highly among the existing tea estates in Bangladesh. In meeting the growing domestic demand for premium black and green tea as well as that of international markets, UTE has secured a position not only as a tea producer but also as a multi-product estate. In particular, it enjoys abundant rainfall, is in an area of elevated altitude above sea level and has acidic and well-draining soils, all combining to make favorable conditions for tea production.

Since plants produce their necessary energy via the use of leaves together with the photosynthesis process, the leaves may have more activity than the other parts of plants, therefore they may contain relatively more radionuclides which are normally taken up by the root system along with other minerals. Moreover, leaves are more exposed to aerial deposition, e.g., radionuclides in dust or radionuclide particles themselves, if there are any artificial phenomena in the surrounding environment or nearby countries. Therefore, tea plants may be subjected to direct and indirect contamination of various radionuclides, and these radionuclides can be distributed in different parts of the plants according to the chemical characteristics and parameters of the plants and soil [33]. Since tea forms the second most popular drink (after water) in all populations in Bangladesh, the presence of a low level of radioactive material in tea leaves may produce a non-negligible health hazard via cumulative exposures. By acknowledging that ingestion doses above permissible levels are harmful for human beings [34], assessment of radionuclides due to the consumption of foodstuffs is important for public health. Furthermore, assessment of any release of radioactivity to the environment is important for the protection of public health, especially if the released radioactivity can enter into the food chain [35].

While studies of natural radioactivity in various foodstuffs are available in the literature, information on the distribution and enrichment of radionuclides in tea leaves is sparse, especially in tea leaves collected from the major tea gardens including the UTE in Bangladesh. The main objectives of the present study are to determine the transfer of radionuclides from soil to tea leaves harvested from the Udalia Tea Estate, further calculating the associated health hazards following the consumption of tea by the populace in Bangladesh. The activity concentrations of marketed tea leaves were also analyzed to observe the effect of manufacturing processes. This study may also help to enrich the radioactivity database on tea, i.e., the most popular stimulating drink.

2. Experimental

2.1. Local Geology of the Udalia Tea Estate

The Udalia Tea Estate (UTE) is located in the hilly region of Fatickchari Upazilla in the Chittagong district of Bangladesh (see in Figure 1). The UTE can be addressed as latitude 22°36′39″–22°39′41″ N and longitude 91°45′6″–91°51′15″ E. The estate is covered by low hill ranges and terraces having an altitude of 30–46 m above sea level and contains an area of 3096 acres (3.9 km × 3.2 km = 12.5 square kilometer) [36]. The climate of this area is tropical monsoon. An average annual rainfall of about 2794 mm is recorded in this area, while July is the wettest month. The soils in this area are mainly yellowish to reddish brown, the texture is mostly clay loam on flat land while the hilly soil is mostly sandy loam to coarse sand, which is characterized by broken shale or sandstone and mottled sand at different depths [37]. The soils are strongly acidic and poor in organic matter and nutrients.

2.2. Sample Collection and Preparation

The samples (soil and tea leaves) were collected from different places in the Udalia Tea Estate (UTE). The sampling location was chosen on a random basis, but the distance between each sampling location was almost 700–800 m. Since the estate consists of ranges of low lying hills, separate hills were chosen as different sampling locations. A total of 5 locations were chosen for collecting the tea leaves and the surrounding soil samples throughout the garden. While an approximate amount of 2.5 kg of fresh tea leaf samples were collected from each selected location around the UTE, a total of 4 (×3) soil samples were also collected from three different depths, 0 to 5 cm, 6 to 12 cm and 13 to 20 cm, of the corresponding locations to the tea leaves samples. More specifically, the soil samples were collected from four different points within an area of 1 m^2 around the tea tree. The 25-year-old tea trees were selected for collection of the tea leaf samples. Usually, the branches of the tea plant are cut and are fertilized twice per year, and in some cases fertilizers are used more than twice for influencing the growth of the tea plant in the garden. At present, the UTE produces approximately 0.7 million tons of tea per year with six separate grades of tea including the export quality one. In the local market and throughout the country, this estate supplies one of the popular tea brands "Mostafa tea". Five marketed tea leaf samples were also collected from the local market, allowing comparison of the measured radioactivity from these with that from the fresh tea leaves collected from the UTE.

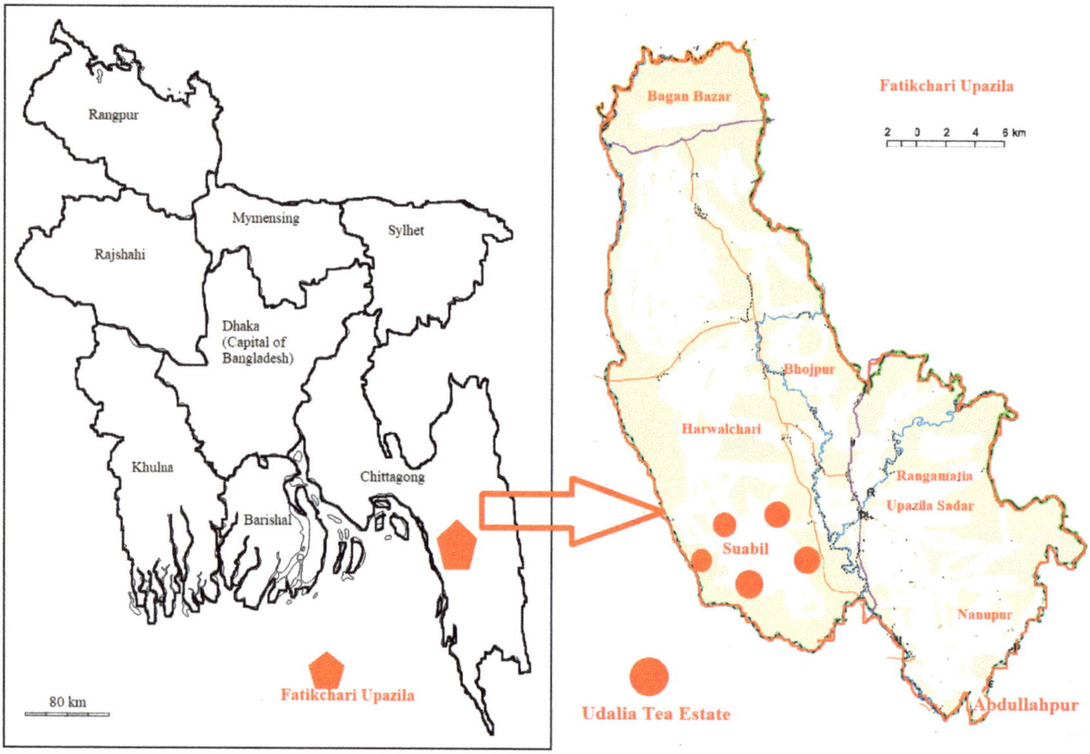

Figure 1. Location of the Udalia Tea Estate at the Fatikchari Upazila of the Chittagong district in Bangladesh.

The procedure for sample collection followed that recommended by the IAEA [4]. The collected samples were stored separately in sealed plastic bags and tagged with an identification number, and with the date and location of sampling. The samples were dried under direct sunlight for several days to allow evaporation of moisture content, subsequently being further dried for a period of 24 h in an oven maintained at 85 °C to remove any remaining moisture. The dried samples were then mechanically crushed into a fine powder, homogenized with a mortar and pestle and filtered through a sieve of 0.395 mm mesh size to obtain similarly sized particles. A constant dry weight was measured out for each sample evaluation. For the determination of activity concentration, the dried sample was transferred to an individual cylindrical container having dimension of 3.5 cm height and 8.5 cm diameter. To settle and obtain a homogeneous mixture of the samples, the containers were simply shaken by hand. The containers were then sealed tightly by using an insulating tape to reduce the possibility of moisture contamination. The samples were then kept undisturbed for 5–7 weeks at room temperature to attain secular equilibrium between short-lived progeny with the respective long-lived parents, ^{226}Ra (from ^{238}U) and ^{228}Ra (from ^{232}Th) [38,39]. It was assumed that ^{222}Rn and ^{220}Rn could not escape from the sealed containers during the period of storage. The samples were then ready for subsequent measurement and analysis by γ-ray spectrometry.

2.3. Measurement of Radionuclides

This study used a co-axial high-purity germanium (HPGe) γ-ray detector (GC2018, CANBERRA, USA), having a relative efficiency of 20%, resolution of 1.8 keV at 1332 keV of peak of ^{60}Co source, to measure the samples and standards obtained from the IAEA. The detector was coupled with a digital spectrum analyzer and GENIE 2000 to acquire the γ-ray spectra emitted from the samples. To ensure a low background environment, a cylindrical lead shielded arrangement (5.08 cm thick) with fixed bottom and movable cover was installed. The efficiency of the detector was measured using the reference samples RGU-1, RGTh-1 and RGK-1 provided by the IAEA [40], with results as presented in Figure 2. The standard sources containing known concentrations of ^{226}Ra, ^{232}Th and ^{40}K were supplied by the Canada Center for Mineral and Energy Technology (CAMET) under a contract with the IAEA. Considering the leaves' texture and density, a radioactive standard with leafy vegetables was prepared by mixing/spiking ^{226}Ra standard source of solid matrices in identical containers to the samples, and using them accordingly. Necessary information on the calibration of the efficiency of the detector is available elsewhere [41]. In this study, each sample was measured for 10,000 seconds to achieve reasonable statistics. The net count rate from the primordial radionuclides originating from the samples was obtained by subtracting the background count from the gross count, both acquired for the same counting time. The activity concentrations of ^{226}Ra and ^{232}Th radionuclides were assessed using the characteristic gamma lines of their short-lived progeny [42,43]. The concentrations of ^{40}K and ^{137}Cs were determined by the gamma ray lines of 1460.77 keV and 661 keV, respectively. For evaluation of ^{226}Ra and ^{232}Th, a weighted mean approach was applied following reference [14].

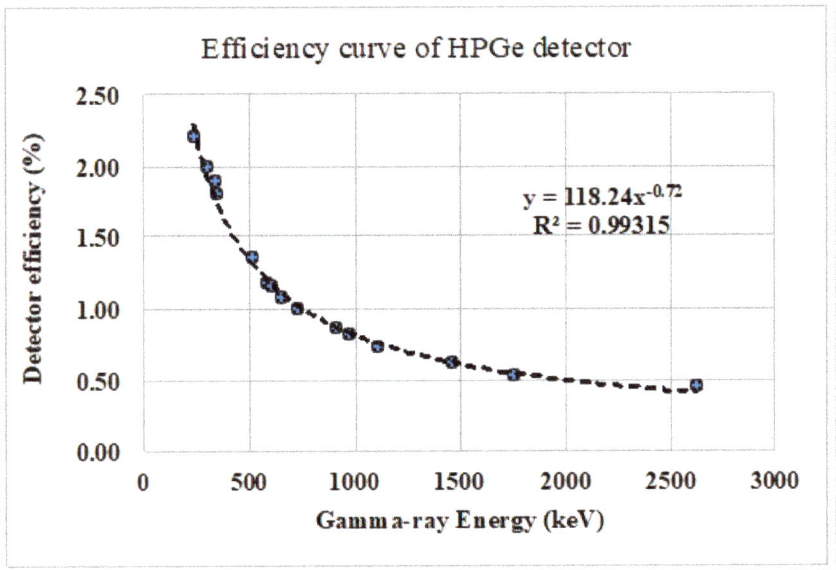

Figure 2. Counting efficiency curve of the HPGe (high-purity germanium) detector.

2.4. Calculation of Activity Concentration

Activity concentrations of radionuclides (Bq kg^{-1}) in surface soil, sub-surface soils and plant samples were calculated using Equation (1) [41]:

$$\text{Activity concentration} = \frac{\text{CPS} \times 100 \times 1000}{\varepsilon_f(\%). \times I_\gamma \times w_s(\text{kg})} \quad (1)$$

where CPS represents the net counts per second, ε_f the efficiency of the detector, $I\gamma$ the branching ratio and W_s the weight of the sample in kg. The statistical uncertainties were expressed in terms of standard deviation ($\pm\sigma$), where σ is expressed in Equation (2) [41]:

$$\sigma = \sqrt{\left[\frac{N_s}{T_s^2} + \frac{N_b}{T_b^2}\right]} \quad (2)$$

where N_s and N_b represent the sample and background counts in time T_s and T_b, respectively. The total uncertainty for each measured sample was calculated taking into account the statistical and other components of uncertainty. The combined uncertainty of the activity was estimated by using the quadratic sum of relevant quantities, which can be defined by Equation (3) [44]:

$$\Delta A = A \times \sqrt{\left(\frac{\Delta N}{N}\right)^2 + \left(\frac{\Delta \varepsilon_\gamma}{\varepsilon_\gamma}\right)^2 + \left(\frac{\Delta \rho_\gamma}{\rho_\gamma}\right)^2 + \left(\frac{\Delta m_s}{M_s}\right)^2} \quad (3)$$

where ΔA is the combined uncertainty of each measured value. The symbols ΔN, $\Delta\varepsilon_\gamma$, $\Delta\rho_\gamma$ and Δm_s represent the uncertainties due to the counting statistics, N (<7%), detection efficiency, ε_γ (4%), gamma ray emission probability, ρ_γ(<1%), and sample weight, M_S (<2%), respectively. The determined radioactivity levels, together with the uncertainties, are presented in Table 1.

Table 1. Concentrations of ^{226}Ra, ^{232}Th, ^{40}K and ^{137}Cs in the analyzed soil and tea leaf samples (both fresh and marketed tea leaves) and calculated transfer factors from soil-to-tea leaf (fresh tea leaves from UTE (Udalia Tea Estate)).

Sampling Location	Sample Type	Activity Concentrations (Bq kg^{-1}) Together with Uncertainty				Transfer Factor			
		^{226}Ra	^{232}Th	^{40}K	^{137}Cs	^{226}Ra	^{232}Th	^{40}K	^{137}Cs
Location-1	Garden Tea	5.7 ± 0.6	4.4 ± 0.5	190 ± 31	<0.4	0.13 ± 0.08	0.09 ± 0.05	0.69 ± 0.39	-
	Soil	45 ± 3	51 ± 1	275 ± 80	8.5 ± 1				
Location-2	Garden Tea	3.6 ± 0.7	3.2 ± 0.4	258 ± 29	<0.4	0.11 ± 0.08	0.05 ± 0.04	0.41 ± 0.13	-
	Soil	34 ± 9	63 ± 5	635 ± 35	7 ± 1				
Location-3	Garden Tea	5.7 ± 1.0	2.4 ± 0.5	175 ± 32	<0.4	0.15 ± 0.06	0.05 ± 0.01	0.55 ± 0.16	-
	Soil	37 ± 7	50 ± 13	391 ± 73	9 ± 1				
Location-4	Garden Tea	3.6 ± 0.6	5.8 ± 0.9	136 ± 22	<0.4	0.10 ± 0.03	0.09 ± 0.02	0.36 ± 0.13	-
	Soil	36 ± 7	65 ± 21	373 ± 73	4 ± 1				
Location-5	Garden Tea	4.1 ± 0.8	5.8 ± 1.1	132 ± 25	<0.4	0.10 ± 0.03	0.10 ± 0.04	0.54 ± 0.12	-
	Soil	42 ± 12	50 ± 19	245 ± 30	3 ± 1				
Mean						0.12 ± 0.08	0.08 ± 0.05	0.46 ± 0.35	-

The lower limit of detection or the minimum detectable activity concentration (MDA) of the measurement system was calculated using Equation (4) [45,46]:

$$\text{MDA} = \frac{k_\alpha \times \sqrt{\beta}}{\varepsilon_\gamma \times \rho_\gamma \times T_s \times M_s} \quad (4)$$

where K_α represents the statistical coverage factor which is equal to 1.64 (at the 95% confidence level), β is the background count in the energy of interest and the other symbols ε_γ, ρ_γ, T_S and M_S represent detection efficiency, gamma-ray emission probability, counting time, and sample weight, respectively. The MDAs for the studied radionuclides ^{226}Ra, ^{232}Th, ^{40}K and ^{137}Cs were calculated to be 0.32 Bq kg^{-1}, 0.60 Bq kg^{-1}, 2.5 Bq kg^{-1} and 0.4 Bq kg^{-1}, respectively.

2.5. Soil to Tea Leaves Transfer Factor (TF)

Within the food we eat, plants are the principal recipients of radioactive contamination, a result of atmospheric or other releases of radionuclides and from naturally occurring radioactivity within the soil [40]. Basically, the transfer factor defines the uptake of radionuclides from soil to plants, which can be calculated by the ratio of the radioactivity per unit dry weight of plant (C_P) to the radioactivity per unit dry weight of soil (C_S) in the rooting zone, using the Equation (5) [47,48]:

$$\mathrm{TF} = \frac{C_P}{C_S} \tag{5}$$

Dry weight analysis is preferred, the amount of radioactivity per kilogram dry weight being subject to much less variability than the amount per unit fresh weight, thereby reducing uncertainties in the measured TF (transfer factor) [49]. The calculated TFs for the studied tea leaf samples are shown in Table 1.

2.6. Annual Committed Effective Dose (ACED)

Following the consumption characteristics of foodstuffs, the committed effective dose due to the ingestion of radionuclides can be calculated using Equation (6), as below [50]:

$$\mathrm{AECD}\ (\mu\mathrm{Sv/y}) = C_r \times \sum_{i=1}^{3} D_{cfi} \times A_i \tag{6}$$

where C_r is the intake of radionuclides through use of the tea leaves, $D_{CF,i}$ are the ingestion dose conversion coefficients of 2.8 × 10^{-7} Sv Bq^{-1}, 2.2 × 10^{-7} Sv Bq^{-1} and 6.2 × 10^{-9} Sv Bq^{-1} for ^{226}Ra, ^{232}Th and ^{40}K, respectively, for an adult [51] and A_i is the measured activity concentration (Bq.kg^{-1}) of each radionuclide. According to the typical statistics, an average of 2 g of tea leaves is needed to prepare a cup of tea and if one person drinks two cups of tea per day, then an amount of some 1.5 kg/year is consumed by an individual. The two cups of tea is a typical tea consumption characteristic for the Bangladeshi population. The C_r is also defined as the consumption rate.

2.7. Threshold Consumption Rate of Tea (kg/y)

The threshold consumption rate (DI_{thresh}) represents a reference dietary level to avoid deleterious health hazards due to the intake of radionuclides via foodstuffs [52]. The particular threshold data due to the drinking of tea can be estimated by using the following Equation (7) [53]:

$$DI_{thresh}\ (\mathrm{kg/y}) = \frac{E_{ave}}{\sum_{i=1}^{3} D_{cfi} \times A_i} \tag{7}$$

were E_{ave} (290 μSv/y) is the threshold committed effective dose due to the ingestion of radionuclides of interest via the consumption of foodstuffs [54,55], A_1, A_2 and A_3 are the activity concentrations of ^{226}Ra, ^{232}Th and ^{40}K, respectively, in the tea leaf samples and D_{cfi} is the activity to dose conversion coefficient for the radionuclides of interest, as before.

2.8. Carcinogenic Risk

The carcinogenic risk for a population is estimated by assuming a linear no threshold, dose–effect relationship as per ICRP practice. For low doses, the ICRP suggest a fatal cancer risk factor of 0.05 Sv^{-1} [56], which indicates that the probability of a person dying of cancer is increased by 5% for a total dose of 1 Sv received during a lifetime. The estimated average annual committed effective dose for tea leaves is used herein to calculate the carcinogenic risk for an adult, made of the following relationship Equation (8):

$$EICR = \mathrm{AECD}(\mu Sv/y) \times R_f\left(Sv^{-1}\right) \times A_{ls}(y) \quad (8)$$

where R_f is the risk factor per sievert of annual effective dose received by the consumption of tea and A_{ls} is the cumulated time of tea consumption by Bangladeshi populace. Considering the local typical tea consumption characteristics, a duration of 50 years was used for both sexes.

3. Results and Discussion

3.1. Activity Concentration in the Tea Garden Doil Samples

The mean activity concentrations of ^{226}Ra, ^{232}Th, ^{40}K and ^{137}Cs in the soil samples collected from five locations of UTE are given in Table 1. The measured values in the investigated soil samples are found in the order ^{40}K > ^{232}Th > ^{226}Ra > ^{137}Cs. ^{40}K dominates over the other nuclides, which is not unexpected. This is because potassium is the seventh most abundant element in the Earth's crust, making up 2.6% of the weight of the earth's crust [57]. The greater activity concentration of ^{232}Th over that of ^{226}Ra may be attributed to the differences in the physical and chemical characteristics in a natural environment. In the earth's crust, both uranium and thorium tend to occur together due to the some inherent characteristics. However, throughout the various superficial processes like weathering and transportation, and the soil characteristics (pH and redox), they become fractionated. In general, thorium possesses low solubility and accumulates on particular phases whereas uranium is chemically more soluble and mobile. Consequently, uranium can easily be redistributed and transported in various environmental matrices compared to thorium [48].

Table 1 shows the mean activity concentrations of ^{226}Ra in soil samples for all locations other than that at location 2 to be greater than the UNSCEAR reported worldwide mean value of 35 Bq.kg^{-1} [49]. Among the three studied layers/depths of soil, the greatest concentration of ^{226}Ra (53 $\pm$ 8 Bq.kg^{-1}) was at location 5, at a depth of 13- to 20 cm, whereas, the minimum concentration of ^{226}Ra (27 $\pm$ 7 Bq.kg^{-1}) was at location 2, at a depth of 6 to 12 cm. This may be correlated to the ambient environment, i.e., the presence of high moisture content in the clay silty sand soil of this location which allows better solubility of ^{226}Ra [58].

For all locations the mean activity concentrations of ^{232}Th are greater than the worldwide mean value of 30 Bq.kg^{-1} [49], while the ^{40}K data for all locations other than location 2 are less than the UNSCEAR [49] reported mean concentration of 400 Bq.kg^{-1}. In respect to the vertical distribution, the greatest concentration of ^{232}Th (82 $\pm$ 11 Bq.kg^{-1}) was in soil from location 4, at a depth of 6 to 12 cm, whereas the minimum ^{232}Th concentration (29 $\pm$ 7 Bq.kg^{-1}) was from soil at location 5, at a depth of 0 to 5 cm. Soil samples from location 5 are clay silty sand and have large carbonate content, a matter correlating with the low ^{232}Th concentration. The data show the level of natural radioactivity forming a similar distribution in the surface and deep layered soils.

The greatest concentration of ^{40}K (672 $\pm$ 81 Bq.kg^{-1}) was shown to be at location 2, at a depth of 6 to 12 cm, whereas the minimum ^{40}K concentration (201 $\pm$ 78 Bq.kg^{-1}) was at location 1, at a depth of 6 to 12 cm. The majority of ^{40}K is a part of a clay mineral component rather than organic matter and its mobility depends on its solubility in the soil [59]. The low concentration of ^{40}K may be correlated to the soil texture, i.e., the presence of more sandy soil. Moreover, use of NPK fertilizer at least two times per year for better yield of leaves may contribute to the higher values of ^{40}K activity [60,61].

^{137}Cs, an anthropomorphic nuclide, as detected in trace amounts in the UTE soil, predominantly in the topsoil layers and less so or otherwise not detectable in sub-surface layers. The greatest ^{137}Cs mean activity concentration, at 10 $\pm$ 1 Bq.kg^{-1}, was found in surface soil at location 1, at a depth of 0 to 5 cm, while the lowest concentration of 3 ± 1 Bq.kg^{-1} was found at location 5, at the same depth. ^{137}Cs in other locations was not detected, the one exception being at location 1, at a depth of 6 to 12 cm. The mean ^{137}Cs concentrations at the different locations were found to be less than the world average value 51 Bqkg^{-1} as reported by UNSCEAR [49]. The ^{137}Cs is a quasi-permanent source of external gamma ray exposure, the activity slowly decaying in accordance with a half-life of some 30 years. The small likelihood that this nuclide will form a significant soil to plant pathway is generally acknowledged, the contaminant for the most part being linked to widely publicized nuclear establishment accidents. When detected, most typically at very low levels, the variation in the activity concentrations of the radionuclides are due to meteorological factors, the difference in sampling depth, physiochemical soil characteristics and the time of deposition.

3.2. Activity Concentration in Tea Leaf Samples

The measured activity concentrations of ^{226}Ra, ^{232}Th, ^{40}K and ^{137}Cs radionuclides in tea leaf samples collected from the Udalia Tea Estate as well as from the local market are summarized in Table 1. The concentrations of radionuclides in tea leaf samples are reported in Bq.kg^{-1} dry weight. The activity concentrations of studied radionuclides in the investigated tea leaf samples were in the order ^{40}K > ^{226}Ra > ^{232}Th > ^{137}Cs. The activity concentrations of ^{226}Ra were found to be greater than that of ^{232}Th in most of the tea leaf samples collected from UTE. One probable reason is that the ^{238}U (^{226}Ra) tends to move towards the outer extremities of the tree and accumulates more greatly in new leaves and sprouts [62].

The greater concentration of ^{40}K in tea leaf samples can be attributed for the most part to the specific metabolic processes of potassium involved in plant growth. Furthermore, for faster plant growth, the extra use of muriate (potassium chloride) of potassium fertilizer may be another factor causing the increase of ^{40}K concentration in the tea leaf samples [60]. It has been reported that about 88–96% of K is taken up by the plant from the soil through the root system [61]. Since plants take up a high amount of potassium and natural potassium contains 0.0117% of ^{40}K, the detection of high level of ^{40}K in plants is not unexpected.

3.3. Transfer Factors (TF) of Radionuclides from Soil-to-Tea Leaf

Soil to tea leaf transfer factor (TF) values from the five different locations are also presented in Table 1, the values depending on soil properties such as nutrient and moisture contents and pH [51]. It can be observed from the results that the TF values for ^{226}Ra, ^{232}Th and ^{40}K lie within the range 1.1×10^{-2} to 1.0, in accordance with values reported by the IAEA [63]. ^{137}Cs in all of the tea leaf samples was found to be below the detection limit, therefore the transfer factors could not be calculated. Note that the IAEA report a TF range of 0.02–3.2 for ^{137}Cs [63]. This indicates that, compared to ^{226}Ra, ^{232}Th and ^{40}K, ^{137}Cs is less efficiently transported from soil to the tea bush, as well as to the leaves

The transfer factors in the studied tea leaves are in the order ^{40}K > ^{226}Ra > ^{232}Th > ^{137}Cs, that for ^{40}K being significantly greater than those of other radionuclides in all samples. It is well known that K is a very essential nutrient for plants metabolism and depending upon the particular metabolism a variable amount of K is taken up by plants from soil. Since elemental potassium is homeostatically controlled by the body (intake and excretion maintaining balance), such amounts of ^{40}K in tea leaves are not to be considered to be of any particular concern as a potential radiation hazard. The actual concentration of radium in plant species clearly depends on the radium content of soil, its uptake by the plants species and also the metabolic characteristics of the plants [14]. Moreover, the chemical factors such as the presence of exchangeable amount of calcium in the soil may influence the absorption rate of radium by the plants [64]. The calculated TF show mean values of less than 1 for all radionuclides. It is worth mentioning that a value of TF >1 is indicative of radiation hazards to human health via the soil–plant–human body pathway.

On the other hand, a TF = 1 would be indicative of a particular species or plant forming a useful natural process for decontamination of soil affected by a nuclear accident or deliberate nuclear device testing.

Table 2 shows a comparison of results from the present study with reported data for tea leaves from the Chittagong region of Bangladesh. Considering the similar geographical conditions, humidity and quality of soil, comparability of data might thus be expected. Within the Chittagong region, there are 17 tea growing estates in Fatickchari, 3 in Rangunia and 1 in Banskhali. In this respect, Table 2 shows the measured radioactivity of ^{226}Ra to be similar to the available literature data, while UTE values for ^{232}Th and ^{40}K show much lower values compared to the reported data in the literature. Moreover, radioactivity in the estate tea leaves from Rize in Turkey also show higher values than the present results from UTE. The activity concentration of the artificial ^{137}Cs radionuclide for the UTE sample is shown to be below the detection limit (<0.4), while a substantial amount of ^{137}Cs was reported in estate tea leaf samples from Turkey. Such a result indicates the contamination of sampling area via some known/unknown nuclear activities. There are no available studies on the radioactivity of marketed tea samples in Bangladesh, therefore studies on marketed tea leaf samples imported from abroad have been chosen for comparison. The average activity concentration of ^{226}Ra and ^{232}Th and ^{40}K in tea leaf samples collected from the local market show greater values than the reported data from Turkey (except ^{40}K) and Serbia. The fact that the concentrations of ^{226}Ra, ^{232}Th and ^{40}K vary substantially across the various regions depends mainly on their concentrations in the bedrock from which the soil originates [65].

Table 2. Average activity concentrations of ^{226}Ra, ^{232}Th, ^{40}K and ^{137}Cs in tea leaf samples from various countries compared with that from present work.

Sample Type	Countries	Activity Concentrations ($Bq.kg^{-1}$) Together with Uncertainties				References
		^{226}Ra	^{232}Th	^{40}K	^{137}Cs	
Fresh tea leaf	UTE, Chittagong, Bangladesh	4.53 ± 0.62	4.31 ± 0.58	178 ± 28	<0.4	Present study
	Chittagong district, Bangladesh	5.34	10.07	429.91	Not measured	[66]
	Ramgarh, Bangladesh	3.20 ± 2.18	4.65 ± 1.76	625 ± 62.37	Not measured	[66]
	Kodala, Bangladesh	3.56 ± 0.69	27.22 ± 3.65	1243 ± 83.91	Not measured	[66]
	Chandpur Belgaon, Bangladesh	5.67 ± 2.16	12.41 ± 2.82	380 ± 62.06	Not measured	[66]
	Rize, Turkey	36.3 ± 6.1	23.1 ± 4.8	688.4 ± 18.3	20.9 ± 3.8	[67]
Market tea leaf	Mostafa-1	3.8 ± 0.4	6.0 ± 0.7	321 ± 35	<0.4	Present study
	Ceylon	4.3 ± 0.3	4.3 ± 0.5	244 ± 31	<0.4	
	Ispahani	4.0 ± 0.1	3.9 ± 0.5	159 ± 32	<0.4	
	Taza	5.4 ± 0.6	2.8 ± 0.6	141 ± 29	<0.4	
	Mostafa	4.3 ± 0.3	5.5 ± 0.8	183 ± 25	<0.4	
	Turkey-2 (Market tea)	0.9	2.7	501	-	[68]
	Serbia-1 (Market tea)	0.6–8.2	1.7–15.1	126–1243.7	-	[69]

3.4. Committed Effective Dose, Threshold Consumption Rate and Carcinogenic Risk

The calculated values of committed effective dose, threshold consumption rate and carcinogenic risk are shown in Table 3. The committed effective dose due to the intake of the studied radionuclides via tea consumption was found to be in the range of 4.7–5.6 μSv y^{-1} with a mean of 5.0 μSv y^{-1}. This compares with an average worldwide ingestion dose of ^{226}Ra and ^{232}Th of 120 μSv y^{-1} and 170 μSv y^{-1} for ^{40}K, making a total annual dose estimate of 290 μSv y^{-1} from the total diet [49]. The annual effective doses from the ingestion of tea leaves were found much lower than the limiting value recommended by UNSCEAR [49]. Note that the estimated 5.0 μSv y^{-1} is contributed to by only a single dietary element (here tea leaf), thus such a low value is not unexpected. However, the radiation risk via the cumulative consumption of tea leaf may not be totally negligible, because tea forms only a minor part of the total dietary habits [70,71]. Figure 3 shows the dose contribution due to the individual radionuclides. Among the studied radionuclides, ^{226}Ra incurred the maximum dose (38%) followed by ^{40}K (33%) and ^{232}Th (29%). Exposure to radioactive materials, especially radium, over a prolonged time may result in an increased carcinogenic risk. In addition, higher doses of radium are found to have links with anemia, cataracts, reduction of bone growth, etc. [72].

Table 3. Calculated hazard parameters due to the consumption of studied tea leaves collected from the UTE, Chittagong, Bangladesh.

Sample	Annual Effective Dose (uSv/y)				Threshold Consumption Rate, kg/y	Lifetime Carcinogenic Risk
	^{226}Ra	^{232}Th	^{40}K	Total		
TL-1	2.4	1.5	1.8	5.6	77	1.40×10^{-5}
TL-2	1.5	1.1	2.4	5.0	88	1.24×10^{-5}
TL-3	2.4	0.8	1.6	4.8	90	1.20×10^{-5}
TL-4	1.5	1.9	1.3	4.7	93	1.17×10^{-5}
TL-5	1.7	1.9	1.2	4.9	89	1.22×10^{-5}
Mean	1.9	1.4	1.7	5.0	88	1.25×10^{-5}

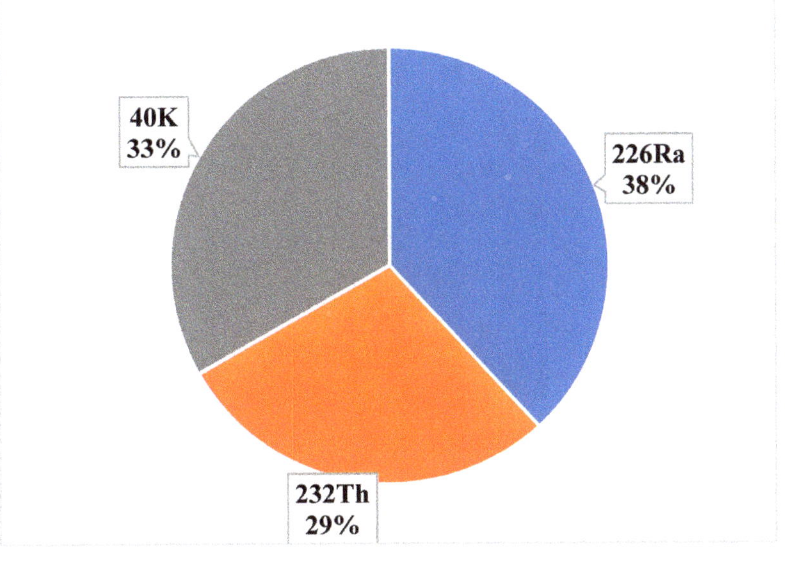

Figure 3. Dose contribution by individual radionuclides due to the consumption of tea leaves.

The estimated mean threshold consumption rate for the studied tea leaf samples was found to be 88 kg/y (equivalent to 241g/d), an untenable value. This parameter indicates that the consumption of tea below the estimated value poses only a negligible radiological health hazard, while a greater rate than the calculated ones indicates enhanced radiological health risk.

Accordingly, the mean cumulative carcinogenic risk via tea leaf consumption (for a period of 50 years) was estimated at 1.3×10^{-5}, significantly lower than the ICRP given cancer risk of 2.5×10^{-3}, based on an annual effective dose limit of 1 mSv for the general population [52].

4. Conclusions

Activity concentrations of ^{226}Ra, ^{232}Th, ^{40}K and ^{137}Cs in soil and tea leaf samples collected from a large tea estate in the Chittagong region of Bangladesh were measured by HPGe γ-ray spectrometry. In addition, the most popular tea brands available in the local market were also measured to observe the effect of production processes. The transfer factor of radionuclides from soil to tea leaves was found to be less than 1, indicating the corresponding uptake by the tea plant to be insignificant. The estimated committed effective dose and the carcinogenic risk all show values far below the limiting ranges as suggested by various international bodies. Thus, the consumption of tea (at 4 g/day/person or two cups/day/person) produced from the Udalia Tea Estate provides an insignificant radiation risk to the health of the local populace. Considering a number of facts such as the non-availability of the literature data on TFs of the UTE tea leaves, the recommended limiting values for total dietary habits and that for the tea arising from this single entity, and that the radiation risk follows the linear no threshold model, the measured data can act as reference values for any future experimental or modelling studies for the protection of human health.

Author Contributions: Conceptualization, N.A., M.M.R., and N.S.; methodology, N.A. and M.M.R.; software, M.I.C. and M.U.K.; validation, M.M.H.M. and J.A.; formal analysis, N.A. and J.A.; investigation, N.A., M.M.R., and N.S.; resources, M.K. and M.I.C.; data curation, N.A. and M.M.H.M.; writing—original draft preparation, N.A. and J.A.; writing—review and editing, M.U.K. and D.A.B.; visualization, A.A.; supervision, M.M.R. and N.S.; project administration, M.R.I.F. and A.A.M.S.; funding acquisition, A.A. All authors have read and agreed to the published version of the manuscript.

Funding: This work was supported by the Taif University Researchers Supporting project number (TURSP-2020/163), Taif 21944, Saudi Arabia.

Institutional Review Board Statement: Not applicable.

Informed Consent Statement: Not applicable.

Data Availability Statement: All data are available in the manuscript.

Acknowledgments: Contributions from the staff of the gamma ray spectrometry laboratory of Atomic Energy Centre Chittagong are greatly acknowledged.

Conflicts of Interest: The authors declare no conflict of interest.

References

1. Allisy, A.; Henri, B. The discovery of radioactivity. In Proceedings of the Becquerrel's Legacy: A Century of Radioactivity, London, UK, 29 February–1 March 1996; Nuclear Technology Publishing: London, UK, 1996; pp. 3–10.
2. Watson, S.J.; Jones, A.L.; Oatway, W.B.; Hughes, J.S. *Ionising Radiation Exposure of the UK population, Health Protection Agency, Centre for Radiation, Chemical and Environmental Hazards, Radiation Protection Division*; Chilton: Didcot, UK, 2005.
3. NCRP. *National Council on Radiation Protection and Measurement. Exposure of the Population in United States and Canada from Natural Background Radiation*; NCRP Report No 49; National Council on Radiation: Bethesda, MD, USA, 1987.
4. Elegba, S.B.; Funtua, I.I. Naturally occurring radioactive material (NORM) assessment of oil and gas production installations in Nigeria. *Int. At. Energy Agency (IAEA)* **2005**, *37*, 256.
5. World Nuclear Association. Radiation and Nuclear Energy. 2011. Available online: http://www.world-nuclear.org/info/inf30.html (accessed on 30 December 2020).

6. Guagliardi, I.; Rovella, N.; Apollaro, C.; Bloise, A.; De Rosa, R.; Scarciglia, F.; Buttafuoco, G. Modelling seasonal variations of natural radioactivity in soils: A case study in southern Italy. *J. Earth Syst. Sci.* **2016**, *125*, 1569–1578. [CrossRef]
7. Dhahir, M.D.; Azhar, S.A.; Ali, A.A.; Hayder, H.H. Assessment of Natural Radionuclide levels for Tea samples in Najaf. *Rap Conf. Proc.* **2019**, *4*, 57–60.
8. Guagliardi, I.; Zuzolo, D.; Albanese, S.; Lima, A.; Cerino, P.; Pizzolante, A.; Thiombane, M.; De Vivo, B.; Cicchella, D. Uranium, thorium and potassium insights on Campania region (Italy) soils: Sources patterns based on compositional data analysis and fractal model. *J. Geochem. Explor.* **2020**, *212*, 106508. [CrossRef]
9. Hassan, Y.M.; Zaid, H.M.; Guan, B.H.; Khandaker, M.U.; Bradley, D.A.; Sulieman, A.; Latif, S.A. Radioactivity in staple foodstuffs and concomitant dose to the population of Jigawa state, Nigeria. *Radiat. Phys. Chem.* **2021**, *178*, 108945. [CrossRef]
10. Khandaker, M.U.; Asaduzzaman, K.; Nawi, S.M.; Usman, A.R.; Amin, Y.M.; Daar, E. Assessment of Radiation and Heavy Metals Risk due to the Dietary Intake of Marine Fishes (Rastrelligerkanagurta) from the Straits of Malacca. *PLoS ONE* **2015**, *10*, 0128790. [CrossRef]
11. Khandaker, M.U.; Nasir, N.L.M.; Asaduzzaman, K.; Amin, Y.M.; Bradley, D.A.; Alrefae, T. Evaluation of radionuclides transfer fromsoil-to-edible flora and estimation of radiological dose to the Malaysian populace. *Chemosphere* **2016**, *154*, 528–536. [CrossRef] [PubMed]
12. Yeltepe, E.; Şahin, N.K.; Aslan, N.; Hult, M.; Ozçayan, G.; Wershofen, H.; Yucel, U. A review of the TAEA proficiency test on natural and anthropogenic radionuclides activities in black tea. *Appl. Radiat. Isot.* **2018**, *134*, 40–44. [CrossRef]
13. Khandaker, M.U.; Shuaibu, H.K.; Alklabi, F.A.A.; Alzimami, K.S.; Bradley, D.A. Study of primordial ^{226}Ra, ^{228}Ra, and ^{40}K concentrations in dietary palm dates and concomitant radiological risk. *Health Phys.* **2019**, *116*, 789–798. [CrossRef] [PubMed]
14. Khandaker, M.U.; Heffny, N.A.; Amin, Y.M.; Bradley, D.A. Elevated concentration of radioactive potassium in edible algae cultivated in Malaysian seas and estimation of ingestion dose to humans. *Algal Res.* **2019**, *38*, 101386. [CrossRef]
15. Schuller, P.; Voigt, G.; Handl, J.; Ellies, A.; Oliva, L. Global weapons' fallout ^{137}Cs in soils and transfer to vegetation in South-centralChile. *J. Environ. Radioact.* **2002**, *62*, 181–193. [CrossRef]
16. Alatise, O.O.; Adebesin, T.C. Assessment of Radionuclides in some Nigerian Cereals and products. *J. Nat. Sci. Eng. Technol.* **2019**, *18*, 128–142.
17. Dufresne, C.J.; Farnworth, E.R. A review of latest research findings on the health promotion properties of tea. *J. Nutr. Biochem.* **2001**, *12*, 404–421. [CrossRef]
18. Zhu, Q.Y.; Hackman, R.M.; Ensunsa, J.L.; Holt, R.R.; Keen, C.L. Antioxidative activities of oolong tea. *J. Agric. Food Chem.* **2002**, *50*, 6929–6934. [CrossRef] [PubMed]
19. Zehringer, M.; Kammerer, F.; Wagmann, M. Radionuclides in tea and their behaviour in the brewing process. *J. Environ. Radioact.* **2018**, *192*, 75–80. [CrossRef] [PubMed]
20. Karak, T.; Bhagat, R. Trace elements in tea leaves, made tea and tea infusion: A review. *Food Res. Int.* **2010**, *43*, 2234–2252. [CrossRef]
21. Zaver, N.T. Green tea and its polyphenolic catechins: Medicinal uses in cancer and noncancer applications. *Life Sci.* **2006**, *78*, 2073–2080. [CrossRef] [PubMed]
22. Phan, L.H.; Le, D.H.; Vu, T.M.; Dang, V.C.; Tran, T.T.; Chau, V.T. Natural and artificial radionuclides in tea samples determined with gamma spectrometry. *J. Radioanal. Nucl. Chem.* **2018**, *316*, 703–707.
23. Friedman, M. Overview of antibacterial, antitoxin, antiviral, and antifungal activities of tea flavonoids and teas. *Mol. Nutr. Food Res.* **2007**, *51*, 116–134. [CrossRef]
24. Korte, G.; Dreiseitel, A.; Schreier, P.; Oehme, A.; Locher, S.; Geiger, S. Tea catechins' affinity for human cannabinoid receptors. *Phytomedicine* **2010**, *17*, 19–22. [CrossRef]
25. Fujita, H.; Yamagami, T. Anti hyper cholesterolemic effect of Chinese black tea extract in human subjects with borderline hypercholesterolemia. *Nutr. Res.* **2008**, *28*, 450–456. [CrossRef]
26. Hamer, M. The beneficial effects of tea on immune function and inflammation: A review of evidence from in vitro, animal, and human research. *Nutr. Res.* **2007**, *27*, 373–379. [CrossRef]
27. Siddiqui, I.A.; Raisuddin, S.; Shukla, Y. Protective effects of black tea extract on testosterone induced oxidative damage in prostate. *Cancer Lett.* **2005**, *227*, 125–132. [CrossRef] [PubMed]
28. Khizar, H.; Hira, I.; Uzma, M.; Uzma, B.; Sobia, M. Tea and Its Consumption: Benefits and Risks. *Crit. Rev. Food Sci. Nutr.* **2013**. [CrossRef]
29. Chow, H.; Hakim, I.; Vining, D.; Crowell, J.; Ranger-Moore, J.; Chew, W.; Celaya, C.; Rodney, S.; Hara, Y.; Alberts, D. Effects of dosing condition on the oral bioavailability of green tea catechins after single-dose administration of polyphenon E in healthy individuals. *Clin. Cancer Res.* **2005**, *11*, 4627–4633. [CrossRef]
30. Mamun, M.; Ahmed, M. Integrated pest management in tea: Prospects and future strategies in Bangladesh. *J. Plant Prot. Sci.* **2011**, *3*, 1–13.
31. Khan, A.; Biswas, A.; Saha, A.; Motalib, M. Soil properties of Lalmai hill, Shalban Bihar and Nilachal hill of greater Comilla district and its suitability for tea plantation. *Tea J. Bangladesh* **2012**, *41*, 17–26.
32. Revitalising the Tea Sector. Available online: https://thefinancialexpress.com.bd/views/revitalising-the-tea-sector-1527864045#: (accessed on 20 December 2020).

33. Khatun, R.; Saadat, A.H.M.; Ahasan, M.M.; Akter, S. Assessment of Natural Radioactivity and Radiation Hazard in Soil Samples of Rajbari District of Bangladesh. *Jahangirnagar Univ. Environ. Bull.* **2013**, *2*, 1–8. [CrossRef]
34. Mst, N.A.; Suranjan, N.D.; Selina, Y.; Mahfuz, S.M.M.; Mizanur, R.A.F.M. Measurement of Radioactivity and Assessment of Radiological Hazards of Tea samples collected from local market in Bangladesh. *J. Bangladesh Acad. Sci.* **2018**, *42*, 171–176.
35. Islam, M.N.; Akhter, H.; Begum, M.; Kamal, M. Comprehensive Review of the Investigation of Anthropogenic and Naturally Occurring Radionuclides in Different Parts of Bangladesh. *Int. J. Adv. Eng. Manag. Sci.* **2018**, *4*, 490–495. [CrossRef]
36. Oodolia Tea Garden, an Untapped Natural Beauty Site at Fatikchhari. Available online: https://www.daily-sun.com/arcprint/details/167547/Oodolia-Tea-Garden-an-untapped-natural-beauty-site-at-Fatikchhari/2016-09-18 (accessed on 12 February 2021).
37. Hossain, M.K.; Azad, A.K.; Alam, M.K. Assessment of natural regeneration status in a mixed tropical forest at Kaptai of Chittagong Hill Tracts (South) forest division. *Chittagong Univ. J. Sci.* **1999**, *23*, 73–79.
38. IAEA. *Intercomparison Runs Reference Manuals*; AQCS: Vienna, Austria, 1995.
39. ICRP. International Commission on Radiological Protection, 1983, Publication 119: Compendium of dose coefficients based on ICRP Publication 60. *Ann. ICRP* **1983**, *41* (Suppl. 42), e1–e130.
40. IAEA. Measurement of radionuclides in food and the environment–technical reports. In *Ser*; IAEA: Vienna, Austria, 1983; Volume 295, pp. 5–27.
41. Abedin, M.J.; Karim, M.R.; Hossain, S.; Deb, N.; Kamal, M.; Miah, M.H.A.; Khandaker, M.U. Spatial distribution of radionuclides in agricultural soil in the vicinity of a coal-fired brick kiln. *Arab. J. Geosci.* **2019**, *12*, 1–12. [CrossRef]
42. Amin, Y.M.; Khandaker, M.U.; Shyen, A.K.S.; Mahat, R.H.; Nor, R.M.; Bradley, D.A. Radionuclide emissions from a coal-fired power plant. *Appl. Radiat. Isot.* **2013**, *80*, 109–116. [CrossRef] [PubMed]
43. Khandaker, M.U.; Jojo, P.J.; Kassim, H.A.; Amin, Y.M. Radiometric analysis of construction materials using HPGe gamma-ray spectrometry. *Radiat. Prot. Dosim.* **2012**, *152*, 33–37. [CrossRef] [PubMed]
44. Asaduzzaman, K.; Khandaker, M.U.; Amin, Y.M.; Bradley, D.A. Natural radioactivity levels and radiological assessment of decorative building materials in Bangladesh. *Indoor Built Environ.* **2014**, *25*, 1–10. [CrossRef]
45. Khandaker, M.U.; Asaduzzaman, K.; Sulaiman, A.F.; Bradley, D.A.; Isinkaye, M.O. Elevated concentrations of naturally occurring radionuclides in heavy mineral-rich beach sands of Langkawi Island, Malaysia. *Mar. Pollut. Bull.* **2018**, *127*, 654–663. [CrossRef] [PubMed]
46. Solak, S.; Turhan, S.; Ugur, F.A.; Goren, E.; Gezer, F.; Yegingil, Z. Evaluation of potential exposure risks of natural radioactivity levels emitted from building materials used in Adana, Turkey. *Indoor Built Environ.* **2014**, *23*, 594–602. [CrossRef]
47. Nurul, A.; Mashiur, R.; Masud, K.; Naziba, S.; Mantazul, I.C. Natural and anthropogenic radioactivity levels and associated radiation hazard in soil of Oodalia Tea Estate at hilly region of Fatikchari in Chittagong, Bangladesh. *J. Radiat. Res.* **2014**, *55*, 1–6.
48. Bajoga, A.; Alazemi, N.; Shams, H.; Regan, P.; Bradley, D. Evaluation of naturally occurring radioactivity across the State of Kuwait using high-resolution gamma-ray spectrometry. *Radiat. Phys. Chem.* **2016**, *137*, 203–209. [CrossRef]
49. UNSCEAR. Exposures from Natural Radiation Sources. *Annex-B* **2000**, *9*, 124–127.
50. James, J.P. Soil to leaf transfer factor for the radionuclides226Ra, 40K, 137Cs and 90Sr at Kaiga region, India. *J. Environ. Radioact.* **2011**, *102*, 1070–1077. [CrossRef] [PubMed]
51. Forkapic, S.; Vasin, I.; Mrdja, D.; Bikit, K.; Milic, S. Correlations between soil characteristics and radioactivity content of Vojvodina soil. *J. Environ. Radioact.* **2017**, *166*, 104–111. [CrossRef] [PubMed]
52. Khandaker, M.U.; Zainuddin, N.K.; Bradley, D.A.; Faruque, M.R.I.; Almasoud, F.I.; Sayyed, M.I.; Sulieman, A.; Jojo, P.J. Radiation dose to Malaysian populace via the consumption of roasted ground and instant coffee. *Radiat. Phys. Chem.* **2020**, *173*, 1–7. [CrossRef]
53. Monica, S.; Jojo, P.J.; Khandaker, M.U. Radionuclide concentrations in medicinal florae and committed effective dose through Ayurvedic medicines. *Int. J. Radiat. Biol.* **2020**, *96*, 1028–1037. [CrossRef]
54. Cember, H.; Johnson, T.E. *Introduction to Health Physics*, 5th ed.; McGraw-Hill: New York, NY, USA, 2009; pp. 1–888.
55. UNSCEAR. *Sources and Effects of Ionizing Radiation. Report to the General Assembly*; Scientific Annexes: New York, NY, USA, 2000.
56. Naturally Occurring Radioactive Materials (NORM IV). In Proceedings of the an International Conference, Szczyrk, Poland, 17–21 May 2004; pp. 1–584.
57. Potassium, Chemical Element. Available online: https://www.britannica.com/science/potassium (accessed on 12 February 2021).
58. OECD. Exposure to Radiation from the Natural Radioactivity in Building Materials. In *Reported by a Group of Experts of the OECD*; Nuclear Energy Agency: Paris, France, 1979.
59. Dragovic, S.; Jankovic, M.L.; Dragovic, R.; Đorđevic, M.; Dokic, M. Spatial distribution of the ^{226}Ra activity concentrations in well and spring waters in Serbia and their relation to geological formations. *J. Geochem.* **2012**, *112*, 206–211. [CrossRef]
60. Khandaker, M.U.; Uwatse, O.B.; Shamsul Khairi, K.A.; Faruque, M.R.I.; Bradley, D.A. Terrestrial radionuclides in surface (dam) water and concomitant dose inmetropolitan Kuala Lumpur. *Radiat. Prot. Dosim.* **2019**, *185*, 1–8.
61. Titus, A.; Pereira, G.N. Potassium Dynamics in Coffee Soils. 2016. Available online: https://ecofriendlycoffee.org/potassium-dynamics-coffeesoils/ (accessed on 12 February 2021).
62. Duc, H.H.; Minh, N.D.; Cuong, P.V.; Anh, L.T.; Leuangtakoun, S.; Loat, B.V. Transfer of ^{238}U and ^{232}Th from Soils to Tea Leaves on Luong My Farm, Hoa Binh Province, Vietnam, VNU. *J. Sci. Math. Phys.* **2019**, *35*, 106–115.

63. International Atomic Energy Agency (IAEA). Handbook ofparameter values for the prediction of radionuclide transfer in terrestrialand freshwater environments. In *Technical Report series No. 472*; IAEA: Vienna, Austria, 2010.
64. Frissel, M.J. Generic values for soil-to-plant transfer factors of radiocesium. *J. Environ. Radioact.* **2002**, *58*, 113–128. [CrossRef]
65. Sabbarese, C.; Ambrosino, F.; D'Onofrio, A.; Roca, V. Radiological characterization of natural building materials from the Campania region (Southern Italy). *Constr. Build. Mater.* **2021**, *268*, 121087. [CrossRef]
66. Jannatul, F.; Munmun, N.N.; Rahman, R.R. Assessment of Radionuclide Concentrations in Tea Samples Cultivated in Chittagong Region, Bangladesh. *Int. J. Life Sci. Technol.* **2018**, *11*, 20–30.
67. Recep, K.; Filiz, K.G.; Nilay, A.; Nazmi, T.O. Radionuclide concentration in tea, cabbage, orange, kiwi and soil and lifetime cancer risk due to gamma radioactivity in Rize. *Turk. Soc. Chem. Ind.* **2011**, *91*, 987–991.
68. Önder, K.; Murat, B.; Sayhan, T.; Yavuz, Ç. ^{232}Th, ^{238}U, ^{40}K, ^{137}Cs radioactivity concentrations and ^{137}Cs dose rate in Turkish market tea. *Radiat. Eff. Defects Solids* **2009**, *164*, 138–143.
69. Milutin, J.; Natasa, L.; Snezana, P.; Milan, O. Radionuclide concentrations in samples of medicinal herbs and effective dose from ingestion of 137Cs and natural radionuclides in herbal tea products from Serbian market. *Isot. Environ. Health Stud.* **2011**, *47*, 87–92.
70. Khandaker, M.U.; Nasir, N.L.M.; Zakirin, N.S.; Kassim, H.A.; Asaduzzaman, K.; Bradley, D.A.; Zulkifly, M.Y.; Hayyan, A. Radiation dose to the Malaysian populace via the consumption of bottled mineral water. *Radiat. Phys. Chem.* **2017**, *140*, 173–179. [CrossRef]
71. Khandaker, M.U.; Wahib, N.B.; Amin, Y.M.; Bradley, D.A. Committed effective dose from naturally occurring radionuclides in shellfish. *Radiat. Phys. Chem.* **2013**, *88*, 1–8. [CrossRef]
72. Health Effects of Radium Radiation Exposure. Available online: https://www.mass.gov/service-details/health-effects-of-radium-radiation-exposure# (accessed on 13 February 2020).

life

Article

Indoor Radon Concentration and Risk Assessment in 27 Districts of a Public Healthcare Company in Naples, South Italy

Filomena Loffredo [1], Federica Savino [2], Roberto Amato [3], Alfredo Irollo [4], Francesco Gargiulo [5], Giuseppe Sabatino [6], Marcello Serra [1] and Maria Quarto [1,*]

1 Advanced Biomedical Science Department, University of Naples, 80131 Naples, Italy; filomena.loffredo@unina.it (F.L.); serraemme@gmail.com (M.S.)
2 LB Business Services srl, 00135 Rome, Italy; savino.federica@gmail.com
3 Occupational Health Service, Public Healthcare "Napoli 3", 34102 Naples, Italy; r.amato@aslnapoli3sud.it
4 Protection and Prevention Service, Public Healthcare "Napoli 3", 34102 Naples, Italy; icratur@libero.it
5 Building Division Office, University of Naples, 80138 Naples, Italy; francesco.gargiulo3@unina.it
6 Advanced Metrological and Technological Services (CeSMA), University of Naples, 80138 Naples, Italy; giuseppe.sabatino@unina.it
* Correspondence: maria.quarto@unina.it

Abstract: Radon is a major source of ionizing radiation exposure for the general population. It is known that exposure to radon is a risk factor for the onset of lung cancer. In this study, the results of a radon survey conducted in all districts of a Public Healthcare in Italy, are reported. Measurements of indoor radon were performed using nuclear track detectors, CR-39. The entire survey was conducted according to a well-established quality assurance program. The annual effective dose and excess lifetime cancer risk were also calculated. Results show that the radon concentrations varied from 7 ± 1 Bq/m^3 and 5148 ± 772 Bq/m^3, with a geometric mean of 67 Bq/m^3 and geometric standard deviation of 2.5. The annual effective dose to workers was found to be 1.6 mSv/y and comparable with the worldwide average. In Italy, following the transposition of the European Directive 59/2013, great attention was paid to the radon risk in workplaces. The interest of the workers of the monitored sites was very high and this, certainly contributed to the high return rate of the detectors after exposure and therefore, to the presence of few missing data. Although it was not possible to study the factors affecting radon concentrations, certainly the main advantage of this study is that it was the first in which an entire public health company was monitored in regards to all the premises on the underground and ground floor.

Keywords: radon; effective dose; workplaces

Citation: Loffredo, F.; Savino, F.; Amato, R.; Irollo, A.; Gargiulo, F.; Sabatino, G.; Serra, M.; Quarto, M. Indoor Radon Concentration and Risk Assessment in 27 Districts of a Public Healthcare Company in Naples, South Italy. *Life* **2021**, *11*, 178. https://doi.org/10.3390/life11030178

Academic Editor: Fabrizio Ambrosino

Received: 18 January 2021
Accepted: 20 February 2021
Published: 24 February 2021

Publisher's Note: MDPI stays neutral with regard to jurisdictional claims in published maps and institutional affiliations.

1. Introduction

Radon is a radioactive noble gas belonging to the ^{238}U radioactive chain, produced by the decay of ^{226}Ra [1]. It is ubiquitous in the earth's crust in a concentration dependent on geology [2]. A fraction of radon produced in rocks and soils, escapes into the outdoor atmosphere where it is quickly diluted while, in confined spaces, it tends to accumulate reaching levels of concentration that are dangerous for health [3]. In 1988, radon was classified as a carcinogenic agent for humans by the International Agency for Research on Cancer (IARC) for evidence of an association between the exposure and onset of lung cancer [4]. Due to its long half-life (3.82 d), radon is almost completely exhaled after inhalation, otherwise, its progenies with short half-life, ^{218}Po and ^{214}Po, being electrically charged, can be attached to dust or smoke particles in indoor air. During the breathing process, they reach the bronchial tissue where they decay emitting radioactive alpha particles capable of damaging the pulmonary epithelium and thereby causing lung cancer. Radon and its progeny contribute more than 50% to the human exposure to natural ionizing radiation [5]. UNSCEAR [6] estimates that the annual effective dose for humans attributable to the radon exposure is about 1 mSv. After the residential exposure, the second important source of exposure to radon and its progenies, is the occupational exposure since people spend about 35% of their daytime in the workplace. Supported by the scientific evidence of epidemiological studies on residential exposure data [7–12] showing a statistically significant increase in the risk of lung cancer, due to the prolonged exposure to radon already at the level of 100 Bq/m^3, the European Union has issued the Directive 59/2013 [13], which established a reference level for indoor radon concentrations in workplaces of 300 Bq/m^3. Despite the European Union regulation, some authors [14–16] have questioned the correlation between the exposure to radon and the increased risk of lung cancer. They believe that the results obtained from the epidemiological studies are conditioned by the dose-response model adopted, the linear no-threshold (LNT). According to this model, the excess risk increases linearly versus the radon concentration but, these studies argue the lack of enough experimental data to support this thesis. Dobrzyński et al. [16] stated, in their meta-analysis, that there is no scientific evidence supporting the thesis that the exposure to radon is significantly correlated to the incidence of pulmonary cancer, at least for concentration values below 1000 Bq/m^3.

In Italy, in July 2020, this Directive was transposed into the current National law [17] for the protection of health of the general population and workers against the ionizing radiation exposure. In addition, it imposes the action level for the home and workplaces of 300 Bq/m^3. Despite this, in Italy, few surveys have been performed in the workplaces, many of them, on a local scale and with different methods of measurement. The aim of this study is to present the results of a radon survey conducted in 2018–2020 in the underground workplaces and on the ground floor at the Public Healthcare Company "Napoli 3" of Campania region, South Italy. Moreover, the annual dose estimation to workers and the excess lifetime cancer risk, were assessed using the data presented in the paper.

2. Materials and Methods

The survey conducted at the Public Healthcare Company "Napoli 3" was divided into four fundamental phases: (1) Sampling plan of the premises to be monitored; (2) exposure of the CR-39 detectors for two consecutive semesters in the sampled premises; (3) chemical etching and statistical analysis; (4) radon risk assessment. These steps are listed below.

2.1. Sampling Plan

The confined spaces to be monitored were selected by the Prevention and Protection Department of Public Healthcare Company "Napoli 3". In this study, regardless of their intended use, all underground workplaces and on the ground floor, were monitored. The sampling plan concerned the 27 districts in which the Public Healthcare Company is divided and, a total of 607 rooms were sampled.

2.2. Monitoring, Detectors Analysis, and Statistical Analysis

The monitoring was conducted during 2018–2020 and involved a total of 1307 solid state nuclear detector CR-39 types. In each monitored workplace, two detectors were exposed for two consecutive semesters to obtain a whole year of exposure. Radon measurements were conducted according to the UNI ISO 11665-4: 2020 standard. CR-39 detectors were positioned about 2 m from the floor and at about 30 cm from the internal wall in order not to record the contribution from the Thoron. The quality assurance was performed by participating in the intercomparison exercise organized by the German Federal Office for Radiation Protection (BfS). Moreover, the laboratory of Radioactivity (lab.RAD, University of Naples Federico II, Naples, Italy) has a certification according to the ISO 9001:2015 standard and accreditation according to the European Standard EN ISO/IEC 17025 for the "integrated measurement method for determining average activity concentration of the radon 222 in the environment air using passive sampling and delayed analysis" (UNI ISO 11665:2020, part 4).

After the exposure, all the CR-39 detectors were chemically etched using a solution of 6.25 M NaOH at (98 ± 1) °C for 60 min. The automatic counting of tracks density was performed by the Politrack system (mi.am s.r.l., Rivergaro, PC, Italy). It consists of an automated microscopic image analyzer equipped with a control computer and track analysis software. Finally, the radon concentration was calculated using Equation (1):

$$C_{Rn} = \frac{\mathrm{N}}{\mathrm{E} \times \mathrm{T}} \tag{1}$$

where N is the track density corrected by the background track density, E is the calibration factor, and T is the exposure time. The background track density was estimated to be 10 tracks/cm^2 and it was determined by counting the tracks of a significant number of unexposed CR-39. The calibration factor was determined by exposing the detectors at a certified atmosphere in the range of exposure from 100 to 3000 Bq h m^{-3} at the National Metrological Institute (ENEA). The detection limit (LLD) of the method was estimated to be 4 Bq/m^3 (with an exposure time of about 4320 h). To obtain the radon annual average concentration, the time-weighted average radon concentration from two consecutive 6-month periods, was calculated using the exposure time as weights, as shown in the following Equation (2):

$$C^{T}_{Rn} = \frac{(\Delta t_1 \times C^{1}_{Rn}) + (\Delta t_2 \times C^{2}_{Rn})}{(\Delta t_1 + \Delta t_2)} \tag{2}$$

where Δt_1 and Δt_2 are the exposure times of two semesters and C^{1}_{Rn} and C^{2}_{Rn} are the integrated measured radon concentrations in the two semesters.

The central tendency of the radon measurements was described using the geometric mean since their distribution was skewed. The evaluation of normality of log-transformed data was tested by the Shapiro–Wilk test. All the statistical analysis was performed using the Stata software (Stata Corp, College Station, TX, USA).

2.3. Risk Assessment

The annual effective dose H to the workers due to the radon and its progeny was calculated as suggested by the Italian Legislation (Decreto Lagislativo 101/2020), using Equation (3) [14]:

$$\mathrm{H(mSv/y)} = \mathrm{C} \times \mathrm{T} \times \mathrm{D} \tag{3}$$

where C is the radon concentration (Bq/m^3), T is the occupancy factor, and D (6.7 × 10^{-9} Sv per Bq h m^{-3}) is the dose conversion factor assuming an average equilibrium factor between radon and its daughters of 0.4 (ICRP 137) [18]. Generally, for a particular radionuclide, the internal dose is evaluated using two pieces of information, the intake and internal dose conversion coefficients called the dose per unit intake. Due to the short half-life of radon progenies responsible for the major contribution to the inhalation dose, it is not possible to perform bioassay measurements to assess the intake. Therefore, ICRP recommends the dose calculation using the activity concentration of radon in the atmosphere and the use of the dose conversion factor.

The excess lifetime cancer risk (ELCR) was estimated using the following Equation (4):

$$\mathrm{ELCR} = \mathrm{H} \times \mathrm{DL} \times \mathrm{RF} \tag{4}$$

where H is the mean effective dose, DL is the average duration of life estimated to 70 years, and RF is the fatal cancer risk per Sievert (5.5 10^{-2} Sv^{-1}) recommended by ICRP 103 [19].

The lung cancer cases per year per million persons (LCC) is estimated using the following Equation (5):

$$\mathrm{LCC} = \mathrm{H} \times \mathrm{RFLC} \tag{5}$$

where RFLC is the risk factor lung cancer induction per million per person of 18×10^{-6} mSv^{-1} y reported in ICRP 50 [20].

3. Results

In this study, the results of a radon survey at the Public Healthcare Company "Napoli 3", are reported. During the survey, 607 rooms were monitored using a total of about 1300 CR-39 detectors. After exposure, the return rate of the CR-39 detectors was 92% for the first semester and 90% for the second semester. In Figure 1a, the frequency distribution of the annual radon concentrations measured, is reported. Although the experimental data appear to show an approximately log-normal distribution, the Shapiro–Wilk test failed to assess normality (p-value < 0.001), as shown also in Figure 1b.

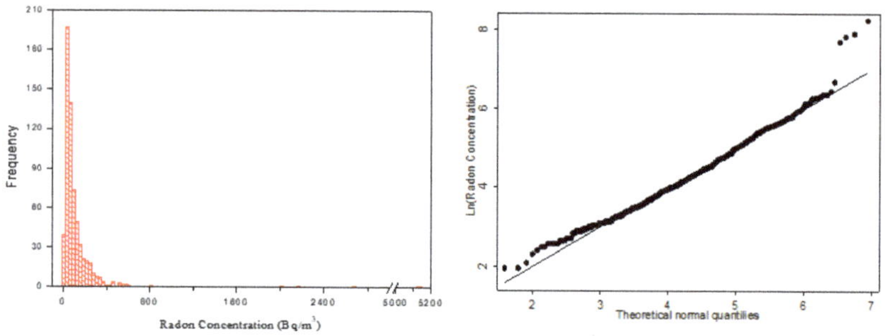

(a) (b)

Figure 1. (**a**) Frequency distribution of indoor radon concentrations in workplaces of the Public Healthcare Company "Napoli 3". (**b**) Q-norm plot of natural log-transformed radon concentration.

The minimum and maximum annual radon concentrations were found to be 7 ± 1 Bq/m^3 and 5148 ± 772 Bq/m^3, respectively, with a geometric mean of 67 Bq/m^3, geometric standard deviation of 2.5, and the median value of 65 Bq/m^3. Of all radon concentration measurements, 87% of the rooms presented radon concentrations lower than 200 Bq/m^3, 8% had radon concentrations between 200 and 300 Bq/m^3, and the other 5% had values greater than 300 Bq/m^3.

The radon concentration varied with the floor where it was measured. In this survey, of all the monitored rooms, 43% were placed on the ground/underground floor, while 57% were placed on the first and second floor underground. The t-Student test on the log-transformed data presents a statistically significant dependence from the floor ($p < 0.001$; Figure 2). Here, it seems that the radon concentration increases with the floor, where the radon concentration in the underground is higher than those measured in the first and second floor underground. It is well known that the concentration of radon decreases as the floor increases, but we do not have an explanation to justify this different trend of radon concentration observed. In this study, no information on the characteristics of constructions, such as age, building material, and presence of forced ventilation affecting the radon concentration, that could explain this trend, were collected. However, our finding agrees with Ruano-Ravina et al. [21] that reported in their study an increase of radon concentration with height.

Risk Assessment

Using Equation (3), the mean annual effective dose received by the workers was estimated to be 1.6 mSv/y, assuming an occupancy factor of 2000 h y^{-1} for a worker [22,23]. Moreover, assuming that the average duration of life is estimated to 70 years and the fatal cancer risk per Sievert is equal to 5.5×10^{-2} Sv^{-1} as recommended by ICRP 103, the mean excess lung cancer risk (ELCR) was found to be 0.62%. This estimate does not involve population-specific adjustments for major factors such as sex, age, and smoking habits. Finally, using the conversion factor for cancer cases per year per million per person of ICRP 50 [20] of 18×10^{-6} mSv^{-1} y, the lung cancer cases per million persons (LCC) was estimated to be 2.1 per million persons.

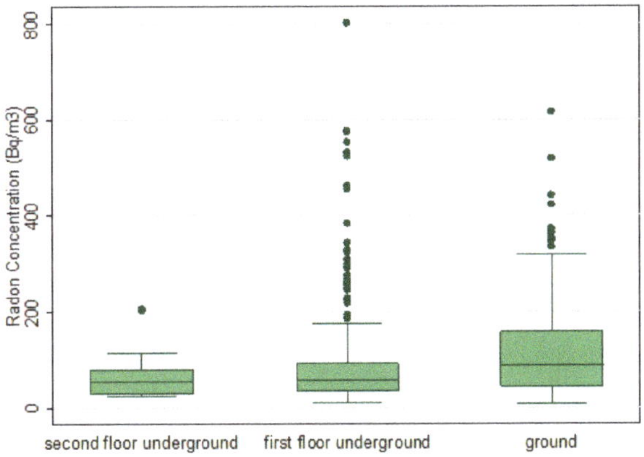

Figure 2. Relationship between radon concentrations and the floor of the monitored workplaces. Radon concentration values higher than 1000 Bq/m^3 are not shown, in order to make the dependence of the average radon concentration appreciable on the floor where the measurements are performed.

4. Discussion

In Italy, the culture of prevention to radon exposure in the workplace is still not widespread, although already in the year 2000, in accordance with the European Directive 96/29/EURATOM, the National Regulation was issued which regulated the protection of workers against the ionizing radiation exposure. Consequently, few surveys have been conducted and most of them have been performed on a local scale with different methods of measurement to determine the radon concentration in workplaces in some Italian regions [24–29]. The lack of mandatory monitoring in homes and workplaces not underground has meant that, although radon is a ubiquitous pollutant, the health risk that it can produce is little as considered by the general population, unlike other risks such as the toxicity of waste and the effects of electromagnetic fields. Studies [30,31] have shown that although people often have a high awareness of the presence of radon in their homes, they do not perceive a real risk and do not consider it necessary to monitor their living and working environments. On 31 July 2020, the Italian government has transposed the Council Directive 2013/59/Euratom into Decreto Legislativo 101 [17], which establishes the value of 300 Bq/m^3 as the annual average of the radon concentration not to be exceeded in all closed environments, thus standardizing the value of the reference level to all work environments and homes. This study was conducted in an area with a high concentration of indoor radon and where building materials such as tuff and pozzolan, are often used. It is the first study conducted in Campania in which various types of workplaces were monitored, such as offices, hospitals, counseling centers, retirement homes for the elderly, regardless of the workers' employment hours. The survey was conducted in the period in which the Italian government was enacting the law and the radon problem was beginning to be of great public interest. This is evidenced by the great participation that the workers of the monitored health company showed and by the high overall percentage of returned detectors during the survey of 97%.

The main results show that indoor radon concentrations were in general within the Italian legislation, in fact, about 95% of them are below 300 Bq/m^3. The mean radon concentration of 118 Bq/m^3 obtained in this study is higher than the national average of

75 Bq/m^3 relative to the national survey carried out in homes [32]. Probably, this higher average value is attributable to the different lifestyle between homes and workplaces. In addition, the highest radon concentrations were found in some rooms that remained closed during the lockdown due to the COVID-19 health emergency. Our findings are in good agreement with previous studies carried out nationwide [33,34]. Moreover, these surveys conducted in different workplaces, spread throughout the Italian region, show an average radon concentration higher than obtained in national survey dwellings.

Numerous studies [1,35,36] report that indoor radon concentrations follow a lognormal distribution. However, our results show a departure from lognormality at a higher concentration level. The reason for this behavior could be attributed to the not too large sample size.

Moreover, the measurements were carried out in an area which was not homogeneous, either for the geology or for the characteristics of the buildings. This may justify the presence of outliers.

In the world, the estimate of the proportion of lung cancers attributable to radon ranges from 3 to 14% depending on the average radon concentration in the country concerned and the calculation methods implemented [37]. In the European pooling [9], the exposure-response relationship appeared to be approximately linear with no evidence for a threshold below which there was no risk. Conversely, Dobrzyński et al. [16], applying three different statistical models to analyze data from 34 radon studies, concluded that no statistical evidence could support the thesis that the linear model best fits the data over low radon concentrations.

In Italy, the total lung cancers attributable to radon is about 10% [38]. The risk of radon-induced lung cancer increases with exposure and the duration of the exposure, so it is very important to evaluate the annual effective dose to the workers. It is well known that the internal exposure to radon can be performed using either the epidemiological method or the dosimetric method. ICRP 126 [39], based on the re-assessment of dose coefficients considering the epidemiological studies on miners and the newest studies on residential exposures, concluded that the results of dose assessments using epidemiological data are similar to those using dosimetric models. In the present study, the assessment of the dose to workers was performed as recommended by the Italian law. This approach is based on the epidemiological method as recommended by ICRP 65. The annual effective dose to the workers due to radon indoor exposure was found to be 1.6 mSv/y. In Italy, national dose data relating to workplaces are not available. However, if we compare our data with exposures in homes, we observe that these values are higher than the Italian national average value which is 1.2 mSv/y [40]. Moreover, the mean annual effective dose is higher than the worldwide average of 1.15 mSv/y, reported by UNSCEAR [6]. The LCC found (2.1 per million persons) results lower than the limit range of 170–230 per million persons recommended by ICRP [22]. Our data show that the impact of radon exposure in these workplaces can be considered modest.

This study has some advantages, including being the first survey conducted in Campania in which all the rooms, at the underground and ground floor of a whole healthcare company, were monitored. Moreover, the radon concentrations measurements were conducted by applying a quality assurance system. The radon concentration data collected in this study could contribute to the radon map of Campania and to the validation of new analysis methods on the correlation between radon concentrations and geology [41,42].

The main limitation is represented by the fact that it was not possible to obtain information on the factors affecting indoor radon concentrations to correlate these to the major characteristics of the buildings, such as age of the construction and building materials.

Author Contributions: Conceptualization, M.Q.; Software, M.S.; Formal analysis, M.Q. and F.L.; Investigation, F.L. and F.S.; Data curation, F.G.; Writing—original draft preparation, M.Q. and F.L.; Writing—review and editing M.Q., G.S. and F.L.; Visualization, A.I.; Project administration, R.A.; Funding acquisition M.Q. and F.L. All authors have read and agreed to the published version of the manuscript.

Funding: This research received no external funding.

Institutional Review Board Statement: Not applicable.

Informed Consent Statement: Not applicable.

Data Availability Statement: The data presented in this study are available on request from the corresponding author. The data are not publicly available due to the non-exclusive ownership by the authors. The data is shared with a third party, "Public Healthcare Company Napoli 3".

Acknowledgments: The authors wish to thank Michele Paduano and Giuseppe Casillo for technical assistance regarding the distribution of detectors in workplaces. This survey was carried out with the kind cooperation of the workers on the monitored premises.

Conflicts of Interest: The authors declare no conflict of interest.

References

1. Gooding, T.D. An analysis of radon levels in the basements of UK workplaces and review of when employers should test. *J. Radiol. Prot.* **2018**, *38*, 247–261. [CrossRef] [PubMed]
2. Kusky, T.M. *Geological Hazards: A Sourcebook*; Greenwood Press: Westport, CT, USA, 2003; pp. 236–239. ISBN 9781573564694.
3. Barazza, F.; Murith, C.; Palacios, M.; Gfeller, W.; Crristen, E. A national survey on radon remediation in Switzerland. *J. Radiol. Prot.* **2018**, *38*, 25–33. [CrossRef] [PubMed]
4. Man-Made Mineral Fibres and Radon. In *IARC Monographs on the Evaluation of the Carcinogenic Risks to Humans Volume 43*; WHO IARC Publications: Lyon, France, 1988; ISBN 978-92-832-1243-0.
5. Collier, C.G.; Strong, J.C.; Humphreys, J.A.; Timpson, N.; Baker, S.T.; Eldred, T.; Cobbi, L.; Papworth, D.; Haylock, R. Carcinogenicity of radon/radon decay product inhalation in rats—Effect of dose, dose rate and unattached fraction. *Int. J. Radiat. Biol.* **2005**, *81*, 631–647. [CrossRef]
6. United Nations Scientific Committee on the Effect of Atomic Radiation (UNSCEAR). *Sources and Effects on Ionizing Radiation*; United Nation: New York, NY, USA, 2000.
7. Lubin, J.H.; Wang, Z.Y.; Boice, J.D.J.; Zhao, Y.X.; Blot, W.J.; De Wang, L.; Kleinerman, R.A. Risk of lung cancer and residential radon in China: Pooled results of two studies. *Int. J. Cancer* **2004**, *109*, 132–137. [CrossRef]
8. Bochicchio, F.; Forastiere, F.; Farchi, S.; Quarto, M.; Axelson, O. Residential radon exposure, diet and lung cancer: A case-control study in a Mediterranean region. *Int. J. Cancer* **2005**, *114*, 983–991. [CrossRef]
9. Darby, S.; Hill, D.; Auvinen, A.; Barros-Dios, J.M.; Baysson, H.; Bochicchio, F.; Deo, H.; Falk, R.; Forastiere, F.; Hakama, M.; et al. Radon in homes and lung cancer risk: Collaborative analysis of individual data from 13 European case-control studies. *Br. Med. J.* **2005**, *330*, 223–226. [CrossRef]
10. Krewski, D.; Lubin, J.H.; Zielinski, J.M.; Alavanja, M.; Catalan, V.S.; Field, R.W.; Klotz, J.B.; Létourneau, E.G.; Lynch, C.F.; Lyon, J.I.; et al. Residential radon and risk of lung cancer: A combined analysis of 7 north american case-controls studies. *Epidemiology* **2005**, *16*, 137–145. [CrossRef] [PubMed]
11. Krewski, D.; Lubin, J.H.; Zielinski, J.M.; Alavanja, M.; Catalan, V.S.; Field, R.W.; Klotz, J.B.; Létourneau, E.G.; Lynch, C.F.; Lyon, J.L.; et al. A combined analysis of North American case–control studies of residential radon and lung cancer. *J. Toxicol. Environ. Health A* **2006**, *69*, 533–597. [CrossRef]
12. Garzillo, C.; Pugliese, M.; Loffredo, F.; Quarto, M. Indoor radon exposure and lung cancer risk: A meta-analysis of case-control studies. *Transl. Cancer Res.* **2017**, *6*, S934–S943. [CrossRef]
13. Basic Safety Standards for Protection Against the Dangers Arising from Expo-Sure to Ionizing Radiation. Council Directive 2013/59/Euratom 17 January 2014. Available online: https://eur-lex.europa.eu/LexUriServ/LexUriServ.do?uri=OJ:L:2014:013:0001:0073:EN:PDF (accessed on 20 February 2021).
14. Scott, B.R. Residential radon appears to prevent lung cancer. *Dose Response* **2011**, *9*, 444–464. [CrossRef] [PubMed]
15. Fornalski, K.W.; Adams, R.; Allison, W.; Corric, L.E.; Cuttle, J.M.; Davey, C.; Dobrzyński, L.; Esposito, V.J.; Feinendegen, L.E.; Gomez, L.S.; et al. The assumption of radon-induced cancer risk. *Cancer Causes Control* **2015**, *26*, 1517–1518. [CrossRef] [PubMed]
16. Dobrzyński, L.; Fornalski, K.W.; Reszczyńska, J. Meta-analysis of thirty-two case–control and two ecological radon studies of lung cancer. *J. Radiol. Res.* **2018**, *59*, 149–163. [CrossRef] [PubMed]
17. DECRETO LEGISLATIVO 31 Luglio 2020, n. 101, in Italian. Available online: https://www.gazzettaufficiale.it/eli/id/2020/08/12/20G00121/sg (accessed on 20 February 2021).
18. International Commission on Radiological Protection. Occupational Intakes of Radionuclides: Part 3. ICRP 137. *Annals ICRP* **2017**, *36*, 1–487.
19. International Commission on Radiological Protection. *The 2007 Recommendations of the International Commission on Radiological Protection*; ICRP 103; Elsevier: Amsterdam, The Netherlands, 2007; Volume 37.
20. International Commission on Radiological Protection. Lung cancer risk from exposure to radon daughters. ICRP 50. *Ann. ICRP* **1987**, *17*, 1–57.

21. Ruano-Ravina, A.; Narocki, C.; Lopez-Jacob, M.J.; Garcia Oliver, A.; de la Cruz Calle Tierno, M.; Peon-Gonzalez, J.; Barros-Dios, J.M. Indoor radon in Spanish workplaces. A pilot study before the introduction of the European Directive 2013/59/Euratom. *Gac. Sanit.* **2019**, *33*, 563–567. [CrossRef]
22. International Commission on Radiological Protection. Protection against Radon-222 at Home and at Work. ICRP 65. *Ann. ICRP* **1993**, *23*, 1–45.
23. ICRP 115, International Commission on Radiological Protection. Lung Cancer Risk from Radon and Progeny and Statement on Radon. *Ann. ICRP* **2010**, *40*, 1–64. [CrossRef] [PubMed]
24. Pugliese, M.; Quarto, M.; De Cicco, F.; De Sterlich, C.; Roca, V. Radon Exposure Assessment for Sewerage System's Workers in Naples, South Italy. *Indoor Built. Environ.* **2013**, *22*, 575–579. [CrossRef]
25. Vimercati, L.; Fucilli, F.; Cavone, D.; De Maria, L.; Birtolo, F.; Ferri, G.M.; Soleo, L.; Lovreglio, P. Radon Levels in Indoor Environments of the University Hospital in Bari-Apulia Region Southern Italy. *Int. J. Environ. Res. Public Health* **2018**, *15*, 694. [CrossRef]
26. Bucci, S.; Pratesi, G.; Viti, M.L.; Pantani, M.; Bochicchio, F.; Venoso, G. Radon in workplaces: First results of an extensive survey and comparison with radon in homes. *Radiat. Prot. Dosim.* **2011**, *145*, 202–205. [CrossRef]
27. Quarto, M.; Pugliese, M.; Loffredo, F.; Zambella, C.; Roca, V. Radon measurements and Effective Dose from Radon Inhalation estimation in the Neapolitan Catacombs. *Radiat. Prot. Dosim.* **2014**, *158*, 442–446. [CrossRef]
28. Panatto, D.; Gasparini, R.; Nenatti, U.; Gallelli, G. Assessment and prevention of radioactive risk due to ^{222}Radon on university premises in Genoa, Italy. *J. Prev. Med. Hyg.* **2007**, *47*, 134–137.
29. L'Abbate, N.; Di Pierri, C.; Martucci, V.; Cianciaruso, G.; Ragone, M. Radon concentrations in Apulian banking workplaces. *GIMLE* **2010**, *32* (Suppl. 4), 248–250.
30. Duckworth, L.T.; Frank-Stromborg, M.; Oleckno, W.; Duffy, P.; Burns, K. Relationship of Perception of Radon as a Health Risk and Willingness to Engage in Radon Testing and Mitigation. *Oncol. Nurs. Forum* **2002**, *29*, 1099–1107. [CrossRef] [PubMed]
31. Loffredo, F.; Savino, F.; Serra, M.; Tafuri, D.; Quarto, M. Cognitive investigation on the knowledge of the risk deriving from radon exposure: Preliminary results. *Acta Med. Mediterr.* **2020**, *36*, 1265–1267.
32. Bochicchio, F.; Campos Venuti, G.; Piermattei, S.; Nuccetelli, C.; Risica, S.; Tommasino, L.; Torri, G.; Magnoni, M.; Agnesod, G.; Sgorbati, G.; et al. Annual average and seasonal variations of residential radon concentration for all the Italian Regions. *Radiat. Meas.* **2005**, *40*, 686–694. [CrossRef]
33. Trevisi, R.; Orlando, C.; Orlando, P.; Amici, M.; Simeoni, C. Radon levels in underground workplaces—Results of a nationwide survey in Italy. *Radiat. Meas.* **2012**, *47*, 178–181. [CrossRef]
34. Rossetti, M.; Esposito, M. Radon levels in underground workplaces: A map of the Italian region. *Radiat. Prot. Dosim.* **2014**, *164*, 392–397. [CrossRef]
35. Kunovska, B.; Ivanova, K.; Badulin, V.; Cenova, M.; Angelova, A. Assessment of residential exposure in Bulgaria. *Radiat. Prot. Dosim.* **2018**, *181*, 34–37. [CrossRef]
36. Bossew, P. Radon: Exploring the log-normal mystery. *J. Environ. Radioact.* **2010**, *101*, 826–834. [CrossRef]
37. World Health Organization. *Handbook on Indoor Radon*; World Health Organization: Geneva, Switzerland, 2009.
38. Bochicchio, F.; Antignani, S.; Venoso, G.; Forastiere, F. Quantitative evaluation of the lung cancer deaths attributable to residential radon: A simple method and results for all the 21 Italian Regions. *Radiat. Meas.* **2012**, *50*, 121–126. [CrossRef]
39. International Commission on Radiological Protection. Radiological Protection against Radon Exposure. ICRP 126. *Ann. ICRP* **2014**, *43*, 1–73.
40. Bochicchio, F.; Campos Venuti, G.; Nucciatelli, C.; Piermattei, S.; Risica, S.; Tommasino, L.; Torri, G. Results of the representative italian national survey on radon indoors. *Health Phys.* **1996**, *71*, 721–748. [CrossRef] [PubMed]
41. Sabbarese, C.; Ambrosino, F.; D'Onofrio, A.; Pugliese, M.; La Verde, G.; D'Avino, V.; Roca, V. The first radon potential map of the Campania region (southern Italy). *Appl. Geochem.* **2021**, *126*, 104890. [CrossRef]
42. Loffredo, F.; Scala, A.; Adinolfi, G.M.; Savino, F.; Quarto, M. A new geostatistical tool for the analysis of the geographical variability of the indoor radon activity. *Nukleonika* **2020**, *65*, 99–104. [CrossRef]

MDPI
St. Alban-Anlage 66
4052 Basel
Switzerland
Tel. +41 61 683 77 34
Fax +41 61 302 89 18
www.mdpi.com

Life Editorial Office
E-mail: life@mdpi.com
www.mdpi.com/journal/life

www.ingramcontent.com/pod-product-compliance
Lightning Source LLC
LaVergne TN
LVHW070401100526
838202LV00014B/1361
* 9 7 8 3 0 3 6 5 6 7 6 7 9 *